Carbon Technocracy

STUDIES OF THE WEATHERHEAD EAST ASIAN
INSTITUTE, COLUMBIA UNIVERSITY

The Studies of the Weatherhead East Asian Institute of Columbia University were inaugurated in 1962 to bring to a wider public the results of significant new research on modern and contemporary East Asia.

Carbon Technocracy

Energy Regimes in Modern East Asia

VICTOR SEOW

The University of Chicago Press
Chicago and London

The University of Chicago Press, Chicago 60637
The University of Chicago Press, Ltd., London
© 2021 by The University of Chicago

Published 2021
Paperback edition 2023
Printed in the United States of America

32 31 30 29 28 27 26 25 24 23   1 2 3 4 5

ISBN-13: 978-0-226-72199-6 (cloth)
ISBN-13: 978-0-226-82655-4 (paper)
ISBN-13: 978-0-226-81260-1 (e-book)
DOI: https://doi.org/10.7208/chicago/9780226812601.001.0001

Library of Congress Cataloging-in-Publication Data

Names: Seow, Victor, author.
Title: Carbon technocracy : energy regimes in modern East Asia / Victor Seow.
Other titles: Energy regimes in modern East Asia | Studies of the Weatherhead East Asian Institute, Columbia University.
Description: Chicago ; London : The University of Chicago Press, 2021. | Series: Studies of the Weatherhead East Asian institute, Columbia University | Includes bibliographical references and index.
Identifiers: LCCN 2021031893 | ISBN 9780226721996 (cloth) | ISBN 9780226812601 (ebook)
Subjects: LCSH: Coal mines and mining—China—Fushun Xian (Liaoning Sheng)—History—20th century. | Energy policy—China—History—20th century. | Energy policy—Japan—History—20th century.
Classification: LCC TN809.C62 F8667 2021 | DDC 622/.334095182—dc23
LC record available at https://lccn.loc.gov/2021031893

For my parents,
Sally Neo and Seow Chuan Bin,
with love and gratitude

CONTENTS

ILLUSTRATIONS

NOTE ON CONVENTIONS

I have transcribed Chinese words and names in *pinyin*, Japanese ones according to the modified Hepburn system, and Korean ones according to the McCune–Reischauer system. East Asian names are given in the customary order in which family name comes before personal name, except when the individuals themselves choose to reverse that order. Several Chinese figures have established forms of their names in English that follow different romanizations or conventions, such as Sun Yat-sen (Sun Zhongshan), H. H. Kung (Kong Xiangxi), and Chang Kia-Ngau (Zhang Jia'ao), and I use those forms when referring to them. I drop the macrons marking long vowels in Japanese for place names that are well known, like Tokyo and Kyushu, and for reign names, like Taisho and Showa. For place names in Taiwan, I romanize them in the way they typically are in English, such as Taipei and Keelung. Throughout the text, I have provided Chinese characters for select individuals, places, institutions, and terms, and I have tried to use the form of the characters appropriate to the time and place. For instance, I switch from traditional to simplified characters for Chinese when writing about the People's Republic of China from the late 1950s onward, after the first simplification scheme. A list of abbreviations for several sources referenced in the notes can be found at the beginning of the bibliography. All translations from Chinese and Japanese, unless otherwise indicated, are my own.

Carbon Technocracy

I came in search of the origins of China's modern industrialization. I found, instead, the beginnings of its end. Before arriving in the coal-mining city of Fushun in the summer of 2011, I had seen old photographs and read historical accounts of its colossal open pit, first excavated by Japanese technocrats almost a century earlier. Pictures of the site showed an expansive industrial landscape molded by the machine: large excavators, electric- and steam-powered shovels, and dump cars hewing rock and moving earth to bring the cavity into being. The Japanese poet Yosano Akiko 與謝野晶子 (1878–1942), who visited Fushun in 1928, described the mine as "a ghastly and grotesque form of a monster from the earth, opening its large maw toward the sky."[1] At first glance, the real thing did not disappoint.

It would have been easy to mistake the gigantic depression in the ground for a natural formation such as a valley were the sides not cut into steps of recognizable regularity: like terrace farming, but for harvesting shale and coal. I had been brought to the pit by a colliery representative eager to show off the sight. As our car trundled down a rocky road into its depths, I could not help but notice that the mine was far less busy than I had anticipated. Along our descent, we passed by a single dump truck loaded with debris. Imposing though it was—its wheels twice the height of our sedan—it appeared to be the only sign of work on site. Overhead, the sky was almost too blue for an industrial city, certainly so for one that

---

1. Yosano and Yosano, "Man-Mō yūki," 140. A prolific poet, writer, and translator, Yosano Akiko was widely regarded as a keen if often controversial observer of social issues. Her visit to Fushun was part of a tour of Manchuria and Mongolia that she took with her husband, fellow poet Yosano Tekkan, in the spring of 1928. We will encounter her again in chapter 2.

I.1. Aerial view of Fushun's Western open-pit mine. This photograph dates to September 29, 2017. By that point in time, much of the work that was being done on site was not mining coal or shale but clearing debris from rockfalls, which often resulted from prior overmining, and putting out fires from exposed coal or shale that had combusted spontaneously. This mine would cease operations less than two years later. (Image courtesy of Imaginechina Ltd./Alamy Stock Photo.)

for decades boasted East Asia's largest coal-mining operations and that was once known as "Coal Capital" (in Chinese, 煤都; in Japanese, 炭都).

Fushun is located in Liaoning, the southernmost of the three provinces that make up China's Northeast—a region formerly referred to as "Manchuria."[2] Sandwiched between layers of green mudstone, oil shale, tuff, and basalt, massive stores of coal lie beneath the city. For the past hundred or so years, this coal has been mined in spades. The South Manchuria Railway Company (南滿洲鐵道株式會社; "Mantetsu" [滿鐵] for short), the Japanese colonial corporation that ran Fushun's coal mines for much of the first half of the twentieth century, developed them into an extractive enterprise of staggering proportions. In 1933, Fushun accounted for almost four-fifths of Manchuria's coal output and more than a sixth of the coal produced in the Japanese metropole and its colonies.[3] It was the pitch-black heart of Japan's empire of energy.

The Chinese Communists continued to exploit Fushun's carbon resources after taking control of the area in 1948. In 1952, this colliery, then

2. On the history of "Manchuria" as a toponym, see Elliott, "The Limits of Tartary."
3. Schumpeter, "Japan, Korea, and Manchukuo," 409, 424.

still China's largest, produced over 8 percent of the country's coal.[4] Decades later, the speed and scale of its extraction have proven unsustainable. Fushun's current annual output is less than three million tons, roughly a third of its 1936 prewar peak and a sixth of its 1960 postwar height.[5] Wasteful mining practices in the past have compromised present and future production. While about a third of the estimated 1.5 billion tons of total coal deposits remain in the ground, mining these reserves risks triggering landslides and subsidence that have caused infrastructure to crack and buildings to sink. According to a 2012 government report, as much as two-thirds of Fushun's urban area rests on unstable ground. As one recent commentator put it, "today the mineral that helped turn the city into a booming metropolis of 2.2 million threatens to bury it."[6]

This book explores how Chinese and Japanese states, in attempting to master the fossil fuels that powered their industrial aspirations, undertook large-scale technological projects of energy extraction that ultimately exacted considerable human and environmental costs. Nowhere is this more evident than in Fushun.[7] Although the former Coal Capital's fortunes may now be flagging, the pattern of fossil-fueled development that enabled its rise persists into the present. As we confront a planetary crisis precipitated by copious carbon consumption, the history of the Fushun colliery offers us a genealogy of our current predicament.

Opened by Chinese merchants at the start of the last century, Fushun's coal mines were occupied by Japan during the Russo-Japanese War (1904–

4. Fushun shi tongji ju, *Fushun sanshi nian*, 21.
5. *FMBN*, 360–61.
6. Dong, "Coal, Which Built a Chinese City."
7. The Fushun colliery has attracted some scholarly attention in the past. On its development as a coal-mining enterprise under Mantetsu, see Chen, *Riben zai Hua meiye touzi*, 29–80, 163–75. On the Japanese management of its largely Chinese workforce, see Murakushi, *Transfer of Coal-Mining Technology*, 48–84; Matsumura, "Fushun meikuang gongren shitai"; You, *Mantetsu Bujun tankō no rōmu kanri*; Teh, "Mining for Difference"; Wang, *"Manshūkoku" rōkō*, 213–319. My book distinguishes itself from and adds to the prior literature in several ways. First, I focus a good deal on the coal itself. It matters, I contend, that Fushun's main product was coal and not some other resource or commodity. Starting from that simple premise, I show how the meanings and materiality of coal had bearing on the extraction, circulation, and production of this fossil fuel and how these processes gave form to particular assemblages of technology, labor, and the environment at this site and beyond. Second, while other works end with the fall of the Japanese empire in 1945, I take the story of Fushun coal up to 1960, which allows us to consider changes and continuities at the colliery as it passed into Nationalist and then Communist hands and, more generally, the legacies of Japanese imperialism on subsequent Chinese states. Third, I place this local and regional history firmly within wider developments in China and Japan, showing how central Fushun was to the vision of fossil-fueled industrial modernity that the Chinese and Japanese states in question shared.

1905) and placed under Mantetsu's management soon after. Following the fall of Japan's empire in 1945, the Fushun colliery was seized first by the Soviets, then the Chinese Nationalists, and finally the Chinese Communists, under whose control it has since remained as a state-owned enterprise. Throughout, operators maximized their hauls by deploying various technologies of extraction. They turned not only to methods like open-pit mining and hydraulic stowage to more completely extract carbon energy from the earth but also to mechanisms like fingerprinting and calorie counting to do the same with human labor. Fushun was the model of modern coal mining in China and Japan, and images of its immense open pit fed fantasies of energy-intensive industrial modernity in Tokyo, Nanjing, Beijing, and beyond.

The coal mined at Fushun and at other sites of energy extraction around the globe catalyzed a distinctive sociotechnical apparatus that presented itself as the epitome of modernity—universal, scientific, inevitable. For all their differences as political regimes, the imperial Japanese, Chinese Nationalist, and Chinese Communist states that controlled Fushun at varying times shared a decidedly technocratic vision in relation to carbon resources. This vision involved marshalling science and technology toward the exploitation of fossil fuels for statist ends. It was further characterized by an embrace of coal-fired development, a focus on heavy industrial expansion, a fixation on national autarky, an interest in labor-saving mechanization, a privileging of cheap energy, and a pegging of economic growth to increases in coal production and consumption. At the same time, that states saw in carbon energy a means to modernity engendered for them tensions between a fear of fuel scarcity and a faith in securing, largely through technoscientific means, a near limitless fuel supply.[8] The emergence of this particularly modern regime of energy extraction that I call "carbon technocracy" is the subject of this book.

The World That Carbon Made

We inhabit a world that carbon made. The intensified exploitation of fossil fuel energy in the form of coal that began in the eighteenth century set off seismic social and material shifts across the globe. Harnessed through

8. Anson Rabinbach observes a similar paradox in his major study of the science of work and the problem of labor, with "concepts of energy and fatigue . . . at once affirming the endless natural power available to human purpose while revealing an anxiety of limits." See Rabinbach, *The Human Motor*, 12.

the steam engine, the considerable power contained in coal helped drive mass industrial manufacturing, altering the fabric of work and patterns of consumption. Coal fueled steam locomotives and steamships, facilitating travel over great distances on both land and water. In so doing, it allowed people, ideas, and objects to more easily circulate. Gas produced from heating coal lit lamps in streets, factories, and domiciles, lengthening the day's activities into the night. Processed into coke, coal fired furnaces for smelting and working the iron and steel used to build the latest machinery and infrastructure, from mining pumps to railway tracks. Beginning in the late nineteenth century, coal burned in thermal power plants generated electricity through which it witnessed even wider application. "Out of this coal and iron complex," social critic Lewis Mumford remarked, "a new civilization developed."[9]

Having shaped our recent past, coal continues to sustain our present. Since 1800, global coal consumption has gone up 450-fold. As of 2019, coal still accounts for about 27 percent of current energy use worldwide, following only oil (around 33 percent) and still leading natural gas (around 24 percent)—both, incidentally, also fossil fuels. Coal remains the main source for electricity generation, at 36 percent.[10] From the onset, coal consumption has been uneven within and across polities, variances in access to carbon energy reflecting and reproducing social and material inequalities. These inequalities have only widened over time, growing with exponential increases in fossil fuel use.[11]

Still, the world that coal brought into being seemed flat in parts. Regions most transformed bore striking resemblances to one another. Manchester, England's coal-fired "Cottonopolis"—characterized by Alexis de Tocqueville in conflicted terms as a "foul drain" from which "the greatest stream of human industry flows out to fertilize the whole world"—was a modern manufacturing hub replicated around the globe, including in East Asia.[12] Osaka, Japan's most industrialized city in the prewar years, was known as the "Manchester of the Orient." An 1896 municipal survey counted as many as 1,370 industrial chimneys that towered above the city, spewing forth clouds of noxious fumes from the combustion of coal that powered the steam engines driving factory operations below. Osaka's urban terrain

9. Mumford, *Technics and Civilization*, 156. For more on Lewis Mumford and his writings about mines and other subterranean spaces, see Williams, *Notes on the Underground*.

10. Ritchie, "Energy Mix"; Ritchie and Roser, "Fossil Fuels."

11. Oswald, Owen, and Steinberger, "Large Inequality in International and Intranational Energy Footprints."

12. Tocqueville, *Journeys to England and Ireland*, 107.

came to be defined by this "forest of smokestacks."[13] Another contender for the title of East Asia's Manchester was Shanghai. The city claimed the "biggest cotton factory in the Chinese empire, and one of the largest in the world."[14] Here, too, "unlike other cities of China, which are noted for their pagodas and temples," one observer declared in 1929, was a landscape marked by "hundreds of smoke-stacks and chimneys."[15] From Manchester to Osaka, Shanghai, and areas afar, mills and machines, steam and smoke marked the arrival of fossil-fueled industrial modernity.

But if coal has been central to the making of the industrial world, then this black rock is also very much complicit in its unmaking. The profligate burning of coal and other fossil fuels has released (and continues to release) into the atmosphere such massive quantities of carbon dioxide and other greenhouse gases as to provoke an accelerated change in the earth's climate unprecedented in human history. The planet's surface and oceanic temperatures, like sea levels, are rising, while snow and ice melt at a rapid rate. The resulting terrestrial transformation "may well threaten the viability of contemporary civilization and perhaps even the future existence of *Homo sapiens*," chemist Will Steffen and his colleagues warn.[16] In China, beyond the growing cities so frequently covered with smog, the glaciers at the headwaters of the Yangtze River are in rapid recession, foreshadowing floods in the short term and water shortages in the long run—each with potentially devastating consequences.[17] We have, as atmospheric chemist Paul Crutzen proposed, entered the epoch of the Anthropocene.

While humankind has always exerted pressures on the environment, this new "age of man" denotes the fact that we, as a species, have become a force of nature rivaling even the most formidable of geophysical forces in our impact on the biosphere—to its detriment and, as such, our own. Aside from climate change, scientists point out that our fertilizer use and other industrial and agricultural practices have altered biogeochemical cycles of elements essential to life such as nitrogen and phosphorous. Humans have modified terrestrial water cycles through damming or redirecting rivers and clearing vegetation, and we are driving a mass extinction event by despoiling

13. "Commercial and Industrial"; Scott, "Changes and Still More Changes." Like Manchester, Osaka specialized in textiles, but it also produced an assortment of other commodities ranging from the mundane to the majestic: glass, matches, ships, locomotives, and more.

14. Browne, *China*, 283–84.

15. Hsia, *The Status of Shanghai*, 118.

16. Steffen et al., "The Anthropocene," 862.

17. Inglis, "Tibetan Headwaters." On regional projects to tap these waters and how they have already started to be undermined by climate change and glacial retreat, see Pomeranz, "The Great Himalayan Watershed."

and destroying habitats.[18] Each of these activities has in one way or another been enabled or expedited through the might of fossil-fueled machines.

The emergence of the fossil fuel economy, momentous as it was, coincided with another historical development of great importance: the rise of the modern state. As in the case of human influence on the environment, the state as a form of political organization predated the modern era but became more pronounced and keenly felt within it.[19] "By the late nineteenth century," historian Charles Maier writes, "states possessed a degree of dedication to governance, of bureaucratic functionality, of at-oneness with fixed territorial space, of belief in their own competitive mission, that was unprecedented." As he contends, this period into the early twentieth century witnessed "a decisive intensification of state ambition and governmental power" that had been facilitated by new communication and transportation technologies—technologies that were, it should be stressed, largely coal driven.[20] Although modern states differ in a variety of ways, from ideology to capacity, the impulse toward expansion of function, powered at its core by carbon, has been common to most.

But what if the two developments in question—the carbon economy's emergence and the modern state's rise—were not just contemporaneous but coconstitutive? That is, what if the fossil fuel economy made possible the modern state and the modern state the fossil fuel economy? This is the premise of *Carbon Technocracy*: that the mutual production of calorific and political power has defined our industrial modern age.[21]

Fossil fuels fed the appetites and ambitions of the modern state. As energy resources undergirded a range of statist preoccupations, from economic production to the waging of war, states came to see ensuring a steady and growing supply as essential to their variegated objectives and ultimately to their survival and extension of power.[22] To obtain sufficient stocks of coal, states turned to some of the latest developments in science and technology, whether in the form of geological surveys to locate subterranean deposits or mining engineering to wrest those riches from the earth. In Fushun, as in other places in China, Japan, and elsewhere, geolo-

18. Steffen et al., "The Anthropocene," 843. See also, Miller, *The Nature of the Beasts*, 9.

19. The literature on the history and theory of the state is vast and divided. One useful survey is Brooke and Strauss, "Introduction."

20. Maier, *Leviathan 2.0*, 11.

21. My thinking here has been influenced and informed by Sheila Jasanoff's idea of coproduction. See Jasanoff, "The Idiom of Co-production," 1–6.

22. On how the British state began to regard coal as essential to the affairs of the realm by the seventeenth century, see Parthasarathi, *Why Europe Grew Rich*, 164–70.

gists observed outcrops and sunk bores in search of coal while engineers oversaw the design and operations of mines for its recovery.[23]

However, in their eagerness to excavate and exploit fossil fuels, many states became committed to efforts at energy extraction that were as extravagant as they were extensive. I refer to this system as "carbon technocracy" because it rested on an unwavering, oftentimes uncritical belief in the superiority of science and technology in the practice of statecraft and relied heavily upon scientific and technological methods to harness the fossil fuels so central to the project of the modern state.[24] Larry Lohmann previously employed the term in critiquing the Kyoto Protocol for underestimating the costs involved in constructing a carbon market for climate mitigation, which he characterized as a "technical fix" based on "the Kafkaesque logic of the carbon technocracy."[25] I define "carbon technocracy" more broadly as a technopolitical system grounded in the idealization of extensive fossil fuel exploitation through mechanical and managerial means, and I use it here to describe a historical process that is concurrently an alternative account of state formation in modern East Asia and a transnational history of technology. Appealing to mechanistic notions of efficiency, those who saw from the standpoint of the state consistently strove to make their mounting extractive endeavors "rational." In hindsight and from the perspective of the planet and of our continued existence, though, these endeavors have been anything but.

It may seem a bit obvious to go looking for the origins of our modern energy regime in a coal mine, as I do here.[26] Coal mines have, after all, produced much of the power for industrial modern life. Mumford would contend that "to be cut off from the coal mine was to be cut off from the source of paleotechnic civilization."[27] But in fueling the world beyond

23. For studies on geology in China during this period, see Shen, *Unearthing the Nation;* and Wu, *Empires of Coal.* For one account of the same in Japan, see Tanaka, *New Times in Modern Japan,* 39–48.

24. One may regard this as a form of "high-modernist ideology," described by James Scott as the "strong, one might even say muscle-bound version of the self-confidence of scientific and technological progress, the expansion of production, the growing satisfaction of human needs, the mastery of nature (including human nature), and, above all, the rational design of social order commensurate with the scientific understanding of natural laws." See Scott, *Seeing Like a State,* 4.

25. Lohmann, "Marketing and Making Carbon Dumps," 230.

26. Other studies in the history of energy that similarly use a focus on sites of production to reveal transformations in larger social and ecological systems include Andrews, *Killing for Coal;* Black, *Petrolia;* Santiago, *The Ecology of Oil;* and Brown, *Plutopia.*

27. Mumford, *Technics and Civilization,* 156.

it, this site of extraction also came to reflect the nature of the system it supported.

I argue that the coal mine exemplifies several key features of the wider industrial modern world: hubristic attempts to tame and transform nature through technology; mechanization and the disciplining and degradation of labor; and, perhaps above all, the tenacious privileging and pursuit of production. These features were terribly visible in Fushun. They manifested in the open-pit mine, in the assorted techniques of labor control, and in the ever-escalating output targets that strained not only the environment from which the coal was extracted but also the workers on whom that extractive process so deeply relied. The Fushun colliery thus provides a panoramic setting from which we may view the tensions in the workings of the modern world that carbon made.

An Archaeology of Addiction

In tracing the modern state's role in the emergence of the carbon economy, this study offers one explanation for how those of us who live in industrial and industrializing societies have become so dependent on fossil fuels. The heightened awareness of anthropogenic climate change in recent decades has cast the question of dependency into sharper relief. At the same time, a growing chorus of voices has acknowledged the sheer extent of our dependence—tantamount, it seems, to a deep addiction.[28] But if this addiction is intense, it was by no means inevitable. The shift to coal and other fossil fuels as society's primary source of power and the subsequent intensification of their use were historically contingent. Without exhuming this past and examining the forces behind that energy transition, we cannot begin to truly face, let alone overcome, our fixation with fossil fuels and the crisis of climate it has created.[29]

The fossil fuel transition happened in different places at different times

28. Huber, *Lifeblood*, x; Canavan, "Addiction."

29. It bears emphasizing that "energy transition," as stated here and evident throughout this book, refers to a shift in society's primary source of power and not the complete displacement of earlier power sources, many of which remain in use and often in increasing quantities. For a nuanced critique of how "energy transition" as a concept can mask the workings of those older power sources within an environment transformed by coal, see Barak, *Powering Empire*, 8–11, 24–116. For an account that uses British-Burmese encounters with fuels along the Irrawaddy to challenge teleological narratives of energy transitions that presume, for instance, a natural flow from wood to coal to oil, see Sivasundaram, "The Oils of Empire." On the limits of "transition" as a concept in the history of technology more broadly, see Edgerton, *The Shock of the Old*.

(and, in some cases, not at all). When it did, the impact was immense. Working from the case of the English industrial revolution, historical demographer E. A. Wrigley has argued that earlier systems of growth tied the thriving of human populations to the amount of arable land available for producing food, feed, and fuel. This "organic economy," as he termed it, encountered physical limits that circumscribed further growth by the early modern era. In the organic economy, almost all energy for human activity was derived from the sun via plant photosynthesis, which converts solar energy into chemical energy, or through wind and water processes driven, at a remove, by insolation. While fossil fuels, as remnants of organic life, also trace their energetic origins back to the sun—one of coal's many monikers is "buried sunshine"—these are pathways that are far less immediate, stretching hundreds of millions of years into the past. Most work in the organic economy, from tilling fields to transporting goods, was done by human or animal muscle power that was in turn sustained by the consumption of plants or of those who ate them. Plants also served as the main fuel for household and industrial purposes, whether as firewood or charcoal.[30]

To Wrigley, this plant-based energy system ran up against a problem of finitude. "As long as supplies of both mechanical and heat energy were conditioned by the annual quantum of insolation and the efficiency of plant photosynthesis in capturing incoming solar radiation," he concluded, "it was idle to expect a radical improvement in the material conditions of the bulk of mankind."[31] Furthermore, humans' reliance on plants as vectors of solar energy limited the supply of energy according to the amount of agricultural or wooded land. Within this context of constraint, coal offered a way out. Coal, insofar as it could be extracted and expended, promised an energy source that was on average at least twice as calorific as biomass, that was comparatively more compact and transportable, and that was unfettered by the seasonality of sunlight and the limitations of land. Across the long nineteenth century, industrializing societies came to rely ever more on coal, moving toward what Wrigley called the "mineral economy."[32]

30. Wrigley, *Energy and the English Industrial Revolution*, 13–17.
31. Wrigley, 17.
32. Wrigley; Wrigley, *Continuity, Chance and Change*. Other important accounts that similarly illuminate the significance of this shift to fossil fuels in light of the system that came before are Sieferle, *The Subterranean Forest*; and Jones, *Routes of Power*. For efforts to tell the breadth of human history through energy and energy transitions, see Crosby, *Children of the Sun*; Debeir, Deléage, and Hémery, *In the Servitude of Power*; and Smil, *Energy and Civilization*. Bruce Podobnik has offered a forceful analysis of shifts in the modern global energy system toward first coal and then oil, focusing on how the convergence of geopolitical rivalries, corporate competition, and social conflict facilitated those shifts. See Podobnik, *Global Energy Shifts*.

But the transition to a carbon-intensive economy did not take place on the merits of fossil fuels alone. It required new forms of knowledge, be this of steam engines or novel iron smelting techniques.[33] More generally, it involved a combination of calculated considerations and aleatory accidents. In his study of energy transitions in the United States' anthracite-rich mid-Atlantic region, energy historian Christopher Jones shows how entrepreneurs invested in large transport systems of canals, pipelines, and wires to carry coal, oil, and electricity to potential users. By facilitating energy producers' ability to deliver cheap, reliable, and abundant energy, these boosters stimulated a marked growth in consumption, giving form to what Jones terms "landscapes of intensification." As he phrases it, "The *roots* of America's energy transitions can be found in the building of *routes* along which coal, oil, and electricity were shipped."[34] Jones and other historians of energy have done much to demonstrate that culpability for our current crisis ranges widely from corporations to consumers, from the forces of capitalism to the actions of individuals.[35] The problem is so pernicious precisely because the vested interests are so pervasive. What appears to have been left in the background in most of these accounts, though, is the state.[36] This is both surprising and not: the modern era is one in which the state has been overwhelmingly influential and invasive yet concurrently insidious and, at times, even invisible.

In writing this book, I have sought to fill out the narrative of the fossil fuel turn by bringing the state back in.[37] The state was a key player in the energy transition to carbon. I have to clarify, though, that my understand-

33. Renn, *The Evolution of Knowledge*, 362–63.

34. Jones, *Routes of Power*, 2 (italics in original).

35. There is now a sizable and growing literature on the history of energy. For a survey of developments in this field over the past few decades, see Miller and Warde, "Energy Transitions as Environmental Events." Much of the scholarship in energy history focuses on the inanimate energy sources that people today tend to think of when they hear the word "energy"—coal, oil, electricity, and the like. Some concentrate on or at least take into consideration animate energy sources like humans and animals. The classic work that traces flows of energy through both inanimate and animate sources is White, *The Organic Machine*. A more recent work that applies this analytic productively is Demuth, *Floating Coast*. Although I look primarily at coal and other types of inanimate energy in this book, I also pay much attention to the animate energy of human labor that, too, undergirded the making of our industrial world. Energy, as a subject of social inquiry, has also attracted much attention among anthropologists. For a useful overview of the anthropology of energy, see Boyer, "Energopower."

36. One exception, which uses the case of Californian oil to foreground the role that local and national governments have played in driving fossil fuel demand and supply, is Sabin, *Crude Politics*.

37. The phrase "bringing the state back in" comes from Evans, Rueschemeyer, and Skocpol, *Bringing the State Back In*.

ing of the state here extends beyond the bureaucratic institutions of government that are often taken to represent it in its entirety. One of the central contradictions of the modern state is that, while it is typically thought of as distinct from society, the boundary between the two is never quite clear. And yet this boundary is meaningful. As Timothy Mitchell contends, "producing and maintaining the distinction between state and society is itself a mechanism that generates resources of power." The state, then, is manifested as an outcome of such interactions in what he refers to as the "state effect."[38] In this vein, the characters who feature in the story I tell range from individuals easily recognized as state agents, such as Japanese geologists with the quasi-public Mantetsu or Chinese engineers in the Nationalist government's National Resources Commission (資源委員會), to others who presumed to speak for or act in what they framed as the interests of the Japanese and Chinese states. These included coal industry leaders agitating for regulation, economists promoting natural resource autarky, and ordinary women and men moderating their own fuel consumption so as to save energy for their country's war efforts. The state and the system of carbon technocracy emerged out of this uncoordinated but collective enterprise.

The modern state regarded coal as useful in several ways—as an economic necessity, a strategic resource, a public good, a marketable commodity, among other things—but coal's use was also conditioned by variances in its material properties. Coal is commonly differentiated by quality, gauged largely by its carbon content, which correlates to its geological age and indicates how hot it burns. In ascending order of how much carbon it packs, we have lignite, subbituminous, bituminous, and anthracite. Alternatively, coal can be sorted by function, taking into consideration its broader composition. This classification system divides coal into two types. The first, steam or thermal coal, is used for firing boilers and generating electricity. This coal, which is usually subbituminous or bituminous, is typically pulverized into a fine powder so that it burns rapidly at high heat. The second, coking or metallurgical coal, is used to make coke for iron and steel production. This coal, which is usually bituminous, contains lower quantities of sulfur and phosphorus and can withstand the high heat required to turn it into coke, a process through which its impurities are burned away, leaving pure carbon behind. Fushun coal consisted of both of these types. The central and western parts of the coalfield produced good thermal coal, while the eastern part yielded coal that, when mixed with

38. Mitchell, "Society, Economy, and the State Effect," 83; Mitchell, "Limits of the State."

other metallurgical coal, was well suited for coking.[39] Fushun coal seemed primed for the widespread use to which it was put.

Apart from its many material applications, coal also became an abstract means by which many states measured their level of development. Historian Dolores Greenberg has shown how for numerous scientists, engineers, politicians, and economists in early nineteenth-century England and America, "the belief that harnessing nonbiological power to new technologies would realize ideals of automation and abundance became integral to assumptions about work, wealth, and human history." Consequently, they partook in a vision of utopia that was subsumed within "a range of formulations that presented escalating energy use as a primary source of change."[40] In early twentieth-century East Asia, political leaders and social commentators alike similarly championed increases in carbon expenditure and placed Chinese and Japanese coal production and consumption figures alongside those of other countries to ascertain where they stood in the world. A Japanese observer, writing in 1919, remarked that the amount of coal used may be thought of as "a barometer measuring the degree of a country's culture."[41] In line with this logic, states competed against each other to see who could more extensively exploit this carbon resource. Ecologically, this proved to be nothing less than a race to the bottom.

Today, coal may be considered a fossil of a dirtier past, but it was until very recently regarded as the fuel of a brighter tomorrow.[42] Part of the earlier optimism toward coal was that it could power modern machinery to greatly augment human effort. To the Chinese thinker Hu Shih 胡適 (1891–1962), writing in 1928, the ability to fashion new devices driven by coal through steam and electricity "for the conquest of nature and for the multiplication of the power to do work" constituted modernity. It was on this point that one might discern, he claimed, "the real difference between Oriental and Western civilizations." Citing an American friend, Hu noted that "each man, woman and child in America possesses from twenty-five to thirty mechanical slaves, while it is estimated that each man, woman and child in China has at his command but three quarters of one mechanical

39. Quackenbush and Singewald, *Fushun Coal Field*, 9–10.

40. Greenberg, "Energy, Power, and Perceptions of Social Change," 695. On related ideas of endless growth that had come and gone since at least the eighteenth century, see Jonsson, "The Origins of Cornucopianism."

41. Kita, "Sekitan kai."

42. Beginning in the interwar period but becoming more pronounced after World War II, futuristic fantasies about fuel have centered around other sources of energy, especially oil. For a study that examines how oil animated powerful visions of development and modernity in Venezuela, see Coronil, *The Magical State*.

slave."[43] Evidently, the servitude of machines and the energy that powered these devices would enable the liberation of human beings.[44] Still, the oppressive working conditions at Fushun and other mining sites across the industrial world are uncomfortable reminders that the freedom that carbon energy brought some segments of society often came at the expense of the freedom of others.[45]

Coal's allure also lay in its abundance. Coal is one of the most commonly occurring rocks in the earth's crust that humans have figured out how to exploit. Amid anxieties over fuel that followed World War I, Yoshimura Manji 吉村萬治 (1882–1969), head of Japan's Fuel Society (燃料協會), regarded coal, rather than oil, as "the primary fuel of the future." His belief was almost entirely based on coal's comparative bounty—the Earth's largest deposits of oil had yet to be discovered, while sizable deposits of coal could be found on almost every continent.[46] The Geological Survey of Canada, which put together one of the first systematic estimates of the world's coal resources in 1913, came up with a figure of 7.4 billion tons.[47] The current proven reserves—the amount of coal in the ground deemed economically recoverable—is 1.1 trillion tons. At today's rate of consumption, that would supposedly last us as much as half a millennium, if not more.[48]

The tendency to think of coal reserves in this way, in terms of several human lifetimes beyond our own, has been around for a long time. It has often fostered an imagined inexhaustibility and a carelessness that comes with believing that to be true. Early into their occupation of the Fushun coalfield, for example, Japanese technocrats spoke of the area's coal deposits as a "treasure house" (寶庫) that was "limitless" (無限) and that "cannot be depleted" (無盡蔵).[49] Subsequent Chinese regimes would all too readily parrot such characterizations. Between its potential for power and its plenty, coal here and elsewhere animated fantasies of a future propelled by near endless energy. Our addiction to carbon rests upon this long-standing reverie from which we must surely soon awaken.

---

43. Hu, "Civilizations of the East and the West," 27–28.

44. On the idea of the "energy slave" and its predecessors in the form of the "electric slave" and "mechanical servant or slave," see Johnson, "Energy Slaves."

45. For one treatment of this theme in the history of American illumination, see Zallen, *American Lucifers*.

46. Yoshimura, "Waga kuni ni okeru nenryō mondai."

47. McInnes, Dowling, and Leach, *Coal Resources of the World*, xviii.

48. Kopp, "Coal."

49. *BJTK*, 72.

Energy Regimes

In this book, I employ the idea of the "energy regime" to capture the inter-dependence of the calorific and the political. Energy regimes are typically thought of as social and economic systems defined by the predominant type or types of energy used. My more expansive employment of the concept incorporates the assemblages of political institutions, technological artifacts, environmental conditions, labor arrangements, market forces, ideologies, and bodies of knowledge and expertise that come together to govern the extraction, transportation, and consumption of energy.[50] This definition has been informed by Gabrielle Hecht's notion of "techno-political regimes." I find the "regime" metaphor helpful for the same reasons Hecht invokes: its reference to governance in terms of the individuals, ideologies, and instruments involved in the exertion of power; the way the term evokes regimen and prescribes normative visions of sociopolitical order; and its conveyance of contestation within and between regimes.[51] Most of all, by adopting the idiom of regime, I underscore the intrinsically political nature of energy—that is, it is both an important objective of and a main means in the exercise of power.[52]

How, then, did the political nature of energy shape the nature of politics? In his study of modern energy networks and the rise of mass democracy, Timothy Mitchell points out how coal brought into being a new sociotechnical system in which energy, consumed at exponentially increasing rates, was transported along narrow channels of water and rail and concentrated in expanding urban centers. The susceptibility of this system to disruptions meant that workers involved in the mining and movement of coal were able to take advantage of weak points through sabotage and strike and seize for themselves several key rights and concessions. Mitchell goes on to argue that these briefly democratic spaces were closed off, however, in the post–World War II era of oil. As flows of energy began to be directed less by human hands and more by carefully calibrated technical structures of pumps and

50. My definition of "energy regime" builds on and extends that of John McNeill, who characterizes it as "the collection of arrangements whereby energy is harvested from the sun (or uranium atoms), directed, stored, bought, sold, used for work or wasted, and ultimately dissipated." See McNeill, *Something New under the Sun*, 297.

51. Hecht, *Radiance of France*, 16–17. Gabrielle Hecht defines technopolitical regimes as "linked sets of people, engineering and industrial practices, technological artifacts, political programs, and institutional ideologies, which act together to govern technological development and pursue technopolitics." My concept of energy regimes also draws upon envirotechnical approaches advanced by scholars such as Sara Pritchard. See Pritchard, *Confluence*, 1–27.

52. Russell et al., "The Nature of Power."

pipes—in part a function of the difference in physical properties between solid coal and liquid oil—so too did workers find themselves increasingly displaced from larger political processes. If the energy regime of coal had, through unintended features of its design, facilitated the rise of modern democracy, then the energy regime of oil was engineered precisely to undermine that promise of participatory politics.[53] The emergence of the fossil fuel regime in East Asia that I trace in this book offers an alternative account of carbon's political economy. From Fushun, we see a path paved by coal that led not so much to democratic possibilities as to technocratic proclivities.

Technocracy is commonly understood as the rule of experts, the governance and control of society or industry by wielders of technical knowledge, skills, and expertise.[54] Technocracy grants science and technology unquestioned primacy in framing and solving the problems of society and encourages the use of statist directives to bring technoscientific methods to bear on problems identified. In this regard, it may be thought of as scientism—the almost religious belief in the power of science—with an operationalization plan.[55] Equating efficiency with efficacy in the name of rationalization, technocracy privileges mechanization and the substitution of objective machine for subjective man and idealizes industrialization as the means by which a society reaches the elusive endpoint that is modernity. In my analysis, technocracy is first and foremost an ideology in the way that, for instance, democracy is, and technocrats are those with technical expertise who hold fast to that ideology, which nevertheless can have wider appeal across populations.[56] While technocracy has never truly been realized in form if one thinks of it purely as a government of engineers, the long twentieth century includes numerous instances when technocratic visions of modernist futures seized the imaginations of states and, in so doing, shaped the societies onto which these visions were projected.[57] In

53. Mitchell, *Carbon Democracy*, 12–31.

54. For a set of cases that together demonstrate the workings of technocracy, even if the concept is everywhere in the text but in name, see Mitchell, *Rule of Experts*. For a Weberian explication of technocratic politics, see Fischer, *Technocracy and the Politics of Expertise*.

55. Richard Olson contends that technocracy has been "one of the most important manifestations of the scientistic tendency during the twentieth century." See Olson, *Scientism and Technocracy*, ix. For a detailed analysis of scientism in the Chinese context, see Shen, "Scientism in the Twentieth Century."

56. For one clear exposition of such a "technocratic mentality," see Putnam, "Elite Transformation in Advanced Industrial Societies," 385–88.

57. Charles Maier has shown how technocracy, alongside scientific management, garnered interest across the political spectrum in interwar Europe because of associated promises of industrial productivity. See Maier, "Between Taylorism and Technocracy."

China and Japan, the ascendency of such technocratic ideals was inextricably linked to the energy regime of coal, materializing in plans and policies to produce and consume increasing amounts of carbon energy and in the mechanical and managerial infrastructures that enabled those efforts.

To Chinese and Japanese states, the impetus for technocratic control over the coal industry that we see so clearly in Fushun came first from the perceived importance of energy resources to their central aims. Many leaders in China and Japan attributed the Western imperial powers' industrial and military successes to their ability to comprehensively exploit fossil fuels.[58] Since the mid-nineteenth century, these powers had leveraged their access to energy to encroach upon the sovereignty of East Asian polities. Partly in response to imperialistic infringements and partly as a way of dealing with mounting local and regional pressures, Chinese and Japanese states aspired toward a kind of industrial modern development—conceived as this was along a teleological trajectory—that would be similar to that of the Western powers in achievement if not approach.

As with many other states that saw themselves as lower down the international pecking order than they would have preferred, their pursuit of development slid into developmentalism, what historian Arif Dirlik has helpfully defined as "an ideological orientation characterized by a fetishization of development, or the attribution to development of the power of a natural (or even, divine) force which humans can resist or question only at the risk of being condemned to stagnation and poverty."[59] Coal may have been driving development, but developmentalism drove an inordinate desire to dominate this carbon resource. Seeking bountiful supplies, Chinese and Japanese states acted to ensure that their domestic coal industries as a whole thrived and, in several instances, involved themselves directly in establishing and running capital-intensive modern coal mines. Their commitment to securing large volumes of coal would deepen with the rise of autarkic thinking during World War I, as the emergent "fuel question" stoked anxieties in East Asia over continued access to carbon resources and incited statist interventions in attempts to quell those fears.

At the same time, if the energy regime of coal was, as Mitchell contends, vulnerable to organized labor, then the desire to ensure a steady stream of that fundamental fuel prompted moves to mitigate those vulnerabili-

58. For an account of how Qing officials came to regard coal as an essential resource through their engagement with Western imperial powers and their embrace of industrialization in the nineteenth century, see Wu, *Empires of Coal.*

59. Dirlik, "Developmentalism," 30–31.

ties. Mechanization presented itself as a compelling technocratic solution. Fossil-fuel-powered machines have a greater capacity for work than human muscle. In theory, adopting them allowed mine operators to reduce the size of the labor force and, in so doing, reduce the risk of unrest and disrupted energy flows. In Fushun, mechanization reached its most visible expression in the open pit. The new environment created by open-pit mining incorporated not only labor-saving heavy machinery, from line dredges to power shovels, but also a workforce that was more visible and, hence, more easily monitored, as miners were brought from the subterranean depths to the surface. The panoptic gaze that Michel Foucault had identified in the regulatory institutions of the modern age extended to this site of extraction, with similarly disciplining effects.[60]

But in the intensification of energy extraction that defined carbon technocracy, states drew much power not only from coal but from people as well. Industrial society's rise depended, after all, on both harnessing fossil fuels and using and abusing human labor in factories and mines.[61] The exploited miner was, in the evocative imagery of George Orwell, "a sort of caryatid upon whose shoulders nearly everything that is not grimy is supported."[62] This was certainly the case at the Fushun colliery. For all the effort they put into mechanization, intended precisely to displace the labor they deemed so volatile, Fushun's managers and engineers nevertheless came to rely on more and more workers as they expanded operations across the coalfield.[63] These technocrats then wielded increasingly invasive and illiberal technologies of control as they attempted to master this growing labor force.

As we proceed into our examination of carbon technocracy's history in East Asia, three qualifications are in order. First, this is not a case of energy determinism. Coal did not inexorably produce a particular political form, be it democratic or technocratic. Rather, I am interested in demonstrating the historical processes through which Chinese and Japanese actors of the not-too-distant past both built institutions to tap coal's carbon energy and directed that energy toward the construction of political possibilities.

60. Foucault, *Discipline and Punish*.

61. The exploitation of labor in industrial society has been examined extensively in both contemporary accounts and historical studies. One of the most penetrating analyses remains Engels, *Condition of the Working Class*.

62. Orwell, *Road to Wigan Pier*, 18.

63. For an account of how nineteenth-century British industrialists embraced coal-fired, steam-powered machines as replacements for human labor and water power, see Malm, *Fossil Capital*, 58–254.

Second, this argument challenges the simple through line that links technocratic governance in East Asia's long twentieth century to enduring discourses of Oriental despotism.[64] There was nothing intrinsic to the cultures of East Asia that made their turn to technocracy inevitable. Instead, I suggest that technocratic tendencies around energy resource management can be found in numerous states across the globe and are, in fact, inherent to the modern condition. Third, although the workings of technocracy are often idealized as that of a well-oiled machine, my account reveals multiple instances of slippage in which even the best-laid plans to control coal's carbon energy were in the end foiled. Carbon technocracy may have taken root in East Asia, but to assume that it bore perfect fruit would be to buy into the fiction of technocracy itself.

East Asia in the Carbon Age

East Asia must be central to the stories we tell of the Anthropocene.[65] This is not merely because China, now the world's second-largest economy, has been for over a decade the world's biggest emitter of carbon dioxide—hardly surprising when we consider that it currently consumes almost as much coal as the rest of the world combined. This is not just a feature of the era of market reform, nor is it simply a continuation of the environmentally devastating mass campaigns of the socialist era that Judith Shapiro has framed as "Mao's war against nature."[66] Rather, as this book shows, the desire to use the transformative power of coal to recast China's fortunes had long been a burning interest of modern Chinese states.[67] If differing access to coal contributed to what historian Kenneth Pomeranz has called the "great divergence" between the economies of China and Europe—which had been up until the nineteenth century progressing on relatively parallel paths—then extensively extracting this resource was one way in which Chinese leaders aimed to close the gap.[68] Similarly, the modern Japanese state saw coal as imperative for the industrial and military development it so desired. Coal fueled Japan's ascendance on the global

64. The representative text of that position is Wittfogel, *Oriental Despotism*.

65. On Asia, more broadly, in the Anthropocene, see Ghosh, *The Great Derangement*; Thomas et al., *"JAS* Round Table on Amitav Ghosh"; and Chatterjee, "The Asian Anthropocene."

66. Shapiro, *Mao's War against Nature*.

67. Robert Marks, Micah Muscolino, and others have argued that the Chinese Communist state's exploitation of the environment should be placed within the longer historical context of developmentalist plans reaching back at least into the early twentieth century if not earlier. See Marks, "Chinese Communists and the Environment"; and Muscolino, "Global Dimensions."

68. Pomeranz, *The Great Divergence*.

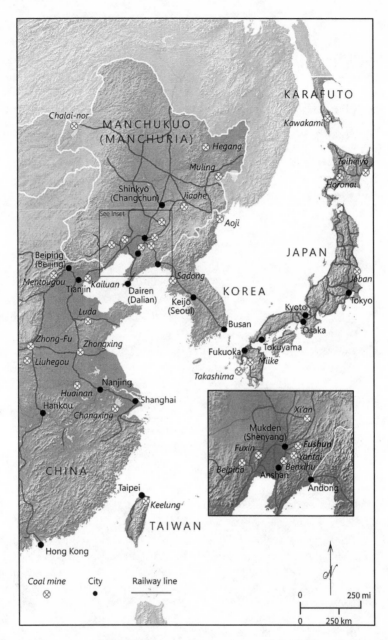

I.2. Major coal mines and railway lines in East Asia, ca. 1935.

stage, reinforcing advances made upon the material and ideological resources of earlier times to engender a "great convergence" between Japan and other imperial powers of the carbon age.[69] It quite literally powered Japan's "total empire."[70] Amid the tumult of the times, coal and the sociotechnical possibilities it drove proved inseparable from Chinese and Japanese visions of industrial modernity.

For China, the first half of the twentieth century was a period of enormous challenge and epochal change. In 1912, the Manchu Qing dynasty collapsed, ending a dynasty that had lasted over two and a half centuries and a dynastic system of more than two millennia. Following the brief presidency of military strongman Yuan Shikai 袁世凱 (1859–1916), China fell into political division and split into regions ruled by individual warlords. In 1928, the Nationalist party under Chiang Kai-shek 蔣介石 (1887–1975), having ostensibly succeeded in reunifying the country by crushing or co-opting the regional warlords, established a national government in Nanjing. Over the next decade, the Nationalist regime undertook ambitious projects of industrial development. Their actions were guided by a technocratic impulse, characterized by historian William Kirby as "an ethos of optimism, not describable or even rational in economic terms, that China could be remade physically, and indirectly economically, by the planned application of international technology under the leadership of homegrown scientific and technical talent."[71]

Two main motivations drove the Nationalists' technocratic endeavors. First, they wanted to complete what was taken to be the unfinished work of industrialization that had begun in the nineteenth century and that was still regarded as the means to the "wealth and power" (富強) necessary for China to survive in a world where only the fittest would. In a manner similar to late Qing "self-strengtheners" confronting the challenges of Western imperial encroachment beginning in the mid-nineteenth century, they understood science and technology as key to this desired industrial transformation.[72] Second, and relatedly, the Nationalists aspired to build up their military-industrial capacity as they prepared to face Japan, which

<hr/>

69. Thomas, "Reclaiming Ground."

70. The term "total empire" was coined by Louise Young. See Young, *Japan's Total Empire*.

71. Kirby, "Engineering China," 152. For other accounts of technocratic development in Nationalist China, see Kirby, "Technocratic Organization and Technological Development"; Pietz, *Engineering the State*, 39–118; Bian, *Making of the State Enterprise System*, 45–100; and Greene, *Origins of the Developmental State*, 14–46.

72. On Chinese engagements with science and technology in the late Qing, especially in relation to industry, see Elman, *On Their Own Terms*, 281–395; Meng, "Hybrid Science versus Modernity"; Wu, *Empires of Coal*, 66–128, 160–79; and Hornibrook, *A Great Undertaking*.

had, by 1932, occupied Manchuria.[73] While the Nationalist regime weathered the storm of the Japanese invasion, it was soon felled by its Communist adversaries, led by Mao Zedong 毛澤東 (1893–1976). In 1949, after a civil war that followed the eight long years of the Second Sino-Japanese War (1937–1945), the Communists brought down Chiang's Nationalist government and declared the founding of the People's Republic of China.

For Japan, the turn of the twentieth century was marked by the anxieties of an ascendant imperial power. Its recent victory in the First Sino-Japanese War (1894–1895) saw it outpace the traditional regional hegemon and suggested that the far-reaching social and political reforms underway since the Meiji Restoration of 1868 could be considered something of a success. Despite this, to Japanese leaders, the shadow of Western gunboat diplomacy and the unequal treaties that had prompted those reforms still loomed large, even as Japan secured another historic win in the Russo-Japanese War and further added to its empire by annexing Korea in 1910. The Japanese approach to imperialism was riddled with the inferiority of one who came late to the table of imperial powers and was made to feel as if they did not quite belong there. This persistent sense of inadequacy spurred the state's ongoing efforts at economic and military development.[74]

As the decades went by, increasing numbers of Japanese came to regard empire and further continental expansion as essential to the country's continued prosperity and security. Manchuria, long valued for its bounty in coal and other resources, became, in popular discourse, a "lifeline" (生命線) that would save Japan from the economic crisis of the late 1920s and early 1930s and supply it with the material foundation for a "national defense state" (國防國家).[75] In 1885, reformer Fukuzawa Yukichi 福澤諭吉 (1835–1901) argued that Japan should "leave Asia" (脱亞), by which he meant that it should part ways with its moribund neighbors, China and Korea, and cast its lot with the "civilized nations of the West."[76] In attempting just that, Japan ended up over time being pulled more assuredly than before onto the Asian continent, from which it seemed increasingly unable to extricate itself until its overextended empire finally collapsed in 1945.[77]

73. Although the Nationalists were also in the late 1920s and early 1930s engaged with fighting resurgent warlords and the Communists, whom they drove underground but did not manage to destroy, their military-industrial development was oriented toward the expected showdown with Japan. See van de Ven, *War and Nationalism in China*, 151–52.

74. Duus, *Abacus and the Sword*, 424–38.

75. Young, *Japan's Total Empire*, 88–95, 130–40.

76. Fukuzawa, "Datsu-A ron."

77. For ways in which Japan's imperial project lived on in East and Southeast Asia after 1945, see Mizuno, Moore, and DiMoia, *Engineering Asia*.

The tectonic shifts of the era completely transformed Manchuria. During the Qing dynasty, this region was designated as the hallowed homeland of the ruling Manchus. It was the "land from whence the dragon arose" (龍興之地), where Manchu emperors worshipped their august ancestors.[78] Although the court strictly regulated Chinese migration to Manchuria, that did not stop thousands from coming and settling in the area.[79] The court relaxed restrictions in the latter half of the nineteenth century in an attempt to prevent tsarist Russia from advancing into this territory on which it had developed designs by further populating it. Over the following decades, millions more would arrive. In the age of empires, Manchuria witnessed, as Owen Lattimore observed, the "invasion of colonists and rivalry of civilizations."[80] Early in the twentieth century, Japan established itself as the dominant imperial power in the region, having grabbed from Russia its leased territory and railway and mining concessions in southern Manchuria. It was at this time that Fushun's coal mines fell into Japanese hands. After the collapse of the Qing, Manchuria came under the control of warlord Zhang Zuolin 張作霖 (1875–1928), a bandit of humble origins who became the region's "uncrowned king."[81] For a spell, Zhang and the Japanese shared a symbiotic relationship. Between them, Manchuria was remade.[82]

The Manchurian economy grew rapidly up to and through the 1920s, as railway and port infrastructure expanded, more land came under the cultivation of cash crops (most notably the soybean), and new factories and mines churned out finished products and raw materials for regional and global markets. Japan was heavily implicated in this increase. By 1932, almost two-thirds of the region's industrial capital was Japanese.[83] Although the Japanese initially relied on Zhang to protect their interests—supplying him with arms and capital in return for his support for their presence in Manchuria—they eventually tired of his appeals to anti-Japanese nationalism and attempts to undermine their economic endeavors by backing com-

---

78. On the processes by which the Qing court made and maintained the ideal of Manchuria as the venerated homeland of the Manchus, see Elliott, "The Limits of Tartary," 607–14; and Shao, *Remote Homeland, Recovered Borderland*, 25–67.

79. Chinese settlers in southern Manchuria during the Qing contributed to the growth of its agrarian sector, which exported grain and soybeans to other parts of China even before the opening of the treaty ports. See Isett, *State, Peasant, and Merchant*, 211–38.

80. Lattimore, *Manchuria*, xi. On Manchuria as a region marked by the intersection of imperial, national, and transnational interests in the modern era, see Tamanoi, *Crossed Histories*.

81. "New Cabinet in Peking."

82. For more on Zhang Zuolin's regime, see McCormack, *Chang Tso-lin in Northeast China*; and Suleski, *Civil Government in Warlord China*.

83. Duara, *Sovereignty and Authenticity*, 48–49.

peting enterprises. On June 4, 1928, the Kwantung Army (關東軍), Japan's garrison force in Manchuria, assassinated Zhang by blowing up his train on the outskirts of Mukden (present-day Shenyang). Later, on September 18, 1931, the Kwantung Army engineered the Mukden Incident that led to Japan's invasion of Manchuria and the establishment, in 1932, of its client state of Manchukuo.

Manchuria was pivotal to the rise of carbon technocracy in East Asia. The region's unparalleled hoard of coal riches fueled Chinese and Japanese developmentalist dreams. To Nationalist China, Manchuria was a territory that had to be recovered, if nothing else than for the coal and other resources that could go toward its industrialization drive. To imperial Japan, Manchuria was a laboratory of empire where various technoscientific projects served as materializations of what historian Aaron Stephen Moore termed the "technological imaginary"—"the ways that different groups invested the term 'technology' (*gijutsu*) with ideological meaning and vision."[84] At the Fushun colliery, scientists, engineers, industrialists, military men, and planners realized several firsts in science and technology for the Japanese empire, including open-pit mining and shale oil distillation, techniques through which the energy that powered Japan's imperial machinery would be more thoroughly exploited. These experiments in extraction bequeathed to the Chinese Communist state some of the means by which it pursued its ideal of progress. In exploring the emergence of the energy regime of coal in Fushun and Manchuria, I highlight the important part that Japan played in the coal-fired industrial transformation of post-1949 China.[85] The presumption that this was necessarily a positive legacy is something this book calls into question.

In the chapters that follow, I track carbon technocracy's development in China and Japan through a mainly diachronic account of Fushun's history. Our narrative begins at the dawn of the twentieth century, when modern mining technologies first cut into "dragon veins" (龍脈) to extract coal from this site. It ends in 1960, at the height of the Great Leap Forward, when it became clear that the Chinese Communist state, like its predecessors, would privilege production over people in its pursuit of fossil-fueled

---

84. Moore, *Constructing East Asia*, 3. Other important studies of Japanese technocracy in Manchuria and the wider empire include Mimura, *Planning for Empire*; and Mizuno, *Science for the Empire*.

85. The parallels here with how Japan served as a primary vector for the rise of hygienic modernity in treaty-port China, as Ruth Rogaski has shown, point to shared ways in which transnational circulations of science and technology were crucial to the making of modernity in its variegated instantiations. See Rogaski, *Hygienic Modernity*.

industrialization. Centered on Fushun and attentive to how imported technologies, migrant labor, transnational expertise, and colonial management came together to give rise to its coal-mining industry, my analysis further situates this remarkable enterprise within the larger Chinese and Japanese energy regimes of which it was a linchpin. This I do in part by tracing how Fushun coal circulated through the systems it fueled, from power plants in Shanghai to textile mills in Osaka, darkening the skies over these urban industrial hubs. I also explore how Fushun often featured prominently in discussions about energy among state planners and metropolitan elites at the centers of calculation, particularly in the context of fuel anxieties that threatened their developmentalist preoccupations.

Grounded in an expansive archive spanning collections across China, Japan, Taiwan, and the United States, with sources that include company records, technical reports, trade journals, literary works, travelogues and diaries, worker oral histories, and the personal papers of mining engineers, these chapters together furnish a connected yet comparative examination of how Chinese and Japanese states staked their fossil-fueled futures on Fushun coal and mobilized multiple extractive technologies toward its exploitation.

While the story of carbon technocracy that I tell in these pages may be set in East Asia, the narrative arc stretches across the history of the wider industrial world.[86] In *The Road to Wigan Pier*, George Orwell writes, "Our civilization . . . is founded on coal, more completely than one realizes until one stops to think about it. The machines that keep us alive, and the machines that make machines, are all directly or indirectly dependent upon coal."[87] But coal did more than just help reorder the material world; it enabled transformations in social and political ones too. If our civilization is indeed founded on coal, we need to consider what kinds of edifices we have built on this coal foundation.

86. My overall approach in this book has been to situate shifting Chinese and Japanese experiences within concurrently evolving global contexts. Although I engage in several broader comparisons along the way, I mainly focus, in those instances, on the differentiation between and within the East Asian polities under study. I am generally more interested in the connected and contemporaneous transnational developments that I regard as having defined the fossil fuel age, carbon technocracy, and the constitution of linear and singular ideas of progress amid the unevenness and inequalities of the industrial modern world. Some of the arguments advanced by historians of science and technology that have informed this approach include Anderson, "From Subjugated Knowledge to Conjugated Subjects"; Bray, "Only Connect"; Fan, "Modernity, Region, and Technoscience"; Raj, "Beyond Postcolonialism . . . and Postpositivism"; and Roberts, "Situating Science."

87. Orwell, *Road to Wigan Pier*, 18.

It may very well be, as Lewis Mumford has suggested, that "the mine is the worst possible base for permanent civilization." His rationale was that "when the seams are exhausted, the individual mine must be closed down, leaving behind debris and deserted sheds and houses." Although Fushun's coal deposits have not been entirely depleted, difficulties in continuing mining operations have brought the former Coal Capital to essentially the same point. "The byproducts are a befouled and disorderly environment," Mumford concluded, "the end product is an exhausted one."[88] As I traveled through Fushun that first summer and in trips back after, my eyes would be repeatedly drawn to the mounds of dusty rubble strewn across this de-industrializing landscape, particularly around the individual mines that have closed down in succession. We who live in this world that carbon made have yet to use up all the "buried sunshine" beneath our feet. But we see similar effects, a result not so much of coal's exhaustion as its un-relenting use. As we now begin to reckon with the harrowing devastation wrought by climate change, Fushun's fate appears, then, to be nothing less than a chilling microcosm of this most pressing of our planetary problems.

88. Mumford, *Technics and Civilization*, 157.

ONE

Vertical Natures

Late in the spring of 1927, Liu Baoyu wrote to the Fushun magistrate seeking redress. Liu had owned two *mu* of land on the eastern hill of the Guchengzi district in southwest Fushun.[1] Acquired by his late father half a century earlier, the land had been set aside as a family burial ground on which four tombs were raised. For several decades, Liu's family visited those graves to pay respects and offer sacrifices to their dead relatives. A few years back, however, the South Manchuria Railway Company (Mantetsu) began buying up property from villagers in the area to convert into mining lands, and Liu sold the plot and the rest of his estate and moved to another village some five miles away.[2]

According to custom, Liu retained the right to worship at the tombs.[3] Although his visits became less frequent, he made a point of returning for special observances. Earlier that spring, Liu traveled back to the tombs for the Qingming festival—the annual day of ancestral remembrance.[4] As he approached, he spotted from afar "lofty coal machine factories standing tall." When he came closer, he saw "not a trace of the graves in sight." Liu asked around and learned that they had all been dug up by the Japanese. A search for the coffins and the remains yielded nothing.[5]

"Just recounting it makes me ache," Liu lamented. "If the colliery wanted to use a piece of land . . . and if there were graves on it, it should

---

1. One *mu* is about a third of an acre.

2. "Jumin Liu Baoyu cheng," 222.

3. Qian Nanjing Guomin zhengfu Sifa xingzheng bu, *Minshi xiguan diaocha*, 28.

4. Liu did not specify that his visit was during the Qingming festival, but his petition to the Fushun magistrate was on May 17, 1927, and the Qingming festival, which is the fifteenth day of the spring equinox, fell on April 4 that year, making it likely that it was so.

5. "Jumin Liu Baoyu cheng," 222.

have noted the number of graves, offered compensation accordingly, and then set a time for those graves to be relocated," he contended, "Instead, it secretly dug up these graves and disposed of the bones without a trace. . . . This is an offense to reason and an affront to justice. How can I, your humble servant, bear it?"[6] The archival record is silent as to how Liu's case was resolved. It is probable that the company paid him a small sum, as it did with a similar instance in 1929, when it offered another Fushun local ten yuan for each grave it destroyed while clearing land for further development.[7]

There are several ways that we can read Liu's case. One is in terms of how the Japanese colonial company—which, by that time, had been operating for two decades—disrupted the lives of ordinary people in this Chinese locale. This reading would be consistent with the now conventional narrative of Japan's empire building in Manchuria.[8] In Fushun, these disruptions extended beyond disturbing the remains of the departed to visiting acts of violence upon the living. Another way that we may read the case is in terms of local Chinese agency. Liu knew well the standard procedure through which Mantetsu was supposed to deal with graves on land that it had acquired. He did not hesitate to call the company out on its failure to follow that procedure. Throughout this period of Mantetsu management, there were numerous episodes in which the Chinese of Fushun attempted to check the oftentimes audacious actions of this colonial enterprise by appealing to terms the company itself had reached in negotiations with Chinese authorities. This effort to contest from below based on rules set above mirrored the experiences of many colonial and semicolonial subjects pushing back against imperialistic incursions.[9] Yet as the years went

6. "Jumin Liu Baoyu cheng," 222–23.
7. Fushun colliery manager to Fushun magistrate (July 22, 1929), in *DBDX*, 27:130–31. The company typically paid out five yuan of "moving fees" per grave.
8. The literature on Japanese imperialism in Manchuria has swelled in the past two decades. An instant classic that inspired many of these studies is Young, *Japan's Total Empire*.
9. See, for instance, Cederlöf, "Agency of the Colonial Subject." Fushun as a site and its Chinese populace as subjects may be considered semicolonial in a basic sense. Japan's jurisdiction in Fushun was technically contained to the concession, which it held with the consent of Chinese authorities. But, as we will see, it had seized and refused to let go of that concession in spite of sustained objections from Chinese officials and frequently reached beyond the concession's boundaries to encroach upon Chinese territory and subjects. That is to say, Japan often behaved like an occupying colonial power in Fushun even if it was not one on paper. Under this framework, the creation of Japan's client state of Manchukuo in 1932, in formally ending Chinese sovereignty in Manchuria, might then be regarded as the point when the semicolonial becomes colonial (even if Manchukuo was ostensibly an independent nation). I think that this simple designation of semicolonial versus colonial is reasonable enough insofar as it does not

by, Mantetsu met fewer remonstrations with restitution, as the company increasingly conducted itself with impunity within and around Japan's concessionary space.

At the same time, unearthing graves to mine coal represented a specifically vertical engagement with the land that marked the emergence of a modern extractive enterprise under the energy regime of carbon technocracy. In his examination of Canadian geological surveys in the late nineteenth century, geographer Bruce Braun argued that state geologists, in probing "the 'inner architecture' of the earth," contributed not only to "such places as Canada's west coast [being] drawn into global circuits of extractive capital" but also to the construction of an intelligible nature with depth that could be subjected to governance.[10] Japanese geologists, occasionally with the help of Chinese surveyors, similarly rendered Fushun's "vertical territory" legible, not only by enumerating the expanse of coal that lay beneath but, just as importantly, by making a case for its value. In so doing, they "opened a space—simultaneously epistemological and geographical—that could be incorporated into forms of political rationality."[11] The depths of the earth were no longer for depositing the dead if there was mineral wealth to be withdrawn.

In this chapter, I explore how Japan came to own and operate the coal mines of Fushun and how this process of establishing a site of energy extraction rested on expressions of verticality both literal and metaphorical. Japan may not have arrived in Fushun on account of coal, but it stayed mainly because of it, raising a technocratic edifice through which it mined the prized black rock. Underlying this edifice were arrangements of land, labor, and locale that would be reconfigured along the way. Beginning with the geological and human history of coal in the region before carbon technocracy, I trace how modern coal mining began in Fushun at the turn of the twentieth century through Chinese, Russian, and then Japanese efforts. I follow, in particular, the practices through which Japanese imperialists determined Fushun's worth and orchestrated its exploitation, from geological surveys to diplomatic negotiations to labor recruitment. Together,

---

detract from the fact that there was often little difference for those subjected to the exercise of colonial power and violence. On colonial violence in semicolonial China, see Hevia, *English Lessons*. For more on semicolonialism as an analytic in understanding China's interactions with imperial powers and its role in constituting the uneven geographies of global capitalism, see Osterhammel, "Semi-colonialism and Informal Empire"; Barlow, "Colonialism's Career"; Goodman and Goodman, "Introduction"; and Karl, *The Magic of Concepts*, 113–40.

10. Braun, "Producing Vertical Territory," 22–24.
11. Braun, 28.

these activities and their impact on the local environment and its inhabitants reflected and reproduced a series of stacked hierarchical relationships: between Chinese and Japanese, labor and management, community and concession. How foundational these variations of verticality were to the rise of carbon technocracy is the central focus of this chapter.

Deep History

Mantetsu established its coal-mining enterprise in Fushun upon a deeper geological and human history, the long past before carbon technocracy. This was a history Japanese empire builders in Manchuria knew little about before they acquired the concession. According to geologist Asada Kamekichi, a member of a small team of experts the army sent to take charge of the nearby Yantai coal mines in the midst of the Russo-Japanese War, they had not even been aware of the riches of Fushun's coal hoard. Hearing rumors of coal in Fushun, the team dispatched several Chinese surveyors to the site. These men returned with a fine coal sample and reported that "the coal in Fushun is of good quality and the seams are more than thirty feet thick." The Japanese experts initially disbelieved the men and "rebuked them for their deceit," assuming that the sample was actually Japanese coal that they had picked up at the nearby port of Yingkou. Follow-up investigations more than verified the Chinese surveyors' claims.[12] As the Japanese began exploring the possibility of exploiting Fushun's resources, they found that they had to familiarize themselves with aspects of local history and the landscape on which that history was inscribed.

Fushun sat squarely in the middle of southern Manchuria, a region centered on the Liao Valley that climbed west to the vast Mongolian Plateau and east to the Changbai Mountain Range, which ran along the border of the Qing empire and Chosŏn Korea.[13] It was more than 450 miles northeast of Beijing, the empire's seat of power, and just under 40 miles to the east of the former Manchu capital and Qing Manchuria's largest city, known at that time as Mukden in Manchu and Shengjing (盛京) or Fengtian (奉天) in Chinese.[14] The territory that made up Fushun was divided by the Hun River, which flowed from east to west, passing Mukden before joining the Liao River and emptying into the Bohai Gulf. To the north of the river lay

---

12. Manshikai, *Manshū kaihatsu yonjūnen shi*, 34–35.

13. Lattimore, *Manchuria*, 14.

14. Zhao, Cheng, and Li, *Fushun xian zhi lüe*, "Shanchuan biao di er" [Second section on landscape], 1a.

1.1. Japanese picture map of Mukden and its environs, with Fushun in the upper right-hand corner. Titled *Hōten meisho zue: fu Bujun meisho* [Picture map of Mukden's famous sites: Including Fushun's famous sites], this map was published in 1930. It depicts both of these urban centers, the railway lines between and around them, several hamlets and villages, and a number of places of general interest, such as the imperial tombs. For instance, to the left of where Fushun is on this map, we have Fuling, the tomb of Nurhaci, the founding father of the Qing empire. (Image courtesy of the Harvard University Library Map Collection.)

relatively flat plains upon which stood Fushun's walled city.[15] On the other side, the topography was more textured, with rugged hills and low mountains that wrapped around to the east.[16] It was in this expanse south of the river that the coal-mining operations that would remake this terrain began. At the turn of the century, the seams of coal were still mostly hidden beneath mudstone and shale, occasionally peeking out in outcrops exposed by weathering and erosion. Dotting the landscape were several villages and hamlets from which the coal mines were to draw their first laborers.[17]

While its physical geography may have provided hints of Fushun's rich subterranean resources and the human settlements the means to their ini-

15. According to the 1911 local gazetteer, the county seat was home to but 371 households or 2,399 people. This was out of the 23,907 households or 165,699 people across Fushun county. See Zhao, Cheng, and Li, *Fushun xian zhi lüe*, "Cuntun hukou biao" [Section on village and hamlet households], 1a, 15b.

16. Zhang and Zhou, *Fushun xian zhi*, 5.

17. Zhao, Cheng, and Li, *Fushun xian zhi lüe*, "Cuntun hukou biao," 1a–15b.

tial excavation, the origins of Fushun coal were buried far deeper under layers of its geological past. This story begins some thirty-three to sixty-six million years ago, during epochs geologists call the Paleocene and the Eocene. At that time, as one account goes, "prehistoric animals such as the saber-toothed tiger, the shovel-mouthed mastodon and the dynosaurus" wandered around this corner of the Eurasian landmass. The area that later became the Fushun coalfield was a large freshwater lake fed by "a torrential stream that flowed through dense forests of pine and spruce." This verdant mix of evergreen and deciduous trees and shrubs thrived in the warm and wet climate of the era. The plants took in energy from the sun and, through photosynthesis, turned this solar energy into the chemical energy they needed for their survival and growth. When these plants died, their remains ("vast quantities of timber, logs, boughs, roots, leaves, pine-cones") fell into the rushing waters and were carried to the lake, where they sank to the murky bottom. Along the way, microbes consumed a portion of this plant matter; the rest reached anaerobic depths, where decomposition continued more slowly in the absence of oxygen. As time passed, the sunken plant material accumulated and became peat, the dark and damp remains of organic life and energy. The layers of peat were further overlain by sedimentary and volcanic deposits. Through increased pressure from the burden above and the accompanying geothermal heat from below, the peat, over millions more years, became coal. Closer to the surface, the Hun River would be all that remained of "the original mighty torrent and lake that co-operated in the laying of the coal deposit."[18]

In contrast to the millions of years it took for Fushun coal to form, its discovery and use by the people of the region was only a matter of millennia. One local legend dates the finding of coal in Fushun to the Han dynasty (206 BC–AD 220) some two thousand years ago. According to this legend, there was once a poor man who had gone to the mountains to collect firewood. There, he chanced upon an injured fox, who immediately took flight, scampering back into its hole. In an attempt to get to the wounded animal, the man began digging around the narrow opening of the burrow, but when he widened it enough to look inside, he could only see a band of black and lustrous rock. The next day, the man went back to the site and started a fire at the mouth of the hole, presumably to smoke the fox out. The rock, much to his surprise, quickly caught fire. From then

18. Powell, "The Marvellous Fushun Colliery"; Johnson, "Geology of the Fushun Coalfield," 221, 225–27; McElroy, *Energy*, 107–9.

on, the man returned regularly to carve out pieces of the rock to burn as fuel. This rock, the story goes, was none other than Fushun coal.[19]

Although this tale of the coal-filled foxhole may be apocryphal, archaeologists have suggested that it was indeed during Han times that Fushun's inhabitants first used coal as fuel. The evidence rests in the remnants of burnt coal found alongside copper coins, clay tiles, and other artifacts in a Han domicile unearthed in the Yongantai district in central Fushun. Coal was certainly used by the people of the Parhae kingdom (698–926) a few centuries later when that precursor of the modern Korean state extended into southern Manchuria. Excavated pottery shards and the remains of kilns and old mining pits dating from the seventh to the tenth century point to coal being mined and burned in Fushun to fire the ceramics characteristic of the regional culture.[20]

The early history of coal use in Fushun is consistent with what we know about the use of coal in premodern China more generally. Tracing the origins of coal use in China has long been a challenge for scholars given the ambiguity of terminology in the textual record: *mei* (煤), the more frequently used of the two characters in the word for coal today, was not commonly recognized as a term for coal until at least the fourteenth century; the other character, *tan* (炭), could, and often did, also refer to charcoal.[21] Archaeological digs have, however, uncovered coal and coal briquettes in the excavations of iron-producing sites dating from the beginning of the first millennium.[22]

Coal was central to the industrial expansion of the Northern Song dynasty (960–1127) a few centuries later, when iron production reached a scale that historian Robert Hartwell argues "probably was not equalled anywhere in the world until the Industrial Revolution of the nineteenth century."[23] The production of iron in the Northern Song had initially relied on charcoal for fuel. However, as the iron industry grew and urban centers, with their heating and construction needs, also expanded, pressures on timber resources mounted. North China faced deforestation, and charcoal fell in short supply. The Northern Song's frontier strategy further compounded this fuel shortage by calling for restrictions on cutting down

19. *FKZ*, 25–26.
20. *FKZ*, 26–27.
21. Ting, *Coal Industry in China*, 1; Read, "Earliest Industrial Use of Coal," 125.
22. Golas, *Science and Civilisation*, 190–94. Archaeologists have also excavated ornaments carved out of lignite that date back to almost 4000 BC, but there is no material evidence of coal having been used as fuel until later.
23. Hartwell, "Cycle of Economic Change in Imperial China," 123.

trees in order to maintain sufficient forest cover to slow down the advance of mounted enemies along its northern border. To keep up the prodigious pace of iron production, smelters switched, in the eleventh century, to firing their furnaces with coal or, more precisely, coke—bituminous coal heated in an oven under low-oxygen conditions to remove impurities so that it burns hotter, faster, and with less smoke.[24]

Beyond the iron industry, coal was used, by the late imperial era, as a fuel in smelting other metals like copper and lead, calcinating lime, making bricks and glass, brewing alcoholic beverages, and boiling brine to produce salt.[25] The seventeenth-century scholar Song Yingxing 宋應星 (1587–1666) detailed many of these industrial processes in his famous 1637 encyclopedia of technologies and crafts, *The Works of Heaven and the Inception of Things* (天工開物). In a section on coal, Song described ways to work coal seams by sinking shafts and erecting subsurface supports. Of note is his assertion that "once the coal is extracted, the well can be filled with earth, and in twenty or thirty years, the coal would regrow—it cannot be exhausted."[26] While one might now scoff at such a claim, the patterns of coal production and consumption into and through the modern era have been predicated on essentially similar notions of inexhaustibility.

Aside from industrial purposes, coal was mined for household use. Northern Chinese homes often burned coal to heat the *kang* (炕), a raised platform built atop a stove that, when warmed, served as a comfortable surface to sit or sleep on in the chill of the long winter months. "These stoves are made of Brick like a Bed or Couch three or four Hands Breadth high, and broader and narrower according to the number of the Family. Here they lie and sleep upon Matts or Carpets; and in the day time fit to-gether either upon Carpets or Matts, without which it would be impossible to endure the great Cold of the Climate," recorded Jesuit observer Gabriel de Magalhães in the late seventeenth century. "On the side of the Stove there is a little Oven where in they put the Coal, of which the Flame, the Smoak and Heat spread themselves to all the sides of the Stove, through Pipes made on purpose, and have a passage forth through a little open-ing, and the Mouth of the Oven, in which they bake their Victuals, heat their Wine and prepare their *Cha* or *Thê*; for that they always drink their tea hot." These pleasures were not without their perils: "the Heat and Smoak

24. Hartwell, "Revolution in the Chinese Iron and Coal Industries," 159–61; Golas, *Science and Civilisation*, 195.

25. Ting, *Coal Industry in China*, 4; Golas, *Science and Civilisation*, 197.

26. Song, *Tiangong kaiwu*, 283. On Song and his encyclopedia, see Schäfer, *Crafting of the 10,000 Things*.

of which are so violent, that several Persons have been smother'd therewith; and sometimes it happens that the Stove takes Fire, and that all that are asleep upon it are burnt to Death."[27] Many, nevertheless, braved those dangers. Their survival through the seasons depended on it.

Although coal saw a range of uses across China before the so-called industrial age, consumption of this fuel was very much bound by place. Difficulties in transportation, due in large part to coal's bulk and weight relative to its market value, meant that coal tended to be the primary fuel only in areas situated near coalfields. Mines close to rivers may have been able to move coal by boat or barge many miles away at somewhat reasonable cost, but overland haulage was extremely expensive. Even in instances when water transport was possible, the coal often had to be carried some distance from the mine to the river. In one account from the mid-nineteenth century, coal from a mine in North China sold at the Yellow River just over thirty miles away for five times its price at the pithead. At a broader level, there was an issue of geographical distribution across the empire. Coal deposits were concentrated in the North, while riverine transport networks were clustered in the South. As historian Tim Wright has argued, "these high transport costs were the single most important constraint on the growth of coal consumption."[28]

Coal did not, however, seem to feature too prominently in the history of Fushun after the Parhae period. During the Ming dynasty (1368–1644), Fushun was part of a defense system along the empire's northeastern border, as reflected in its name, which it acquired during this period. "Fushun" comes from combining two characters from the words "to pacify" (fusui 撫綏) and "to guide" (shundao 順道) in the phrase "to pacify the frontier, to guide the barbarian peoples" (撫綏邊疆, 順道夷民). The "barbarian peoples" (夷民) here referred to the Mongols, who ruled over the preceding Yuan dynasty (1234–1368) and still exerted a measure of influence in the region into the 1380s, and to the Jurchens, a Tungusic people who had long inhabited the fields and forests of Manchuria.[29]

The naming of Fushun may have been aspirational, but it foreshadowed early Ming success at quelling challenges to its authority along that border. The walled city of Fushun was built in 1384; three years later, the Ming launched a military expedition that resulted in the surrender of the Mon-

---

27. de Magalhães, *New History of China*, 10–11. See also Handler, *Austere Luminosity*, 166–67.

28. Wright, *Coal Mining in China's Economy and Society*, 9.

29. *FKZ*, 2; Wang, "Ming xiu Fushun cheng."

gol chieftain Nayaču, who had been gathering troops and expanding his control in southern Manchuria. Other Mongols and Jurchens would submit to the Ming soon after.[30] Fushun became one of the designated sites where the Ming Chinese traded with the Jurchens. Officially, these sites hosted horse markets in which the Ming cavalry obtained mounts from Jurchen breeders in exchange for ironware and textiles, but they also drew Han Chinese merchants interested in purchasing from Jurchen traders all manner of "wild things," from ginseng and mushrooms to sable pelts and freshwater pearls.[31]

In the early seventeenth century, a Jurchen chieftain by the name of Nurhaci (1559-1626), who had enriched himself by this lucrative trade, rose up against the Ming. Having unified the various Jurchen tribes—whom his successor Hong Taiji would in 1635 collectively rename "Manchus"— Nurhaci began his campaign in 1618.[32] The first Ming outpost Nurhaci's forces seized, attacking under the pretense of a scheduled horse market, was Fushun.[33] Fushun's city walls were destroyed in that assault; when they were rebuilt in 1657, the flags that flew from the ramparts were those of the Manchu Qing dynasty.[34]

The historical record does not reveal the extent of coal mining in Fushun over the Qing period. The court did, however, ban all mining activity in the area during the reign of the Qianlong emperor (r. 1735-1796), citing geomantic reasons. Fushun was located near three imperial mausoleums, including that of Nurhaci, who had died before the conquest of China and was posthumously recognized as the dynasty's founder. The court claimed that digging for coal would sever the "dragon veins" pulsing power between the sacred Changbai Mountain Range and these imperial tombs. That the Daoguang emperor (r. 1820-1850) had to issue another prohibition against mining in Fushun in 1838 suggests, though, that such activity had persisted at least to some degree.[35] Even when the Qing government opened Manchuria to mineral exploration and extraction in

30. Rossabi, "Ming and Inner Asia," 258-59.

31. Rawski, Early Modern China and Northeast Asia, 76. Manchuria continued to be a region from which natural resources were extracted for monarch and market during the Qing dynasty. See Schlesinger, World Trimmed with Fur.

32. On the renaming of the Jurchens, see Elliott, The Manchu Way, 71-72.

33. On the fall of Fushun, see Wakeman, The Great Enterprise, 59-62; and Swope, Military Collapse of China's Ming Dynasty, 11-13.

34. Zhao, Cheng, and Li, Fushun xian zhi lüe, "Jianzhi lüe" [Brief account of construction and establishment], 1b.

35. FKZ, 27.

1896, it stipulated that sites along the "dragon veins"—like the Fushun coalfield—remained out-of-bounds.[36] If there was mining in Fushun at the end of the nineteenth century, it was done surreptitiously and on a small scale by locals who had to turn to the black market to peddle the little coal they unearthed.[37]

Japanese accounts of Fushun's deep history placed particular emphasis on this last period, when the Qing banned mining because of concerns over graves and geomancy. In his 1908 survey of Japanese coal mining, industry chronicler Takanoe Mototarō referred to this matter of mines and mausoleums as one of "superstitions of *fengshui*" (風水の迷信).[38] Such a characterization of prior practice suggested that the Chinese of the recent past had failed to make full use of Fushun's valuable coal resources because of backward thinking, shackled as they were by superstitious beliefs.[39] Japan's ongoing development of the coalfield, in contrast, was cast as right, rational, and scientific. Charges of under- or misutilization of lands and their resources were a common tactic by which colonizing powers justified their reach into other territories.[40] This was certainly so with Japan in Fushun. In this way, then, Japanese imperialists placed themselves in a vertical history of largely untapped potential that they were now all too ready to exploit.

Geo-Politics

Modern coal mining in Fushun had, however, preceded the arrival of the Japanese, beginning under Chinese auspices a few years earlier at the turn of the twentieth century. This development was a product of both the broader expansion of coal mining in China that started around the mid-nineteenth century and the more specific opening up of southern Manchuria following the Boxer Uprising (1899–1901). Soon after their establishment, Fushun's coal mines attracted Russian investment. This became the basis for Japan to claim these mines as part of its spoils of battle following the Russo-Japanese War. Insofar as they acknowledged this period of Fushun's history, Japanese accounts stressed the Russian backing of the nascent Chinese enterprises so as to defend Japan's occupation of the mines

36. "Moukden: Official Sanction of Mining."
37. Yang, "Fushun meikuang kaicai," 2.
38. Takanoe, *Nihon tankō shi*, 444.
39. Such a characterization appeared in non-Japanese writings too. For example, see "Coal and the Manchu's Ghost."
40. See, for instance, Whitt, *Science, Colonialism, and Indigenous Peoples*, 10–14.

even as Japanese imperialism began to remake the region and created the conditions for the materialization of carbon technocracy.

On October 8, 1901, Mukden General Zeng Qi 增祺 (1851–1919), the highest ranking official in Manchuria, submitted a memorial to the throne requesting that mining be permitted in Fushun. He had been approached by Weng Shou 翁壽 (1859–1925) and Wang Chengyao 王承堯 (1865–1930), two merchants independently interested in excavating coal mines in Fushun. Both men were also expectant officials—holders of degrees obtained through examination or, just as likely, purchase who were technically eligible for appointment in the imperial civil service—which suggested status and wealth.[41] Zeng was writing to back their petition. Recognizing that the tombs might still be a point of concern, the Mukden general tried to assure the court that he had looked into the matter and that Fuling, the founding emperor's mausoleum, was over forty *li* away from the proposed mines.[42] This, he noted, was farther than the thirty *li* stipulated as the minimum distance between the tombs and the railway lines being laid in the region. As such, there was no danger of disturbing the imperial resting place. The opening of the mines, Zeng reasoned, would be in the state's best interest. During the recent Boxer Uprising, the "flames of battle" had swept through the province, and now, in its wake, the local government needed funds for reconstruction. Having merchants like Weng and Wang pay sizable service fees to open and operate businesses in the area presented a way to help finance these efforts.[43]

At the time, coal mining was an important part of China's small but growing industrial sector. With the establishment of the treaty ports, Chinese officials, often together with local businessmen in "government-supervised, merchant-operated" (官督商辦) arrangements, raised capital to open large, mechanized mines to supply the mounting demand for coal. The most prominent of these was the Kaiping colliery, established in 1877 by elder statesman and Viceroy of Zhili Li Hongzhang 李鴻章 (1823–1901) and merchant Tang Tingshu. At the same time, beginning with the German seizure of Jiaozhou Bay in 1898, foreign powers started more aggressively acquiring mining concessions and opening mines in their expanding spheres of influence across China. Foreign involvement in the industry also came in the form of mining enterprises that were jointly owned and oper-

41. On the sale of office in late imperial China, see Kaske, "Price of an Office"; and Zhang, "Legacy of Success."

42. One *li* is about a third of a mile.

43. *BJTK*, 9–12.

ated by Chinese and foreign business interests. Because of their greater access to capital and the latest in mining technology and expertise, mines that were wholly or partially foreign owned tended to be more productive. Between 1898 and 1906, a national movement to recover mines and mining rights from foreign powers gained traction, driven largely by the nationalist sentiments of local gentry, who bought up a number of foreign and Sino-foreign mines with Chinese capital.[44] Still, even as China's coal-mining industry grew over the successive decades, foreign and Sino-foreign mines ruled the roost. Fushun and its closest competitor, the Sino-British Kailuan Mining Administration (formed between Kaiping, which came under British control around the turn of the century, and the Chinese Luanzhou coal mines), together accounted for at least half of China's annual coal output from large mines in the period before the Second Sino-Japanese War.[45]

While transportation challenges had impeded the growth of the coal sector in the past, the construction of railway networks allowed it to flourish from the late nineteenth century onward. What is typically regarded as China's first railway was built in 1881 to transport coal from Kaiping to the nearby Lutai canal, where it was then ferried over water to markets downstream.[46] By all accounts, lowering transport costs through the railway was one major factor in the coal industry's expansion. Moreover, coal was not only one of the most important freight goods the railway carried but also the very fuel on which its engines ran. The railway and the energy regime of coal were so interdependent that the foreign powers on Chinese soil typically secured railway and mining concessions in one bundle, the right to lay tracks often coming with the permission to excavate mines.[47]

Such was the case with the Russians in Manchuria. In 1896, the Russian government, represented by finance minister Sergei Witte and foreign minister Aleksey Lobanov-Rostovsky, concluded a secret agreement with Li Hongzhang, who had by that time become the Qing empire's most in-

---

44. Li, *WanQing de shouhui kuangquan*; Lee, "China's Response to Foreign Investment."

45. Wright, *Coal Mining in China's Economy and Society*, 78, 117–18. On the British takeover of Kaiping and the establishment of Kailuan, see Carlson, *The Kaiping Mines*, 47–113. For a history of Kailuan until just before the Second Sino-Japanese War, see Yun, *Jindai Kailuan meikuang yanjiu*.

46. The first railway was actually built in 1876, and it ran between Shanghai and Wusong. However, the Viceroy of Liangjiang, Shen Baozhen, saw this as a sign of foreign encroachment, took over the line, and had it dismantled in 1877. On the development of railways in China across the long twentieth century, see Köll, *Railroads and the Transformation of China*.

47. The Jiaozhou convention of 1898, for instance, gave the Germans the right to mine coal within thirty *li* of the railway lines that extended from their leased territory in southern Shandong. See Ting, *Coal Industry in China*, 13.

fluential diplomat. This agreement allowed Russia to construct a railway line through Manchuria in exchange for the promise of Russian support against Japan, to whom China had just lost a war. The trunk line of this Chinese Eastern Railway ran almost a thousand miles from the Siberian city of Chita to the port of Vladivostok, cutting across northern Manchuria along its way.[48] In 1897, Russia seized Lüshun and Dalian in southern Manchuria, secured a lease from the Qing government for these two ports, and successfully negotiated the rights to build a north–south branch line.[49] This endeavor became the seven-hundred-plus-mile South Manchuria Railway that linked the Chinese Eastern Railway to the newly leased territory in the Liaodong Peninsula. As part of these arrangements, Russia also acquired timber and mining concessions in areas adjacent to the railway lines through which it obtained materials and fuel for the construction and operation of the train network.[50] The opening of Fushun's coal mines coincided with the extension of Russian interests into southern Manchuria, and Saint Petersburg, too, became embroiled in the coming contest over Fushun coal.

By early December 1901, Wang Chengyao and Weng Shou heard back from Zeng Qi. The court agreed to allow mining development across Fengtian Province, including in Fushun.[51] After paying ten thousand taels in service fees to Zeng's office, each man set up his own company, with the Yangbaipu River dividing the Fushun coalfield between them. To the west of the river were the mines of Wang's Huaxingli Company (華興利公司); to the east, those of Weng's Fushun Coal Mining Company (撫順煤礦公司). Both men secured multiple investors, including Russian ones, to help finance their capital-intensive undertakings.[52]

The two companies, as competitors in close proximity, started out on hostile terms. Soon after work commenced, Weng disputed the demarcation line between their properties and dispatched men to forcibly occupy two of Wang's mining pits. Their workers clashed, which resulted in several casualties before local authorities stepped in to break up the fighting and force Weng to return the seized pits to Wang. Following this episode, Wang

48. Paine, *Imperial Rivals*, 185–90.

49. Paine, "The Chinese Eastern Railway," 19. Russia claimed to have taken over these two ports as a countermeasure to assist China given that Germany had just seized Jiaozhou Bay. In actuality, Russia and Germany had colluded to support each other in these territorial grabs.

50. Paine, *Imperial Rivals*, 190–94. For the agreement regarding the building of the South Manchuria Railway, see "Agreement Concerning the Southern Branch of the Chinese Eastern Railway."

51. *BJTK*, 11.

52. *BJTK*, 13, 18–19, 22–23.

attributed Weng's boldness to the backing of Russians who had invested in the latter's business. Fearing that conflicts down the road would not turn out in his favor, Wang decided to bring the powerful Russo-Chinese Bank on board as a major shareholder.[53] In early 1903, the Russian Far Eastern Forestry Company took interest in Weng's Fushun Coal Mining Company and purchased it for fifty thousand rubles.[54]

Russia's advances in Manchuria worried Tokyo. To Japanese leaders, this presented a potential threat to neighboring Korea, which they took to be of much geostrategic significance, having come to see the peninsular kingdom as "a dagger pointed at the heart of Japan."[55] Diplomatic attempts to settle this issue of colliding spheres of influence followed, but when talks seemed like they were going nowhere, Japan decided to strike first. The Imperial Japanese Navy launched a surprise attack on the Russian Pacific Fleet at Lüshun on February 8, 1904. Soon after fighting commenced, Russian forces built a feeder line between the Sujiatun station on the South Manchuria Railway and Fushun's Laohutai (老虎臺) mine, with the intention of having the coal flow more easily through the widening rail network and hence better fuel their war effort.[56] Wang had halted Huaxingli's operations in the spring of 1904, but when the provincial capital started to face fuel shortages as winter fell, he reopened the mines for work. On February 16, 1905, Russian troops took over three of Wang's five mines and seized the products from the other two to meet their energy needs.[57] Less than a month later, on March 10, the Japanese Yalu River Army entered Fushun, drove the Russian units north, and occupied the area. The war, however, dragged on for another six months, ending on September 5.

In the age of steam power, coal's carbon energy proved pivotal in the fighting of the Russo-Japanese War. The Russians, for one, had to secure coal to fuel their Baltic Squadron's eighteen-thousand-mile trip by sea from its base in Kronshtadt to Vladivostok, where it had planned to launch an offensive against the Japanese fleet. Each requiring about three to five tons of coal a day, the Russian warships together consumed around half a million tons of coal on this journey. In order to make it, the Baltic Squadron had to refuel at coaling stations operated by other powers along the

53. *BJTK*, 12–13; *KWD*, 2158, Zeng Qi to Board of Foreign Affairs (BFA) (August 29, 1903).

54. *BJTK*, 27; Fu, *Zhong-Ri Fushun meikuang an*, 14–15.

55. This popular metaphor for Korea had originally been coined by a German military adviser to Japan. See Duus, *Abacus and the Sword*, 49.

56. *KWD*, 2162, BFA to Russian ambassador (1904); Fushun shi difang zhi bangongshi, *Fushun shi zhi*, 557.

57. *BJTK*, 21–22, 33.

way. This required some careful negotiation on the part of Russian representatives, for the British, French, and Germans they would approach were afraid of compromising their neutrality in the ongoing war by assisting Admiral Zinovy Rozhestvensky's naval force. Although there were some delays, the Baltic Squadron managed to obtain the coal supplies it needed from these neutral parties and finally made its way to Northeast Asia. There, it was almost completely destroyed in the Battle of Tsushima Strait, May 27–28, 1905.[58]

On the Japanese side, there had been concern over coal supplies for the navy in the years before the war. Japanese collieries in the home islands, particularly in Hokkaido to the north and Kyushu to the south, mined considerable volumes of coal. Much of this coal was, however, bituminous of rather low quality. Aside from its lower calorific value, which made it almost unsuitable for powering battleships, this coal also gave off plumes of black smoke when burned, placing vessels at a strategic disadvantage by rendering them visible to adversaries from afar.[59] Japanese naval planners thus looked abroad for alternatives and began purchasing British coal from Cardiff for their fleet. Between stockpiles of Cardiff coal and fresh imports from Britain as part of the Anglo-Japanese Alliance of 1902, the Imperial Japanese Navy had sufficient amounts of this key resource for what would be its decisive naval campaign.[60]

Meanwhile, back in Fushun, the occupying Japanese troops took steps to revive coal mining in the area. As they fled, Russian forces, in an attempt to hamstring production, had torched buildings, wrecked machinery, and flooded the Laohutai and Yangbaipu (楊柏堡) mines. They rightly recognized that coal mines were assemblages whose worth lay not only in extant deposits but also in the equipment put in place for their extraction. Fushun's carbon resources were far too valuable to just leave to the Japanese for easy taking. The Japanese military in Fushun started by mining the old pits of Qianjinzhai (千金寨), then drained Laohutai's inundated pits and opened several new pits at both mines. Within a few months, daily output reached three hundred tons, a modest revival but a promising start nonetheless.[61]

Japan's victory over Russia marked a turning point in the geopolitics of the region and in the character of Japanese imperialism. Russia lost the

58. Cecil, "Coal for the Fleet," 990–92, 1001, 1005.
59. Evans and Peattie, *Kaigun*, 66.
60. Evans and Peattie, 66–67.
61. *BJTK*, 37; Manshikai, *Manshū kaihatsu yonjūnen shi*, 50.

ground it had been gaining in southern Manchuria even as Japan established a firmer foothold on the continent. At the same time, the war and this victory in particular prompted many Japanese politicians, bureaucrats, military officers, and ordinary citizens to become more sympathetic toward the promise of empire and the prospect of expansion.[62] As one of the terms in the Treaty of Portsmouth that brought the war to a close, Russia ceded to Japan the portion of the South Manchuria Railway that ran from Changchun to Dalian, a stretch of over four hundred miles. In addition, Russia also handed over all the coal mines in the area that were "worked for the benefit of the railway."[63] It was under these stipulations that Japan would stake its claims to the coal mines of Fushun.

Surveying the Prize

Japan's administration of its new acquisitions in Manchuria was marked from the start by a convergence of the colonial and the scientific. To manage the railway and mining concessions, the Japanese government set up Mantetsu as a "semi-public, semi-private" (半官半民) corporation in 1906. This massive colonial joint-stock company boasted an initial capitalization of two hundred million yen, with the government as majority shareholder by mandate.[64] Gotō Shinpei 後藤新平 (1857–1929), a medical doctor turned colonial bureaucrat, was appointed its first director. As civilian governor of Taiwan, Gotō had promoted "scientific colonialism"— colonial governance grounded in extensive information about the specifics of population and place gleaned through systematic research.[65] Even though he now ran a corporation and not a colony, he would ensure that this mode of management became central to Mantetsu as well. This was most evident in the company's Research Section, which Gotō established as part of Mantetsu at its founding and which would carry out thousands of socioeconomic surveys across Manchuria and beyond over the next few decades.[66] It could also be seen in other research agencies that Mantetsu set up in the years that followed, such as the Geological Research Institute (地質研究所) and the Central Laboratory. Furthermore, scientific colonial-

62. Matsusaka, *Making of Japanese Manchuria*, 54–59.

63. This was Article VI of the Portsmouth treaty.

64. Matsusaka, "Japan's South Manchuria Railway Company," 37–38.

65. On Gotō Shinpei and his ideas about "scientific colonialism," see Peattie, "Japanese Attitudes toward Colonization," 83–85.

66. On Mantetsu's Research Section, see Kobayashi, *Mantetsu Chōsabu*; and Young, *Research Activities of the South Manchurian Railway Company*, 3–34.

ism guided Mantetsu's industrial operations, including at the Fushun colliery, which it officially took over on April 1, 1907, and which thereafter existed as a subsidiary company under the larger railway enterprise.

One of the most pressing tasks Mantetsu had to undertake after arriving in Fushun was conducting geological surveys. It needed, after all, a vertical view of the territory to take stock of the coal that lay underfoot. Such surveys entailed several steps. First, surveyors walked around the terrain observing outcrops, looking, in particular, for rocks belonging to a coal-bearing age such as sandstone, limestone, or shale. In the event that they deduced the possible presence of coal, the surveyors then worked to produce a topographical map of the area in order to determine the potential shape and structure of the formation, determining through triangulation the coordinates of such a map. After this was done, they could then sink bores to establish whether there was actually coal present and, if so, to retrieve samples for analysis.[67]

Mantetsu's surveys built on earlier attempts to size up the Fushun coalfield. As Japanese troops crossed Manchuria during the Russo-Japanese War, they were trailed by geologists and mining engineers tasked with assessing the earth and its material prospects.[68] It was in this context that the probe by Chinese proxy mentioned earlier took place, and Japanese empire builders came into knowledge of Fushun coal—even if they were initially skeptical of its excellence or extent.[69] After the war, the government of the Kwantung Leased Territory, the concession at the southern tip of Manchuria that Japan acquired from Russia, immediately dispatched teams to surrounding areas to assess production in agriculture, forestry, commerce, fisheries, and mining. Those in charge of surveying the region's mining industry went off in pairs consisting of a geologist and a mining engineer—the former would ascertain the presence of minerals; the latter

67. Matsuda, "Shisui"; Colliery Engineer Company, *Elements of Mining Engineering*, 12.1–12.52. For more on prewar Japanese prospecting and mapping of coalfields, see Kuboyama, *Saishin tankō kōgaku*, 1:77–137. On the late nineteenth-century development of mine mapping in the United States, which would be a global leader in mining engineering and, on this front, a reference point for Japan, see Nystrom, *Seeing Underground*, 17–52. On the broader American coal consulting enterprise in the same period, see Lucier, *Scientists and Swindlers*, 11–140.

68. Matsuda, *Mantetsu chishitsu chōsajo*, 10–11.

69. Interestingly, the Fushun coalfield seemed to have escaped the attention of Japanese geologists who conducted surveys in southern Manchuria slightly under a decade earlier during the First Sino-Japanese War. One reason could be that their exploration may have been guided by the work of German geologist Ferdinand von Richthofen, who did not mention this coalfield in his report on the region. The record shows that they were most certainly aware of his research. See Inouye, "Geology of the Southern Part," 1–2. On Richthofen's geological surveys in China, see Shen, *Unearthing the Nation*, 29–35; and Wu, *Empires of Coal*, 33–65.

the economic means for their recovery. The geologist in the pair bound for Fushun was Ogawa Takuji, a graduate of Tokyo Imperial University's geology department. Ogawa, who previously authored a topographical study of Taiwan, had been working for the Imperial General Headquarters, Japan's military joint chiefs of staff, since the start of the war. His mining engineer partner, Ōhashi Takichi, had also been serving the wartime state, having completed a series of mine surveys across Manchuria for the military in the past year.[70] The months that the teams were in the field, from the late fall into the winter, were not optimal for investigative work, as Ishizuka Eizō, then head of the Kwantung Leased Territory's civil administration, readily admitted. The ground was "entirely locked under hard ice and harsh snow," which meant that drilling for core samples was difficult.[71] Further complicating matters, the teams often lacked proper prospecting tools. Despite these difficulties, Ogawa and Ōhashi nevertheless completed a report that laid the foundation for future studies of Fushun's geological features and the characteristics of its coal.[72]

The general shape and scale of the Fushun coalfield came into view through such surveys. The coal was located in a relatively narrow east–west trending sequence of rocks that dipped toward the east.[73] Topographical maps that took into account the geological texture of the landscape replaced earlier schematic drawings. This was not, however, merely about creating representations on paper that were truer to the dimensions of nature. Rather, the maps served as guides to the considerable extractable resources below. Although estimates of Fushun's coal deposits varied, surveyors agreed on the immensity of volume. Ogawa and Ōhashi had furnished a conservative figure, placing possible yearly production at one or two million tons for several decades.[74]

Most authoritative among the early surveys was the one by geologists from Mantetsu's Geological Research Institute, whose very first assignment had actually been to calculate the Fushun coalfield's deposits.[75] Led by Kido Chūtarō 木戸忠太郎 (1874–1960)—who previously worked for the Japanese government's Mining Inspection Office and a Japanese iron mining

70. Matsuda, *Mantetsu chishitsu chōsajo*, 11–14.

71. Ishizuka Eizō, preface to Kantō totokufu, *Manshū sangyō chōsa shiryō (kōzan)*, 5–6.

72. "Ogawa gishi no Bujun tankō."

73. *BJTK*, 49. See also Johnson, "Geology of the Fushun Coalfield," 217.

74. Kantō totokufu, *Manshū sangyō chōsa shiryō (kōzan)*, 112.

75. *BJTK*, 47. See also Kido, "Bujun tandan chishitsu chōsa." Other geological surveys from that time include, for example, Ishiwata, "Bujun tankō"; and Matoba, "Bujun tankō ni tsuki."

company in Hubei—the geologists carried out half a year of survey work before arriving at a staggering estimate of eight hundred million tons.[76] Fushun overshadowed Japan's largest known coalfield at that time, a region in the northern island of Hokkaido with an estimated six hundred million tons.[77] In Kido's words, this was nothing less than a "treasure house."[78]

The staggering figures that experts like Kido produced circulated widely across official and mass channels. Often, these were accompanied by claims that Fushun coal could be mined for hundreds of years, its eventual exhaustion so far in the future as to have no bearing on behavior in the present. Over time, the size of the reserves and their projected productivity prompted pundits and planners to talk about this coal as effectively "limitless" or "inexhaustible."[79] This rhetoric fostered an idea among many Japanese that continued access to Fushun's carbon resources was a matter of national interest. It also had implications for the scale of operations that Mantetsu brought to bear on its coal seams.[80]

The value of Fushun coal lay not only in its quantity but also in its quality. According to the report by Ogawa and Ōhashi, Fushun coal, like Japanese coal, belonged to the Tertiary period. Its outward appearance was jet black and shiny; it was hard and did not break easily; and it contained few inorganic impurities. All served as indicators of its high quality.[81] Fushun coal was put through additional tests that measured, among other things, water content, volatility, proportions of sulfur and ash, and calorific capacity. These tests were conducted at a variety of venues, from the Miike colliery in the home islands to the laboratory of a Dr. Prof. Gottschhe from the Natural History Museum in Hamburg, Germany, to the boiler room of Mitsui Bussan's *Fujisan Maru* as the ship sailed between the Manchurian port of Niuzhuang and the Japanese port of Moji late in the spring of 1908. The tests further revealed the character of Fushun coal. It caught fire easily, "with a slight crackling noise," and burned almost completely, giving off little smoke and leaving behind minimal amounts of clinker—the residue that could damage boilers if allowed to build up. In the test onboard the *Fujisan Maru*, for instance, the engine grates needed to be changed only

76. *BJTK*, 70–71.
77. "Mining in the Far East."
78. *BJTK*, 72.
79. There are numerous mentions of this, but for one example, see Shinozaki, *Bujun tankō Ōyama saitanjo*, 1.
80. Drawing from the American experience, Martin Melosi has detailed ways that the idea and reality of energy abundance have encouraged excessive extraction and needless waste. See Melosi, *Coping with Abundance*.
81. Kantō totokufu, *Manshū sangyō chōsa shiryō (kōzan)*, 112–19.

once every six hours when Fushun coal was used as opposed to every four hours when Tagawa coal, one of Japan's best, was burned. Coupled with its high calorific value, Fushun coal was, these tests concluded, particularly suitable as fuel for trains, steam vessels, and power plants, surpassing most Japanese coal on these fronts.[82] In advertisements for Fushun coal, Mantetsu would market it as "the best steam coal in the Far East."[83]

Japanese empire builders may have seemed particularly excited about the volume and virtue of Fushun coal, but this was because they had already come to regard coal as an important resource in fueling Japan's rise in the modern era. According to Scottish engineer Henry Dyer, writing in 1904, "The people of Japan were not slow to recognise the fact that the Britain of the West in great part owed its predominant position among the nations of the world to its abundant mineral resources, and especially those of coal and iron, which enabled it to carry on all kinds of manufacturing industries, and they determined to develop the mineral resources of their country as rapidly as possible."[84] Dyer had spent almost a decade in Japan in the late nineteenth century assisting the Japanese government in establishing a modern engineering curriculum that Meiji leaders saw as essential for, among other things, the development of coal and other mineral resources. Coal drove the expansion of the Japanese industrial sector, closely linked as this process was to the manufacturing of munitions and the military buildup that the Meiji state had so greatly prioritized.[85] With the ascendency of steam power in warfare, coal propelled Meiji Japan to significant naval victories over not only tsarist Russia but also the Qing empire earlier in 1895. When Mantetsu took over the coal mines of Fushun in 1907, Japanese coal output in the home islands was already three times more than it had been just a decade before.[86]

The work of geologists, mining engineers, and other experts in those initial years proved essential to establishing Japan's coal-mining enterprise in Fushun. The research they carried out and the recommendations

---

82. *BJTK*, 72–98; "Fushun Coal Mines," 438–41. Japanese geologists, engineers, and fuel experts, in working out the specificities of the site and its resources while holding the Japanese home islands and Japanese coal as constant points of reference, bear resemblance to the "imperial-royal scientists" of the Austrian empire, who, as Deborah Coen has contended, established their "public authority on a capacity to set local details in appropriate relation to a synthetic overview, to attend to minutiae without losing sight of the coherence of the whole." See Coen, *Climate in Motion*, 63–91, quotation on 64.

83. "Fushun Coal."

84. Dyer, *Dai Nippon*, 157.

85. Samuels, *"Rich Nation, Strong Army,"* 84–88.

86. Walker, *Toxic Archipelago*, 177–79.

they made formed the basis for the colliery's subsequent development. In certifying the quantity and quality of Fushun's coal deposits, such studies helped advance the case that this was a concession worth holding on to. There was an undeniable attractiveness to the abundance. As they calculated the sheer extent of the underground holdings, Japanese earth scientists put forward figures that excited the imaginations of state planners and the wider public back in the metropole, many of whom came to equate with Fushun a promise of inexhaustible energy that was so central to the vision of carbon technocracy.

Rights to Resources

Japanese imperialists, desirous as they were of Fushun coal, thought they well deserved access to it. This access was, they contended, part of the spoils of war. On an affective level, it was Japan's reward for the sacrifice of its subjects. "The Fushun coal mines were redeemed by the blood of the people of our great Japan," one Japanese engineering student wrote in 1912 after finishing a few months of practicum at the colliery.[87] Japanese claims on the Fushun coal mines did not, however, go uncontested. Wang Chengyao, whose mines were first seized by the Russians during the war and then taken over by the Japanese thereafter, sought help from the Qing government in recovering his property. The diplomatic back-and-forth that ensued reveals how the creation of this site of extraction rested upon tensions between the international legal norms on which the parties based their arguments and the hierarchies of power that allowed Japan to dominate discussions and, ultimately, to yield little while maintaining its claims to Fushun's carbon resources.

On April 24, 1906, the Qing Board of Foreign Affairs issued a diplomatic note to Japanese ambassador Uchida Kōsai requesting the return of Wang's mines. The note began with the premise that these mines, which lay to the west of the dividing Yangbaipu River, belonged to a Chinese merchant and had "nothing to do" with the Russian-owned Fushun Coal Mining Company on the eastern side, to which Japan could stake claim in accordance with the Portsmouth treaty. As such, Japanese troops had no business in occupying the mines and should, the note went on to urge, swiftly hand them over.[88] There was no reply from Uchida, who was reassigned less than a month later. The board proceeded to send a few more

---

87. Kubo, *Bujun tankō*, 1.
88. *KWD*, 2168, BFA to Uchida Kōsai (April 25, 1906).

diplomatic notes regarding this matter to the Japanese ambassadors who came after, first to Abe Moritarō, who held the position for just over two months, and then to Hayashi Gonsuke 林権助 (1860–1939), the Japanese diplomat to Korea who had signed the 1904 protocol that placed the peninsula kingdom in a subordinate position to the Japanese empire and paved the way for its colonization.[89]

By the time of its December 20 note to Hayashi, the board had a more elaborate argument. Aside from insisting that the mines had been an entirely Chinese enterprise before the occupation, it also appealed to the Sino-Japanese agreement on Manchuria signed in Beijing on December 22, 1905. This agreement included a clause stating that all Chinese public and private property seized by Japanese forces under "military necessity" should be restored to the original owners upon the withdrawal of troops or the cessation of need.[90] The board also pointed out that the Fushun coal mines were located more than thirty *li* away from Mantetsu's main trunk line and thus fell beyond the permissible range for mining operations that were part of railway concessions. It repeated its request for Japan to return the mines. At that time, the Japanese were also occupying the nearby Yantai coal mines, which, though smaller in scale, had a story similar to Fushun: established in the late Qing, supported by Russian capital, and taken over by Japanese troops during the war.[91] In this note and in most discussions thereafter, the board treated the two cases of Fushun and Yantai as one, seeking common resolution to both these challenges to Chinese sovereignty.[92]

Hayashi's response, sent months later, on May 11, 1907, was fiery. The Fushun coal mines, the Japanese ambassador maintained, had been under Russian management before the war, and there was "clear evidence" to support this fact. He argued that Russian investment in the mines yielded a "monopoly of true managerial authority" and that the Russians exercised this authority by placing troops in the area and constructing the feeder line between the Sujiatun station and the Laohutai mine for the express

89. *KWD*, 2172, BFA to Abe Moritarō (June 16, 1906); and *KWD*, 2181 and 2185, BFA to Hayashi Gonsuke (October 1 and December 20, 1906).

90. "Treaty and Additional Agreement Relating to Manchuria, December 22, 1905." This was Article IV of the additional agreement to the treaty. It states, "The Imperial Government of Japan engage that Chinese public and private property in Manchuria, which they have occupied or expropriated on account of military necessity, shall be restored at the time the Japanese troops are withdrawn from Manchuria and that such property as is no longer required for military purposes shall be restored even before such withdrawal."

91. Yu, "Liaoyang Yantai meikuang," 66–68.

92. *KWD*, 2185.

purpose of obtaining coal for their Chinese Eastern Railway. Claiming that the mines had been under the railway concession, Hayashi concluded that they now belonged to Japan per the terms of the Russo-Japanese Portsmouth treaty and that according to the 1905 Sino-Japanese treaty, which the Qing Board of Foreign Affairs itself had cited, China had consented to all transfers of holdings from Russia to Japan.[93] As for the point about the thirty-*li* range, Hayashi contended that there was already a precedent of the Chinese Eastern Railway mining beyond that stipulated limit. He went on to accuse the Qing government of "acting arbitrarily" and attempting to use the treaties to restrict Japan's rights.[94]

As these diplomatic exchanges were taking place, Wang wrote repeatedly to Qing officials, pressing them to push forward with his case.[95] On May 20, 1907, the Board of Foreign Affairs informed the Mukden general that the Japanese foreign minister, Hayashi Tadasu, had advised that Wang meet directly with Mantetsu president Gotō Shinpei to discuss the resolution of this affair.[96] Wang was notified accordingly but then faced difficulties in getting Gotō to the negotiating table. When he was finally given a meeting on July 26, it was with a Mantetsu representative Gotō sent in his stead. Wang would describe the encounter as one that "began in ambiguity and continued into elusiveness" (始而含混, 繼而支吾).[97] Nothing was determined at that meeting. Although Wang continued to demand that the mines be returned, he was swiftly sidelined in the discussions that followed.

Negotiations between Qing and Japanese officials over Fushun and Yantai went on for a few more years. Apart from the issue of rights to resources, several other matters arose: Mantetsu's purchasing of land around the mines, its stationing of police in the vicinity, and—most pressing from the Japanese point of view—its payment of taxes to the Qing government on the coal produced. Fushun coal was taxed by Chinese customs three taels per ton exported, and Japanese negotiators agitated for lowering the

93. Article I of the 1905 Sino-Japanese treaty states, "The Imperial Chinese Government consent to all the transfers and assignments made by Russia to Japan by Articles V and VI of the Treaty of Peace above mentioned [that is, the Portsmouth treaty]."

94. *KWD*, 2192, Hayashi Gonsuke to BFA (May 11, 1907).

95. Wang wrote to the Qing court a total of eleven times between December 27, 1905, and October 5, 1909. See Fu, *Zhong-Ri Fushun meikuang an*, 227–34.

96. *KWD*, 2193, BFA to Fengtian general (May 20, 1907). For the Japanese correspondence calling for such a meeting, see Japanese consul in Fengtian to Hayashi Tadasu (June 8, 1907), in *NGB*, vol. 40, no. 2, 230–32.

97. *KWD*, 2196, Wang Chengyao to BFA (August 30, 1907).

per ton rate to one tael.[98] Talks between the two parties regarding this assortment of issues culminated first in a general agreement in 1909 and a more specific set of regulations in 1911.

An agreement on Manchurian mines and railways, signed on September 4, 1909, settled the statuses of Fushun and Yantai. In accordance with its third clause, the Chinese government formally recognized the Japanese government's right to work those two sites. The Japanese, in turn, agreed to pay all duties on the export of Fushun and Yantai coal, "respecting the full sovereignty of China," although these were to be set "upon the basis of the lowest tariff of coals produced in any other places in China."[99] Regulations concerning Fushun and Yantai, which were drawn up almost two years later on May 12, 1911, provided elaborations on that third clause. In particular, these regulations made provisions for not only the export tariff but also a tax on coal at the mouth of the mine. This mining tax was set at 5 percent of the coal's initial value for daily output exceeding three thousand tons, or one tael per ton in instances when volume fell below that figure. The regulations also called for Mantetsu to confer with local Chinese authorities should it want to purchase private property within the vicinity of the mines or extend the existing railway line.[100]

During the course of the Sino-Japanese negotiations, there arose the issue of paying Wang Chengyao a compensatory sum. In a telegram Hayashi sent the governor-general of the Kwantung Leased Territory on May 20, 1907, the foreign minister proposed that since Wang had invested capital into the business of the mines, Mantetsu, which had taken over the operations, could award him some money for his losses. Hayashi suggested 100,000 taels.[101] When the issue was tabled in Sino-Japanese negotiations later, it was decided that the sum would be determined by the amount that Wang had initially invested. After more deliberation, both sides agreed that Wang should receive 205,000 taels. In presenting this sum to him, the Japanese representatives were careful to refer to it as "relief" (救恤) rather

98. Japanese consul in Niuzhuang to Hayashi Tadasu (April 30, 1908), in *NGB*, vol. 41, no. 1, 588–89. See also *KWD*, 2005, Tax Affairs Bureau to BFA (June 17, 1908).

99. "Agreement Concerning Mines and Railways in Manchuria, September 4, 1909." See also Fu, *Zhong-Ri Fushun meikuang an*, 220–21.

100. "Detailed Regulations for Fushun and Yentai Mines, May 12, 1911." See also Fu, *Zhong-Ri Fushun meikuang an*, 222–23. For an earlier draft of these regulations, see Japanese consul in Mukden to Japanese foreign minister (December 11, 1909), in *NGB*, vol. 42, no. 1, 569–72.

101. Hayashi Tadasu to Kwantung Leased Territory governor-general (May 20, 1907), in *NGB*, vol. 40, no. 2, 230.

than "reparations," giving the transaction an appearance of charity as opposed to that of atonement.[102]

It is testament to the larger pattern of changing global norms, with firmer ideas about the sanctity of state sovereignty, that Japan was so intent on making sure that its rights to Fushun's coal resources were not only established by force but also grounded in legal legitimacy.[103] Still, power differentials weighed heavily on the outcome of Qing and Japanese negotiations over Fushun. Although the parties involved appealed to treaties signed and the letter of the law, the Japanese interpretation—that the mines had formerly been under Russian control and therefore now belonged to the Japanese as part of the spoils of war—held fast.

Division of Labor

The emergence of Fushun's coal-mining industry, like that of the energy regime of coal more generally, depended on more than conducting geological surveys or securing mining rights. It also required the labor of thousands who came and worked the depths of its deposits. Most of these miners were migrants hailing from other parts of Manchuria, North China, and, to a lesser degree, Korea and the Japanese home islands. Few left written accounts of their experiences. Yet if we know very little about these miners as individuals, we at least have a good sense of them as a collective. In keeping with its commitment to scientific colonialism, Mantetsu kept close records on these men that only got more detailed over the years. The work that Fushun's Chinese miners did and how the company attempted to exercise control over them will be covered in the next chapter. Our interest here is in their recruitment and living conditions in the growing mining town, especially in contrast to those of Mantetsu's Japanese employees.

When Mantetsu took over the Fushun coal mines in 1907, there were

102. The term *kyūjutsu* (救恤) is a little difficult to render directly into English. *Kyūjutsukin* (救恤金) refers to a monetary contribution that one gives to someone who is destitute or a victim for relief. Unlike "reparations" (賠償), it does not necessarily implicate the giver in the circumstances of the receiver. Upon learning that Wang had received money from Mantetsu, the Russian ambassador to China started sending notes to the Qing Board of Foreign Affairs to request that a cut of that sum be given to Russian shareholders of the Russo-Chinese Bank, which had been an original shareholder. This resulted in more negotiations that carried on past the fall of the Qing dynasty in 1912 and into the early years of the Republic of China, and the case apparently only came to a close with the Russian Revolution. See Fu, *Zhong-Ri Fushun meikuang an*, 173–77.

103. This is consistent with the argument put forward by Alexis Dudden regarding Japan's legitimization of its colonization of Korea. See Dudden, *Japan's Colonization of Korea*.

2,589 employees, of whom 1,843 were Chinese. As the enterprise grew, so, too, did its workforce, with the Chinese maintaining their majority. From 1913 onward, no less than 90 percent of the workers at Fushun were Chinese. The decision to employ mainly Chinese workers had been from the start largely about cost. "Using Chinese is more cost effective," a 1906 Japanese report reasoned, "No matter what, the level of their basic needs (clothing, food, and shelter) is entirely lower than those of our miners and workers. Thus, the wages paid to them and the fees for their accommodation can be lower."[104] This logic that Chinese workers could be paid less than their Japanese counterparts because they had fewer basic needs was consistent with social Darwinist ideas of Chinese racial inferiority that had pervaded Japanese discourses, particularly after China's loss in the First Sino-Japanese War.[105] At Fushun, these ideas provided the rationale for a division of labor in which "our countrymen are put in positions like foreman or supervisor . . . and Chinese should be used mostly for lowly work."[106] While there were some Japanese who worked as miners, most held managerial, clerical, or engineering jobs. The mining town also attracted Korean laborers, but almost all of them were engaged in agricultural work in the surrounding fields.[107]

The Chinese who worked in Fushun mostly hailed from the provinces of Hebei and Shandong, across the Bohai Gulf. Although the colliery's first workers were locals from the surrounding communities and other parts of Manchuria, the company soon had to look beyond the region for its labor needs. In 1930, 25 percent of Fushun's Chinese miners came from Hebei and 57 percent from Shandong.[108] As in most other mines in China, labor was procured through a system in which a labor contractor (把頭 or 苦力頭) was in charge of recruiting workers and supervising their work.[109] (More on these contractors, the abuses associated with them, and the changes in their function in Fushun will be said in the following chapter.)

---

104. Kantō totokufu, *Manshū sangyō chōsa shiryō (kōzan)*, 141.

105. Kingsberg, *Moral Nation*, 3.

106. Kantō totokufu, *Manshū sangyō chōsa shiryō (kōzan)*, 141.

107. Out of the sixty-nine registered Koreans in Fushun in 1913, one was a merchant, two were workers, and the remaining sixty-six were all farmers. See the report by the Fushun county office (July 24, 1913), in *DBDX*, 42:32–38. Also, according to a 1923 mining practicum report, there were seventy-four Koreans in fourteen households in the area around the open-pit mine. See Wang, *Bujun rotenbori*, 35. On Korean agricultural migrants to Manchuria, see Park, *Two Dreams in One Bed*.

108. Mantetsu rōmuka, *Minami Manshū kōzan rōdō jijō*, 30.

109. Torgasheff, "Mining Labor in China," 400.

Mantetsu also hired workers directly.[110] In order to better attract the labor it needed, the company set up a recruitment station in Zhifu, Shandong, in March 1911.[111] This was followed by several more in North China and Manchuria, from inland cities like Ji'nan and Chaoyang to bustling ports like Qingdao and Dalian. At these sites, would-be workers could sign up and then arrange passage to Fushun.[112]

In 1913, the magazine published by Mantetsu's Colliery Club put out an article titled "Confessions of a Coal Mining Coolie." Although it is fictional—a ventriloquization for a mostly Japanese readership—this piece, written in the first person, presents one picture of the recruitment process. The essay begins with the author, who was out of work, visiting a friend in another village. As he walks by the local Niangniang temple, he spots a piece of red paper on one of its doors that, upon closer inspection, turns out to be an advertisement recruiting labor for the Fushun coal pits. Since his "money pouch had become light" and he had also heard that jobs in the coal pits were relatively stable and "did not change with the seasons," he decides to follow the directions to the recruitment center. There, a recruiter serves him and several others tea and tells them many things about Fushun. Amazed by what he hears, he decides to go and, leaving his family behind, crosses the Bohai Gulf by steamship. Days later, "without expending any effort or suffering any injury," he arrives in Fushun, where everything, from the dormitories to the wages (fifteen to sixteen yuan a month), seems to be "like out of a dream."[113] The reality was most certainly less rosy, but it is hard to deny the enticement of gainful employment as expressed here.

Prospective work prompted many from Shandong to "charge east of the pass" (闖關東), as their migration to Manchuria was colloquially called. (The "pass" refers to the Shanhai Pass, the easternmost opening in the Great Wall where mountain meets sea.) Shandong was geographically small relative to its population, which, by the late nineteenth century, put great pressure on the land, resulting in widespread poverty, hunger, and death, a situation that was further exacerbated by natural and man-made disasters. Droughts and floods, bandits and soldiers seemed to work together to take from many in Shandong the little that they had. Migration to Manchuria served as a survival strategy and allowed them to search for

110. *BJTK*, 226.
111. *FGY*, 5.
112. Bunkichi, *Manshū ni okeru kōzan rōdōsha*, 57–77.
113. Cai, "Saitan kūrii no jihaku," 23–24.

a better life.[114] Travel to Fushun for work was part of this larger pattern of migration from North China in which twenty-five million made the trip and eight million settled between the late nineteenth and mid-twentieth century.[115]

Most Chinese workers who came to Fushun stayed for just a few months or years. They then took their savings and returned home to buy land and get married. Some left during major holidays like the New Year or during harvest time, when their labor was most needed back in their villages. Unless they were accompanying family or worked outside the coal mines, they were always men. One outcome of this pattern of labor migration was a severely skewed sex ratio in Fushun. Looking at the number of people treated at the local hospital in 1922 and considering that men would have accounted for most of the mine-related injuries and conditions since only men were employed in the mines, the disparity is still stark: there were 19,417 male Chinese patients and only 54 female Chinese patients that year.[116]

During the recruitment process, prospective workers had to undergo a short interview and a medical examination. A set of bilingual self-study textbooks for Japanese supervisors learning Mandarin Chinese, put together by the Fushun colliery's general section bureau, allows us to reconstruct this process. Organized by topic areas from labor affairs to transport matters, these textbooks consisted of numerous practice dialogues between a Japanese superior and a Chinese laborer, with the phonetically annotated Chinese text printed above and the Japanese translation below. Included in the first volume was a telling recruitment scenario involving a prospective Chinese worker being interviewed by a Japanese recruiter.[117] The recruiter asks the worker for basic information such as his name, age, and hometown and inquires about his prior work experience, family situation, and willingness to work for an extended period of time—questions that, on the surface, might seem relatively innocuous.[118]

---

114. Honda, "Santō kūrii ni tsuite." This strategy was, however, more often than not open only to those with at least some financial means and personal connections. See Gottschang and Lary, *Swallows and Settlers*, 4–8. On how the ecological limits of North China, including Shandong, gave rise to other survival strategies in the form of rural violence and a rebellious tradition that hindered later revolutionary mobilization, see Perry, *Rebels and Revolutionaries in North China*.

115. Gottschang and Lary, *Swallows and Settlers*, 2. The number of Chinese migrants staying on in Manchuria actually exceeded that of those who settled in Southeast Asia during this period. See Amrith, *Migration and Diaspora*, 52–53.

116. Miyakuchi, *Bujun tankō Ryūhō kō chōsa*, 13.

117. *NMKY*, vol. 1, *Jimu rōmu hen* [Office and labor affairs volume], 99–115.

118. *NMKY*, 1:99–104.

How the textbook presented the exchange over the worker's age is representative of its general tone and reveals notable aspects of the recruitment process. Mantetsu ostensibly had stipulations that workers in its mines should be no younger than eighteen and no older than forty. In response to the question about age, the worker in the textbook states that he is thirty-five. Here, the interviewer refutes him: "You are so old that your hair is all white and you still want to say that you are thirty-five. I do not believe you." After further interrogation, in which the interviewer asks him for his zodiac sign and uses that to calculate his year of birth, the worker admits that he is forty-two.[119] Although this scenario is fictional, it is still telling. First, it shows that the Japanese management harbored much distrust for the Chinese workers it employed. Even fabricated scenarios for language learning defaulted to depicting Chinese workers as deceitful and were designed to train Japanese recruiters in how to point the accusatory finger and call them out on their supposed chicanery. Second, it suggests that Chinese workers may indeed have had a sense of what the company was looking for and were willing to tell recruiters what they wanted to hear, whether it was claiming to meet the age requirement or, speaking directly to company fears about labor turnover, that they were not ones to "flit around from job to job" and, "if hired, would work for at least a year."[120]

If they got through the interview, workers were then sent for a medical examination to see whether their bodies were up to the exacting task of mining labor. Recruits were brought together, lined up, and called forward one by one. They were asked to strip, and their weight and height would be taken, grip strength tested, eyesight and hearing checked, and chest circumference measured—the last was an indicator of "vital capacity," or lung function, before spirometers became commonplace in collieries.[121] The recruits also gave urine samples to test for sugar—too much pointed to health issues for which one could be turned away—and lifted a heavy weight to gauge upper body strength.[122]

In addition to screening for other ailments, the Japanese doctors visually inspected the recruits for signs of drug use.[123] Although substance abuse was rampant in early twentieth-century Manchuria across all classes and among both Chinese and Japanese, many Japanese observers seemed

---

119. *NMKY*, 1:104–5.
120. *NMKY*, 1:103.
121. Braun, *Breathing Race into the Machine*, 140–60.
122. *NMKY*, 1:109–10.
123. On drug use in Manchuria, see Kingsberg, *Moral Nation*; and Smith, *Intoxicating Manchuria*.

to have trained their eyes to see only the figure of the wretched Chinese "coolie" addict.[124] The textbook provided several denunciatory phrases, from "Why is your face so pale? Is it because you have been snorting cocaine?" to "What are these black spots all over your body? You must be shooting morphine."[125] Drug use may have resulted in some Chinese being denied employment, but it would also be a means by which many who ended up getting employed later came to relieve the aches and pains of hard labor.[126]

Recruits who passed the various tests and were hired were encouraged to move into company dormitories. While the colliery usually permitted those who lived nearby to stay in their own homes, it remained apprehensive of this arrangement, believing that "all those who are not in company dormitories frequently skip work, choosing not to come on a whim if it rains heavily or if they are busy with agricultural work."[127] Chinese workers' living spaces were typically crowded and dingy, the air a potpourri of sweat, human and animal waste, and smoke from coal, wood, and sorghum stalks. "I once tried to go to the Chinese quarter, and my nose was assaulted by such an array of stenches," a young Japanese visitor to Fushun wrote in 1914.[128] Tightly packed company dormitories housed seventy to eighty men to each unit, which consisted of little more than an earthen floor and an elevated sleeping area, a *kang* that wrapped around the room and that was heated from beneath by coal or other combustibles. These spaces were often dirty, in part because of the nature of mine work and the coal dust and debris workers inadvertently tracked back from the pits and in part because of the concentration of people living in such close quarters. Either way, this did not stop more than several Japanese observers from deriding the workers for what they took to be the "filthiness characteristic

124. Kingsberg, *Moral Nation*, 44. The term "coolie," which is said to be of South Asian origin, was widely used, by the nineteenth century, to refer to Asian indentured labor, "largely a product," Moon-Ho Jung contends, "of European expansion into Asia and the Americas, embodying the contradictory imperial imperatives of enslavement and emancipation." See Jung, *Coolies and Cane*, 13. The Chinese and Japanese words *kuli* and *kūrii* (both share the same characters 苦力, literally "bitter strength") appear to be phono-semantic matches. Even if their etymological roots were actually different from that of "coolie," *kuli* and *kūrii* came to take on a similar derogatory and racialized meaning among Chinese and Japanese by the late nineteenth and early twentieth century. Japanese colonial authorities in Manchuria would, for instance, only use the term *kūrii* to refer to Chinese workers. See Tucker, "Labor Policy and the Construction Industry," 33.

125. *NMKY*, 1:109–10.

126. Driscoll, *Absolute Erotic, Absolute Grotesque*, 48.

127. *NMKY*, 1:111–13.

128. Kado, *Bujun tankō chōsa*, 5.

1.2. Inside a dormitory for Chinese workers. Dormitory buildings were about seventy-two by twenty-one feet and divided into four sections, each section holding fifteen to twenty workers. Interiors were sparse, consisting of little more than a dirt floor and the *kang* on which the men sat and slept. This photograph is of a dormitory at the Wandawu mine and was taken in 1922.

of all Chinese."[129] In the conflation of cleanliness and civilization that historian Ruth Rogaski has termed "hygienic modernity," there was often too little room for considerations of class.[130]

This environment bred illness. Although Fushun seemed to have been only lightly hit by the Manchurian plague of 1910–1911, bouts of infectious diseases like typhus and diphtheria still periodically made their rounds.[131] In 1913 alone, the local hospital treated 1,795 cases of infectious diseases.[132] In 1919, a typhus epidemic became severe enough to make a sizable

129. Ishiwata, "Bujun tankō ni tsuite," 4–5.

130. Rogaski, *Hygienic Modernity*.

131. The containment of the plague in Fushun may owe to the robust sanitary supervision that both Japanese and Chinese authorities established at the site. This did, however, lead to clashes between the two constabulary forces as Japanese police restricted the access of Chinese police to disinfection stations. See Nathan, *Plague Prevention and Politics*, 33. On the Manchurian plague of 1910–1911, see Gamsa, "Epidemic of Pneumonic Plague"; Summers, *Great Manchurian Plague*; and Lei, "Sovereignty and the Microscope." On epidemics and Sino-Japanese contestations over meanings of public health in Manchuria, see Rogaski, "Vampires in Plagueland."

132. Inoue, *Bujun tankō*, 8.

dent in Fushun's production. The company, noting that "lice are a powerful medium of carrying the disease," began asking Chinese workers to cut off their queues (many continued to keep them even though the Qing dynasty that mandated this hairstyle had fallen). According to one plan, it would pay two yen to any worker who agreed to "cut off his hairy appendage." Given that "about 20,000 still retained their queues," this would have incurred a hefty expense for the colliery. We do not know whether the incentivized queue clipping came to pass, but the company later gave free haircuts, most probably with the fear of typhus-carrying lice in mind.[133]

The crowded and crummy conditions in which the colliery's Chinese migrant laborers lived appeared even more severe when viewed alongside the accommodations that were mostly available only to Japanese employees. As one travel guide admired, these houses were "built in accordance with sanitary principles."[134] They were outfitted with the latest amenities, such as gas and electric lighting, running water, sewage disposal, and modern heating systems that included, by the early 1920s, radiators.[135] The Japanese quarter, where the colliery also had its administrative offices, was built in 1908 on one thousand *mu* of former sorghum fields the company had acquired to the west of Qianjinzhai. Its wide avenues and boulevards laid out in a grid stood in contrast to the narrow, snaking streets of the neighboring old Chinese town.[136] When Mantetsu decided to demolish much of Qianjinzhai to make way for the expanding open-pit mine excavated on its western edge, it planned and built a new elevated district in Yongantai to the east that boasted, as poet Yosano Akiko admired when she visited, the "orderly and beautiful" buildings and broadways of a "European-style Japanese town."[137]

For Mantetsu's Japanese employees living in this modern space, perhaps one of the most poignant reminders that they were on Chinese soil was the requirement that they learn Mandarin Chinese. In 1924, the company stipulated that all new employees had to study Mandarin "to facilitate their communication with the Chinese," as Yamaguchi Yoshiko 山口淑子 (1920–2014) recalled. Yamaguchi, better known by her Chinese stage name Li Xianglan, was a Japanese actress who launched her career by playing Chinese characters in wartime propaganda films. Born in Yantai, she had grown up in Fushun, where her father, Yamaguchi Fumio, taught Man-

133. "Eruptive Typhus at Fushun."
134. Imperial Japanese Government Railways, *Unofficial Guide to Eastern Asia*, 114.
135. Koshizawa, *Zhongguo Dongbei dushi*, 112.
136. Koshizawa, 100.
137. Yosano and Yosano, "Man-Mō yūki," 140.

1.3. Mantetsu employee residences in Yongantai, with the colliery manager's residence displayed in the upper right-hand corner of the image. These residences, which were equipped with the latest amenities and conveniences, were mostly occupied by Mantetsu's Japanese employees and their families. The English text in the bottom left-hand corner of the image, almost too faint to see, reads, "Fushun is an international coal-mine and a busy manufacturing district in the East." Japan's empire builders in this concessionary space as elsewhere would often use "international" to mean "Western" and "modern" and equate "busy" with "productivity" and "progress." (Image courtesy of the Harvard-Yenching Library, Harvard University.)

darin to Mantetsu employees. As a young girl, she attended his classes, as "the only child, and the only female at that." One memorable episode that she later recalled was of a session in which her father was trying to teach the pronunciation for the aspirated *qi* (e.g., seven)—not to be mixed up, Yamaguchi cautioned, with the unaspirated *ji* (e.g., chicken). Her father had gotten his students to stick tissues to their noses with saliva and to repeat *qi*, which, if said properly, should cause the tissues to flutter. "The sight of all these full-grown adults chirping 'qi' in unison with thin strips of tissues hanging from their noses might strike others as ludicrous," Yamaguchi reflected, "but Father's students, myself included, could not have been more serious."[138] The men, at least, had incentive to be so. Mantetsu gave temporary salary boosts to those who passed language proficiency examinations.[139]

138. Yamaguchi and Fujiwara, *Fragrant Orchid*, 4–5.

139. Increments lasted for two years after the successful proficiency examination. Incidentally, this incentive existed for those employees learning Russian or Japanese as well. For

1.4. The Fushun Colliery Club. Built against the western slope of the hill on which the Yong-antai district sat, this three-story construction included a dining room, a card room, a reading room, a salon, and a number of guest rooms, for this club also doubled as a hotel. Around the facility's sizable grounds was a wall that served to protect the privilege the club embodied. This photograph dates to 1937.

While learning Mandarin was meant to make it easier for Japanese to speak to Chinese, life outside of work in Fushun seemed structured to min-imize such interactions. Aside from residing in areas separate from where most Chinese lived, Japanese employees and their families also generally went about their daily activities within their insular community. They patronized Japanese businesses, attended Japanese schools, and amused themselves in recreational facilities meant mainly for Japanese, such as the club for upper-level employees that opened as the Fushun Yamato Hotel in Qianjinzhai in 1910, was renamed the Fushun Colliery Club a year later, and was then rebuilt in Yongantai in 1924 (where it still stands today as the Coal Capital Hotel).[140] It may have been because she was a child at that time, but Yamaguchi's early memories of Fushun before the Mukden Incident as "just a quiet community of rolling hills where the enchanting

Chinese employees, however, the increments were smaller than those for Japanese ones. See Minami Manshū tetsudō kabushiki gaisha, *Minami Manshū tetsudō kabushiki gaisha dainiji jūnen shi*, 153.

140. Xiao and Jin, *Wangshi jiu ying*, 133.

poplar trees coexisted in perfect harmony with the distant view of mountain valleys and open mines" seem to suggest that the colliery town was so
successfully segregated as to shield her from many of the less idyllic sights
of the extractive enterprise that were otherwise daily fare for the Chinese
workers who labored there.[141]

Contentious Grounds

As Mantetsu established itself in Fushun, the verticality visible in the relationships between the Japanese company and the Chinese in its employ
was reproduced in its dealings with the wider Chinese community in and
around the mining town. Fushun's Chinese residents regularly tried to resist the colonial company's heavy-handedness, taking their grievances to
the local magistrate for him to bring before the colliery's management. In
some instances, they even went to Mukden to seek redress from the provincial authorities. This arrangement persisted beyond the fall of the Qing
dynasty in 1912. The new Republican government in China faced the question of what to do with the "unequal treaties" that its predecessor had
signed with foreign powers. While its officials repeatedly appealed to international laws and conventions in efforts to effect revisions, for the most
part, China abided by prior agreements through the 1910s and 1920s.[142] In
Manchuria, China's transition from empire to nation had been accompanied by the rise of the warlord Zhang Zuolin, who came to extend control
over this region. Zhang, having cultivated a mutually beneficial relationship with the Japanese, would be committed to the status quo. In return
for military and financial support, he recognized imperial Japan's concessionary presence across southern Manchuria, including in Fushun.[143]

Property was the primary point of conflict between concession and
community in the colliery town. Through the geological surveys that it
conducted, Mantetsu quickly realized that the coalfield extended far beyond the boundaries of the mines that Japan had seized. Unlike oil, whose
fluidity meant that it could be extracted at multiple points above the subsurface reservoir where it was found—prompting many states to apply the
principle of the "rule of capture" in which whoever gets the oil out of the
ground owns it—coal, in its solid state, did not move and had to be ex-

141. Yamaguchi and Fujiwara, *Fragrant Orchid*, 2.
142. Wang, *China's Unequal Treaties*, 35–62.
143. McCormack, *Chang Tso-lin in Northeast China*.

tracted directly from where it lay.[144] Moreover, in terms of mining rights, Chinese title deeds typically stipulated that "the owner owns the land from the sky down to the lowest point reachable below the surface."[145] This seemed to be the case with Fushun as well.[146] Although the Japanese in Fushun, like mine operators elsewhere, periodically crossed into neighboring lands from below when they mined along coal seams, they recognized that they needed to acquire the property above if they wished to legally expand their exploitation of the coalfield beneath. Distinguishing themselves from the Russians before them—whom they characterized as "forcefully occupying" lands in "injurious" acts that "harmed the sentiments of the locals"—Japanese empire builders in Fushun claimed to follow due process in negotiating land purchases with local Chinese authorities as stipulated in the 1911 regulations.[147] When Mantetsu took over operations at Fushun, the area it controlled was about 950 acres. By 1916, the company had bought up an additional 2,500 acres, including land used for the new town it constructed.[148] This territory only grew over the next few decades as the colliery continued its purchases.[149] According to company figures from 1930, Mantetsu occupied over 14,500 acres in Fushun, making this by far the largest expanse of property that the company held in the railway zone outside of the Kwantung Leased Territory.[150]

The Chinese of Fushun did not always feel like they had a say in such transactions. In 1912, fifty-eight Fushun residents wrote to officials in Mukden to complain that the Japanese company had "used the excuse of coal tunnels to forcibly buy up a great deal of property, seizing food sources (since the land had been used for crop cultivation) and driving people away, all this as part of a scheme for fat profits."[151] At times, residents claimed the colliery had underpaid those whose lands it acquired. In 1917, several villagers from the southwest approached the magistrate be-

144. For one account of the rule of capture that emerged out of English common law, see Frank, *Oil Empire*, 53.
145. Ting, "Mining Legislation and Development in China." For more on Chinese mining law in the late Qing and early Republic, see Wu, *Empires of Coal*, 129–59.
146. "Fengtian jiaoshe shu gei Fushun xian zhishi de xunling."
147. *BJTK*, 264–66.
148. *MT10*, 542–44.
149. *MSZ*, vol. 4, no. 1, 157.
150. Young, *Japanese Jurisdiction*, 142–43. For comparison, Mantetsu held at the time about 5,000 acres in Anshan and 2,700 acres in Mukden, which were the next two largest areas it administered.
151. "Fushun jumin Chen Rong deng."

cause they felt that the colliery had taken a good portion of their "rich and fertile" land for the extension of the railway while paying them an "unfair sum" in return. They had asked around and found out that the ninety-four yen they received per *mu* was much less than what the company had previously paid residents of Qianjinzhai and Yangbaipu for parting with their property.[152] In still some other instances, residents accused the company of taking over their lands without any compensation at all.[153]

The mining activities of the Fushun colliery also brought about environmental changes that negatively affected the lives of locals, even those residing in areas outside of lands the company owned or claimed. One persistent problem was subsidence. In 1926, the local agricultural and business associations contacted Mukden officials to complain that the colliery had been mining beneath the Qianjinzhai business district adjacent to the coal mines, even though, as they pointed out, the company had not purchased property on the surface of those spaces and hence "should not mine as they please" below. The problem here was that "this mining left behind cavities" and "no one knew if and when the ground would cave in." In fact, multiple businesses in the area were already noticing "sudden cracks in their foundations that were many and deep without measure." This was accompanied by "*kangs* sinking up to more than a foot, walls toppling and falling over, buildings going askew, and tremors all day long." To these men, "the danger had already reached its peak." They had cause for concern. In 1920 the abode of Laohutai resident Chen Qingshan had collapsed entirely into the tunnels the colliery had been digging under it. Chen's family was injured. Chen himself and an acquaintance were crushed to death.[154]

Flooding caused by mining activities also bedeviled local Chinese. In 1925, villagers from Xiaopiaotun to the southwest of the mines petitioned the Fushun magistrate to stop the company from building embankments along the river that separated them. This was a bid, village representatives argued, for the company to further enlarge their lands, as the river shifted from construction work and inundated areas belonging to Xiaopiaotun.[155] It was not the first time that something like this had happened.

152. Fushun magistrate to general affairs section, Fushun colliery (August 4, 1919), in *DBDX*, 17:447–49.
153. See, for example, Fushun magistrate to general affairs section, Fushun colliery (June 12, 1920), in *DBDX*, 18:402–3.
154. "Fengtian jiaoshe shu gei Fushun xian zhishi de xunling."
155. Fushun magistrate to Fushun colliery manager and people of Xiaopiaotun (July 30, 1925), in *DBDX*, 23:228–33.

The company had taken over a piece of land from another village, Wulao-tun, in 1921 and began irrigation work along the adjacent river to supply its mines with water. Downstream, the dikes it had built as part of this process triggered the "flooding of the river as though it had been raining relentlessly."[156] Although this was probably unintentional, the company also caused similar floods when it started dredging for gravel to be used in its underground operations. As a result of these extractive efforts, the dikes of the river by the excavation site weakened and were breached, and the flooding that ensued dealt considerable damage to both agriculture and business in the area.[157]

Apart from the acquisition of land and changes wrought to it through mining activities, a major source of conflict between the concession and the community were episodes of violence perpetrated by Japanese police stationed in Fushun. Upon the initiative of its first governor-general, Ōshima Yoshimasa 大島義昌 (1850–1926), the government of the Kwantung Leased Territory established and oversaw a constabulary charged with safeguarding Japanese interests throughout the leasehold and the railway zone.[158] In Fushun, the Japanese police were responsible for keeping the peace in the coal mines and the Japanese district. They occasionally clashed with local Chinese police along the edges of their respective jurisdictions. A disagreement over Chinese passage into the Japanese concession in 1911 escalated into skirmishes that resulted in the deaths of a Japanese sentry guard and a Chinese policeman.[159] More frequent were the episodes in which the Japanese police brought down the iron boot on ordinary Chinese civilians in their attempts to "maintain order." In 1926, for instance, an altercation between the police and a Fushun resident resulted in the latter's death. The man had been riding the electric tram toward Qianjinzhai when he got into an argument with the conductor over discrepancies with the ticket he was issued. The police were called in, and as they dragged the man off the tram, he hit his head and died.[160] The 1930 case in which a Japanese policeman bludgeoned Chinese resident Meng Litou was not uncommon. It was remarkable in that Meng was but six years old. The policeman had discovered the boy pilfering loose coals on the outskirts of the

156. Wulaotun residents to Fushun magistrate (July 12, 1920), in *DBDX*, 20:368–69.

157. Fushun magistrate to Fushun colliery (September 12, 1920) and Fushun colliery to Fushun magistrate (September 17, 1920), in *DBDX*, 19:15–22.

158. Myers, "Japanese Imperialism in Manchuria," 109.

159. "The Fushun Disturbance"; Hu, "Quarantine Sovereignty," 312–13.

160. Fushun magistrate to Fushun superintendent (December 6, 1926), in *DBDX*, 24:249–51.

mines and chased him back to the Chinese quarter, where the man pushed the youngster into the mud and hit him until "his clothes were covered in blood."[161] Japan's empire builders were often fond of framing their project as rational and benevolent.[162] There was, however, a deep senselessness and cruelty that all too frequently surfaced in its operations.

Conclusion

Fushun's metamorphosis into a mining metropolis through Mantetsu's machinations began with Japan's "geological vision" that beheld value in its "vertical territory." Geological forces over tens of millions of years may have created the huge accumulations of coal, but it was the onset of the carbon age and the emergence of geological sciences in the service of the state that determined their worth. The work that Kido Chūtarō and other geologists and mining engineers carried out in Fushun mirrored that of geological experts around the world at the time. They scoured the earth's strata in search of power. Bruce Braun has noted how late nineteenth-century Canadian geologists frequently "slid freely between the visual language of 'geology' and the speculative language of 'value.'" Kido and his peers made such shifts as well, as they moved back and forth from predictions of reserves grounded in direct observation and core sampling to heady pronouncements celebrating the immense wealth of the "treasure house." Here in Fushun, as it was in Canada and elsewhere, "this geological language of probability speaks in the tongue of an economic and political language of possibility."[163] As soaring figures of Fushun's carbon deposits permeated public discourse alongside bold arguments for Japan's rights to those "limitless" or "inexhaustible" resources, many Japanese came to see in Fushun coal a natural and even necessary bond with the home islands, tying, as they did, Japan's fuel future to this ancient Manchurian rock.

But the rise of the fossil fuel economy was more than just a matter of detection and discourse. It involved a transformation of site that reflected an invasiveness seemingly intrinsic to the exploitation of carbon resources. In her study of the oil industry's emergence in the Huasteca, Mexico, historian Myrna Santiago maintains that "fossil fuel extraction entails the creation of an entirely new ecology." This "ecology of oil," as she calls it,

---

161. Fushun security bureau chief to Fushun magistrate (June 6, 1930), in *DBDX*, 28:1–4.

162. On "rationality," see Moore, *Constructing East Asia*, 7. On "benevolence," see Kingsberg, *Moral Nation*, 5.

163. Braun, "Producing Vertical Territory," 24–25, 28.

included reconfiguring how land was held and used and how local society was composed and structured.[164] Although centered on solid coal and not liquid oil, similar changes took place in Fushun. They started with Chinese entrepreneurship and Russian investment and accelerated under Japanese occupation as Mantetsu made moves to secure land and labor for its budding enterprise. The next chapter will explore how the company's introduction of new technologies of extraction, too, would drastically alter Fushun's social and environmental landscapes.

To some degree, this transformation was not completely unique to Fushun or even to fossil fuel extraction more generally. Mining for minerals in the modern era has often been marked by the scale at which it remade sites of production and their surroundings.[165] The distinctiveness of coal mining lay in the extent to which physical and social power were entwined in the process. In delving into the depths to unearth the coal that powered industrial society under the carbon-energy regime, Mantetsu and mine operators elsewhere relied heavily on the exercise of social power to create the conditions for their extractive endeavors.[166] In Fushun, this dynamic played out along a hierarchical axis between the Japanese colonial company and the Chinese who lived and labored in the mining town.

This verticality of social relations gave rise to conflict. It also largely dictated its resolution. The clashes between colliery and community, including the case of Liu Baoyu's disinterred graves with which this chapter opened, often resulted from Mantetsu's unilateral actions. In these instances, Chinese authorities in Fushun or Mukden attempted to get Mantetsu to either cease its infractions or offer appropriate compensation. The archive holds numerous entreaties of this sort, though few responses from the company were either received or retained. However, that Chinese officials would continue to send such petitions does suggest that they were not completely ineffective. From the replies that we do have, it appears that Mantetsu did occasionally offer money as restitution, especially if the sum was small enough or the protestations seemed irrefutable. Still, when the company paid off those affected by its actions, this would be in lieu of reversing any

164. Santiago, *The Ecology of Oil*, 4.

165. For a discussion of mining's environmental impact in the North American context that has resonances elsewhere, see Vrtis and McNeill, "Introduction."

166. On energy as the source of physical and social power, see Russell et al., "The Nature of Power." As scholars such as Timothy Mitchell and Thomas Andrews have shown and as we will explore in the next chapter, those who worked in those mines, too, attempted to seize social power, though not always with success. See Mitchell, *Carbon Democracy*, 18–27; and Andrews, *Killing for Coal*.

of the damages. We witness this clearly in the case of Wang Chengyao, to whom the company awarded 205,000 taels of "relief" for Japan's takeover of his mines instead of returning these mines to his possession. If conflict was resolved, it was almost entirely on Mantetsu's terms. The contentious grounds on which carbon technocracy took root were notably uneven.

TWO

Technological Enterprise

In late May 1928, Yosano Akiko embarked on a month-long trip across Manchuria and Mongolia. She had come upon Mantetsu's invitation and was traveling at its expense. The colonial company, in a bid to win publicity for its ongoing work, had long sponsored tours of the region for influential Japanese cultural figures.[1] As a noted poet and social commentator, she well fit the bill. Among the sites Yosano would visit was the Fushun colliery.[2] There, her guides took her to the famous open pit, where she was initially overwhelmed by what she regarded as the absolute monstrosity of the mine. After she had a chance to survey the spectacle, though, she came to appreciate its grandeur, "feeling that it was several times more majestic than a large coliseum from Roman days." To Yosano, this colossal cavity "immediately challenged the notion of what a coal mine was." Instead of "an excavation of horizontal and vertical tunnels dug deep into the ground," it was a site of extraction where "one only need peel off the top layer of oil shale, thirty to forty feet thick," to uncover "all of the coal beneath." Reflecting on this spectacle, Yosano concluded that "human be-

---

1. Arguably the most famous of Mantetsu's invitees was the writer Natsume Sōseki 夏目漱石 (1867–1916), who traveled through Manchuria and Korea in the fall of 1909. Like Yosano, he stopped by the Fushun colliery, where he had the chance to go down into a mine. He wrote a travelogue of his experience, which was serialized in the *Asahi shinbun* as "Here and There in Manchuria and Korea" (満韓ところどころ). On this trip and the travelogue, see Yiu, "Beach Boys in Manchuria."

2. For more on Yosano Akiko and this trip to Manchuria and Mongolia, see Fogel, "Yosano Akiko and Her China Travelogue." On the broader phenomenon of Japanese colonial boosters promoting tourism as a means of folding the territorial edges of the empire into the space of the national center, see McDonald, *Placing Empire*, 50–80.

ings who would use nature in such a way were like an intelligent species of ant."[3]

Few could describe the mine in quite so dramatic terms. Many would nevertheless express a similar sense of wonder. Part of it was novelty. This was Japan's first modern open-pit mine, a product of its Manchurian laboratory of empire without equivalent back in the home islands. Part of it was the sheer scale of this engineering feat. Fushun's new urban district, which sat atop the thickest part of the coal seam, had been moved three miles eastward to make way for the excavation. Around the time of Yosano's visit, the cut in the earth was already an "enormous gash" some two miles in length.[4] The visual of Fushun's open-pit mine evoked what Leo Marx, David Nye, and others have termed the "technological sublime"—the mix of amazement and apprehension that can overtake us when we encounter landscapes molded by machines powerful enough to alter the natural world, our original source of subliminal experiences.[5] Carbon technocracy, in urging for the extensive extraction of energy by technoscientific means, manifested in expressions of the technological sublime. Here was a terrain transformed for the exploitation of coal through a technical assemblage of excavators, shovels, and freight cars powered at its core by coal. The pit that emerged was impressive in both dimension and depth. According to figures from 1931, "the quantity of earth removed from the surface of the coal-bed was three times that excavated from the Panama Canal."[6]

So compelling an image was the open pit that it came to symbolize Fushun and, more broadly, Japan's technologically driven economic endeavors in Manchuria. In the numerous reports the Japanese imperial state produced over the decades to tout alleged successes in its development of Manchuria, Fushun's open-pit mine often served as an illustration. Photographs of this site of extraction circulated across a range of media, from newspapers and magazines to pamphlets and postcards. These pictures frequently centered on the expanse of the operations: the huge depression in the ground dotted by the pieces of equipment that facilitated its creation. What one generally had to squint to see were the mine's laborers, who appeared, when they did, dwarfed by the visual of the monstrous excavation. This was striking given that Fushun's workforce numbered in the tens of

3. Yosano and Yosano, "Man-Mō yūki," 140.

4. "Japan Moves Town." At the same time, large-scale efforts at engineering the environment were certainly not unknown to Japanese society, as Philip Brown has demonstrated through the case of riparian construction in the Echigo Plain. See Brown, "Constructing Nature."

5. Marx, *Machine in the Garden*; Nye, *American Technological Sublime*.

6. Powell, "The Marvellous Fushun Colliery."

THE GRAND SIGHT OF THE OPEN WORKING, FUSHUN COLLIERY.
觀壯の堀天露・庫寶の盡無　(偏炭爐撫)

2.1. "The Grand Sight of the Open Working, Fushun Colliery." The Japanese text includes an additional phrase, "inexhaustible treasure house" (無盡の寶庫), appealing to the common perception of Fushun's coal resources as limitless. This image, like a number used in this book, comes from a postcard. Countless Japanese postcards in the early twentieth century depict sites of industry like Fushun, pointing to an industrial modern aesthetic that seemed to be in vogue at the time. (Postcard from the author's collection.)

thousands. And yet, the relative invisibility of miners in depictions of the Fushun colliery was consistent with carbon technocracy's vision of such extractive enterprises as defined more by the working of machines rather than by the labor of humans.

Fushun's Japanese technocrats pursued mechanization as a way of reaching the higher yields they so craved while decreasing their dependence on Chinese laborers, whom they often denounced as incompetent and unreliable and whose occasional turn to collective action hamstrung the workings of the extractive enterprise. Timothy Mitchell has contended that the global transition from coal to oil after World War II represented a conscious effort to curb worker activism and limit democratic possibilities. In the past, workers involved in mining and moving coal managed to wrest rights and concessions by disrupting or threatening to disrupt energy flows through sabotage and strike. One of the main promises, then, of oil—whose extraction and transportation relied more on technical apparatuses like pumps and pipes and less on human labor—lay in curtailing worker power.[7] Mechanization in this Manchurian mine was geared

7. Mitchell, *Carbon Democracy*, 12–31.

toward similarly illiberal ends. However, these efforts failed to shrink the size of the workforce, which ended up growing as production expanded, prompting the colonial company to experiment with new technologies of control in the management of its swelling ranks. At the same time, as Mantetsu enlarged its operations in Fushun, it invariably subjected more and more miners to the deepening dangers of an increasingly engineered environment.

Developmentalist Plans

The technological development of the Fushun colliery had been one of Mantetsu's priorities from the start. Through engineering expertise and the might of the machine, the company hoped to capitalize on the productive potential of its subterranean storehouse. The expectation was that coal output would reach three thousand tons a day within a year and a half, enabling annual profits of around 3.8 million yen. According to one account, the fact that the promising Fushun coal mines were to be included in Mantetsu's property was responsible for the "extreme popularity" of the company's shares when it was raising initial capital.[8]

In order to deliver on Fushun's promise, Gotō Shinpei sought a colliery manager with experience and vision. The man he identified for the job was veteran mining engineer Matsuda Buichirō 松田武一郎 (1862–1911). A graduate of Tokyo Imperial University, Matsuda joined the Mitsubishi company right out of college in 1883. His first posting was to the Takashima colliery, which Mitsubishi had acquired two years prior, early in its process of diversifying from its primary business in shipping. Takashima was Japan's first modern coal mine. There, Matsuda worked under English engineer John Stoddart, who taught him how to use some of the latest mining technologies, most notably the diamond boring drill, which allowed engineers to probe and evaluate the site's underlying strata. With his knowledge of boring procedures, Matsuda was engaged to prospect for several other collieries, including Miike, Meiji Japan's biggest coal-mining complex. A few years later, Mitsubishi assigned Matsuda to the Chikuhō coalfield, where he headed first the Shinnyū colliery and then the Namazuta colliery. Under his leadership, Namazuta came to boast the largest and most mechanized coal-mining operations in the region.[9] Matsuda was

8. "Fushun Coal Mines."
9. On Mitsubishi's development of Namazuta, which involved not only the influx of capi-

at the peak of his career with Mitsubishi when Gotō invited him to come and take the reins at Fushun. Matsuda initially refused, purportedly proclaiming, "I am a Mitsubishi man." The colonial bureaucrat was, nevertheless, persistent and repeatedly approached Mitsubishi president Iwasaki Hisaya about this matter. It was finally agreed that Matsuda be seconded to Mantetsu until the Fushun colliery completed its planned expansion, after which he could return to Mitsubishi. Early in 1908, Matsuda set sail for Manchuria to begin his appointment at Fushun.[10]

Once he got there, Matsuda seemed quite excited about the work to be done, viewing the development of Fushun's coal mines as part of a larger project of resource extraction on the continent. "Given the territorial vastness of the Qing state, its subsurface wealth may be said to be limitless," he wrote, "Be that as it may, in order to obtain this wealth, we will need to employ the latest scientific knowledge [最新式の科學] and exhaust the greatest mechanical might [最大の機械力]." To him, Fushun was a "vanguard" (嚆矢) in these efforts. He held a technocratic vision in which "Fushun's current barren wasteland will be transformed" through science and technology, "becoming a bustling intersection with clouds of smoke filling the air and the rafters of tall buildings so close that they touch." Such a development would be, Matsuda mused, "akin to the sight of the Big Dipper's brilliance lighting up a corner of the sky."[11]

Concretely, Matsuda's plan for Fushun involved enlarging extractive operations and expanding supporting infrastructure. Under his direction, the colliery refurbished the three original mines—Qianjinzhai, Yangbaipu, and Laohutai—expanding and outfitting them with more machines to allow for increased production. It also opened two new mines. Named Ōyama (大山) and Tōgō (東郷) after Ōyama Iwao and Tōgō Heihachirō, these mines stood as monuments to state power—Ōyama was the field marshal and Tōgō the fleet admiral in the Russo-Japanese War. They also served as reminders of how Fushun had entered Japanese hands in the first place. The colliery equipped these mines with the most advanced mining technologies Japan could procure from abroad, from Walker indestructible ventilation fans to Whitmore automatic brakes for winding engines. The Ōyama and Tōgō mines were such large undertakings that more than half

---

tal but also the movement of engineers and miners from Takashima, see Murakushi, *Technology and Labour*, 53.

10. Matsuda, "Matsuda Buichirō shōden"; Hisada, "Chikuhō tankō hatten ki," 11.

11. *BJTK*, 1–2.

2.2. "Before" and "after" photographs of the Qianjinzhai mine's eastern pit. The top one is from the time of the Russo-Japanese War. The bottom one is presumably from sometime just before 1909, the year that the Fushun colliery put out the volume about its operations in which these photographs appeared. This pair is one of a number included in the volume to illustrate the rapidity and extent of Fushun's industrial transformation under Mantetsu's management. Minami Manshū tetsudō kabushiki gaisha Bujun tankō, *Bujun tankō* [The Fushun coal pits] ([Dairen]: n.p., 1909), unpaginated between 230 and 231. (Image courtesy of Hathitrust.)

of the almost 9.2 million yen invested into Fushun in its first five years under Mantetsu management went toward their excavation.[12] Matsuda's plan also saw the completion of surface facilities that supported mining activity—such as a power plant, a machinery factory, and water and gas services—and of amenities for the colliery town—including a hospital, a new urban area, company quarters, and schools.[13] In accordance with Matsuda's first development plan (as it came to be called after the fact), the colliery grew considerably, its output increasing sixfold from 230,000 tons in 1907 to almost 1.4 million tons in 1911—roughly 8 percent of the total volume of coal produced in Japanese mines that year.[14]

In the early days of the Fushun colliery, coal was mined through the pillar-and-stall method. Miners sunk shafts into the ground, opening up passages between the surface and the coal seam. They then drove tunnels, or headings, outward from these shafts directly into the seam. These tunnels, which were typically around ten feet wide and seven feet high, branched out into other tunnels, clusters of which ran parallel to one another and were connected at intervals by cross tunnels.[15] The result was one or more grids of "stalls," the excavated tunnels, and "pillars," the coal left behind to hold up the roof. Considerations of structural integrity set limits on the amount of coal that could be removed. If the stalls were made too wide and the pillars correspondingly too narrow, the miners ran the risk of the roof collapsing. By a similar logic, such a method could not be applied at great depths, where the weight of the unmined coal and overburden above put more pressure on tunnel cavities below, or in seams or parts of seams where the coal was too soft. According to one estimate, the pillar-and-stall method at Fushun meant that "not less than 70 percent of the total quantity of coal would have to be left underground." Furthermore, the extent of exposed coal in these workings proved hazardous, for the coal was prone to slacking—deterioration due to changes in the atmosphere—which could lead to the roof or the sides giving way. The exposed coal was additionally susceptible to spontaneous combustion. One way that Fushun's colliers attempted to mitigate these dangers was to line the walls of shaft bottoms and main tunnels with brick, a costly endeavor that begged for a more efficient solution.[16]

Matsuda was, however, unable to oversee further improvements. Taking

12. *MT10*, 491, 525–26.
13. *MT10*, 491.
14. Murakushi, *Transfer of Coal-Mining Technology*, 33; Murakushi, *Technology and Labour*, 21.
15. *MT10*, 492–93.
16. Yamaoka, "Manchurian Plant," 899–900.

2.3. Miners in a subsurface working at the Yangbaipu mine. The miners are in the process of excavating a main tunnel. The two men on platforms wielding pickaxes and the one bent over with a spade are Chinese workers. The two men holding canes and mining lamps are Japanese overseers. This photograph dates to 1921.

ill in 1910, he returned to Japan for treatment that fall. Writing to Iwasaki Hisaya, his former boss at Mitsubishi, Matsuda appeared cautiously optimistic: "Given the doctor's warning, I understand my case to be serious and resolve to take the necessary care of myself."[17] He passed away before the next spring, succumbing to throat cancer.[18] There is no knowing if this cancer was a result of his decades in the mines, but the excavated environment did indeed pose perils to all who worked within it. For some, as may have been the case with Matsuda, the damage was incremental, dealt over years or even decades by exposure to dust and other features of the work space—a slow assault on the body and its respiratory system.[19] For others, as we will see in greater detail later, it was more immediate, inflicting, often in a flash, disability or even death. To commemorate Matsuda's short but

17. Matsuda Buichirō to Iwasaki Hisaya (September 17, 1910), in Hisada, "Chikuhō tankō hatten ki," 57.
18. Matsuda, "Matsuda Buichirō shōden."
19. McIvor and Johnston, *Miners' Lung*, 50–60.

significant tenure, Mantetsu commissioned the famous sculptor Asakura Fumio to fashion a bronze bust of him, which was then placed at the Yamato Park in Fushun's Qianjinzhai district.[20]

Matsuda was succeeded by Yonekura Kiyotsugu 米倉清族 (1863–1931), another seasoned mining engineer of the same educational pedigree. Finishing at Tokyo Imperial University a few years after Matsuda, Yonekura established his expertise through his work with various collieries in Hokkaido over two decades.[21] He took up the position at Fushun in July 1911 and wasted no time in launching a second development plan that continued and extended what his predecessor had begun. Over the next few years, the colliery excavated three more subsurface mines across the eastern half of the coalfield at Wandawu (萬達屋), Longfeng (龍鳳), and Xintun (新屯). It also further expanded its auxiliary industries. Machine and tool building, in particular, saw advances under the plan. Although Mantetsu still imported much of the colliery's equipment—from German Allgemeine Elektricitäts-Gesellschaft motors to French Citroën gears made in England—its machine and tool factories in Fushun were now able to produce simple devices and proved more than capable of making "alterations and renewals of parts on the various pieces of machinery employed, such as hoists, turbine pumps, ventilating fans, screens, excavators, electric machines, etc."[22] Together, these developments deepened Fushun's industrial character, not only increasing the scale of extraction but also advancing the enterprise's technological self-sufficiency and providing the foundation for more mechanization.

Under Yonekura's plan, the way that coal was mined, too, underwent a transformation. While Japanese boosters may have celebrated Fushun's abundance, this bounty presented a problem for Japanese mining engineers who had "no prior experience whatsoever working such a thick coal seam."[23] The pillar-and-stall method they were familiar with left behind too much coal for their liking, and they sought a more complete recovery of the resource. After some investigation, Fushun's engineers decided to switch to hydraulic stowage (灑砂充填採掘法), through which they hoped to more efficiently and exhaustively excavate the seam. In this method, miners began, as they did in pillar-and-stall, by sinking a shaft into the ground from which they drove tunnels outward into the coal seam. When

20. Yadian, "Youguan Fushun di yi ren kuangzhang."
21. Ōshima, "Ko Yonekura Kiyotsugu."
22. Yamaoka, "Two Strip Pits," 948; Murakushi, *Transfer of Coal-Mining Technology*, 36–37.
23. *MT10*, 492.

these tunnels reached the next seam of shale or basalt, the miners then stopped advancing and started working the sides, widening them to form a room that was typically six feet high, eight to twelve feet wide, and fifty to one hundred feet long, but could go as high as twenty feet and as wide as sixteen feet if geophysical conditions permitted. When the room was completed and the excavated coal removed, the miners closed up the space with wooden boards or mats woven from sorghum stalks and pumped in a sand and water mixture through pipes. After the room was filled, the water was drained, and the miners then went on to work on an adjacent room where the process was repeated. Once they finished mining one level, the miners would cut a new level above the existing one and proceed to excavate rooms on top of the now sand-packed spaces below.[24]

To employ hydraulic stowage, the colliery introduced an assortment of supporting technologies that had impacts on environments beyond the subterranean sites. It installed two dragline excavators, a steam shovel, and a bucket excavator on the banks of the Hun River as well as multiple steam shovels in the sandstone hills by the Yangbaipu and Ōyama mines. Together, these massive machines dredged and dug out sand, which electric locomotives carried to the individual mines. There, the sand was sifted through screens over bins, a process that removed the larger stones. What was left behind was then flushed with water into mixing pans, from which the mixture could be pumped through pipes hundreds of feet into the ground to the workings.[25]

Fushun's engineers first introduced hydraulic stowage at the Yangbaipu mine in 1912. When that appeared successful, they applied it to all the other mines in the colliery.[26] Later, they further adapted the method with longwall mining, in which a seam was worked in lengthy single slices. The resultant method—almost too literally termed "diagonal long wall slicing with hydraulic stowage" (累段傾斜長壁昇払)—would be used in underground mining at Fushun for decades to follow.[27] By packing excavated areas with sand, hydraulic stowage reduced risks of subsidence and fire.

24. Yamaoka, "Manchurian Plant," 901. While there may have been even earlier instance of this practice, the modern version of using water to carry sand and gravel into mines to fill excavated cavities originated in Germany at the turn of the century, and this was what Fushun's system was based on. See Kubo, "Bujun tankō ni hattatsu seru shase jūten saikutsu hō," 1–2.

25. Yamaoka, "Manchurian Plant," 900–901.

26. Yamaoka, 901.

27. Murakushi, Transfer of Coal-Mining Technology, 41–42; Kubo, "Hydraulic Stowage Mining System," 221; Manshikai, Manshū kaihatsu yonjūnen shi, 87–89.

In addition, it also cut down on the amount of timber used as pit props to further support the roof in the pillar-and-stall method. When it began operations, the colliery initially imported Sakhalin fir wood from the home islands as the area around the mines was apparently "desolate as far as the eye can see." This proved prohibitively expensive, and a more local source was sought and found in the pine forests lining the railway between Andong, on the border with Japan's colony of Korea to the east, and Mukden. However, the timber from along the An–Feng line soon needed supplementing as the mining enterprise expanded and the demand for pit props increased. Aside from purchasing supplies from farther afield, such as Manchurian red pine from Korea, the colliery also began cultivating trees around the mines, setting aside almost a thousand acres for afforestation in 1919. With hydraulic stowage, fewer pit props were needed for support. In addition, used logs could be recovered from sand-packed rooms when working the area above and then reused accordingly.[28] Moreover, and perhaps most importantly from the perspective of the company, because hydraulic stowage meant that fewer pillars of coal had to be left in place for structural support, colliers were able to mine as much as 70 to 80 percent of the seam. It thus allowed for a more complete extraction of the coal as was consistent with Mantetsu's productivist imperative.[29]

Through the Japanese technocrats' two developmental plans, Fushun's mechanization and productivity grew in tandem. The colliery's output increased more than tenfold, from around 233,000 tons in 1907 to almost 2.8 million tons in 1919. The initial goal of 3,000 tons a day had been exceeded more than twice over. By 1920, Fushun churned out at least 10,000 tons daily, and Mantetsu made between two and three million yen each year from this enterprise—less than originally estimated but still a significant proportion of the company's profits.[30] Opening new subsurface mines, updating existing ones, and adopting hydraulic stowage could not by themselves claim all the credit for the increase, though. After all, this period also witnessed the beginnings of what would be Fushun's greatest engineering achievement: the excavation of its open-pit mine.

28. *MT10*, 498–502; Yamaoka, "Two Strip Pits," 948. On Japanese imperial forest management in colonial Korea, particularly in regard to afforestation, reforestation, and forest regeneration, see Fedman, *Seeds of Control*.

29. *MT10*, 493.

30. "Fushun Colliery Development," 516; "South Manchuria Railway Company." In 1920, for instance, total profits were twenty-seven million yen, which meant that Fushun accounted for about a tenth of earnings.

Excavating the Open Pit

That open-pit mining proved viable in Fushun was in part a product of geology. On the western end of the coalfield, the seam tilted upward, so much so that it was a mere thirty or forty feet from the surface. In modern open-pit mining, colliers extracted desired deposits by using machines to strip off in steps the overburden—the layers of soil, sand, and, in Fushun's case, shale rock above. The closer the seam was to the surface, the easier it was to mine. This method allowed for a more complete exploitation of the seam, but it was also highly capital intensive, requiring significant initial outlay for excavators and other extractive equipment.[31]

According to one account, Gotō had instructed Fushun's management to explore the possibility of open-pit mining as early as 1909.[32] After several years of prospecting to determine the site's suitability, the colliery began excavating an open-pit mine at Guchengzi (古城子) in April 1914. Mantetsu's Japanese engineers, having no previous experience with modern open-pit mining, looked to Germany and the United States, the global leaders in mining engineering and technology, where this method was being used to extract coal and other resources.[33] At the very beginning, the colliery relied solely on human labor, the topsoil dug out with sweat and spade. As mechanization progressed, it soon employed excavators to remove the rest of the overburden and steam shovels to mine the shale rock and coal beneath. Loaded onto dump cars manually, the resources and tailings unearthed were then sent onward to processing plants and disposal sites. In November 1917, the colliery started work on a second open pit in Qianjinzhai, just west of the shaft mine, and it was here a few years later that plans for a colossal excavation took shape.[34]

The scaling up of open-pit mining in Fushun began under the direction of Inoue Tadashirō 井上匡四郎 (1876–1959), who took over as colliery manager after Yonekura stepped down in 1919. Like the men who came before him, Inoue was a product of the engineering department at Tokyo Imperial University. Unlike his predecessors, however, he spent most of his career not in industry but in academia. Receiving a silver watch from the

---

31. Hartman and Mutmansky, *Introductory Mining Engineering*, 182–91.

32. Koshizawa, *Zhongguo Dongbei dushi*, 116.

33. For open-pit mining in the United States, see LeCain, *Mass Destruction*, which looks at copper mining outside of Salt Lake City, Utah; and Leech, *City That Ate Itself*, which also looks at copper mining, but in Butte, Montana.

34. *MT10*, 532–35; "Kongo no Mantetsu"; "Manshū ni okeru Mantetsu"; "Japan Moves Town."

emperor when he graduated in 1899—an honor conferred for topping his class—Inoue worked as a professor for two decades, first at his alma mater. In 1901, the university sent him to Germany and the United States for further studies. Upon his return three years later, he taught at the Osaka College of Advanced Industry, then at Kyoto Imperial University, and then back at Tokyo Imperial University, where he was put in charge of the lectures on mining. In 1910, Inoue, who had inherited the title of viscount from his adoptive father, Meiji statesman Inoue Kowashi, was raised to the peerage by election.[35] An active member of the House of Peers, Inoue sat on a number of special commissions, such as the one for the 1918 Munitions Industries Mobilization Law, which made provisions for the Japanese state to take over industries deemed militarily strategic in times of war.[36] Although he had never run a mine or a mill before, Inoue, who bore both scientific acumen and hereditary rank, was identified by Mantetsu as an expert and leader who would be able to spearhead the next stage of development for both the Fushun colliery and the company's newly opened ironworks in nearby Anshan.[37] He was swiftly appointed to head these two enterprises.[38] The expectation was for him to oversee their transformation into industrial behemoths. In the case of Fushun, his plans for further development pivoted on the expansion of open-pit mining.[39]

A cosmopolitan technocrat, Inoue drew upon transnational expertise for his efforts, tapping into networks that he most likely formed during his years abroad. In the summer of 1921, he invited a team of American geologists and engineers to Manchuria. This team consisted of William R. Appleby, dean of the University of Minnesota's School of Mines and a metallurgist by trade; two professors of geology, one from the University of Minnesota and the other from the University of Wisconsin; and three Minnesotan mining engineers. The men, whose experience "covered the wide range of exploration, development, mining and beneficiation of ores," had been selected from Minnesota and Wisconsin, Appleby explained, "because of the similarity of the mining problems of this region to those of Manchuria."[40] Minnesota boasted some of the largest mineral

35. "Shishaku giin hoketsu senkyo."

36. "Dōin hōan iinkai."

37. On the history of the iron and steel works at Anshan, see Matsumoto, "*Manshūkoku*" *kara shin Chūgoku e*; and Hirata, "Steel Metropolis."

38. Minami Manshū tetsudō kabushiki gaisha, *Minami Manshū tetsudō kabushiki gaisha dainiji jūnen shi*, 588; Kokugakuin daigaku toshokan chōsaka, *Inoue Tadashirō monjo mokuroku*, 1.

39. "Fushun Colliery Development," 513.

40. "Manchuria Mine Survey Completed"; "Frank Hutchinson Entertains."

deposits in the United States, and its iron-rich Mesabi Range was where open-pit mining was first introduced in the late nineteenth century. Taking advantage of the comparative softness of the ore and its proximity to the surface, open-pit mining helped Minnesota become the country's biggest iron producer—the Mesabi Range alone accounted for more than half of American iron output over the course of the twentieth century.[41] The team spent almost two months surveying both Fushun and Anshan and drafted recommendations on how to best develop them. In Fushun's case, the two main concerns they sought to address were how to extend the open pit in the most economical manner and how to relocate the adjacent urban district that now stood in its path.[42]

Inoue seemed satisfied with their evaluation and was eager to put their expertise to further use. Not long after the trip, he invited L. D. Davenport, one of the team's mining engineers, to return to Fushun as a consultant to help direct work on the open-pit mine. As chief engineer of the Oliver Iron Mining Company, Davenport had been the "engineering brains" behind moving the town of Hibbing, Minnesota, to access the iron deposits beneath it. Inoue hoped he would pull off a similar feat in Fushun.[43] In Duluth, Davenport gathered a crew of fifteen men—steam shovel runners, locomotive drivers, cranesmen, firemen, and brakemen as well as a dump foreman, a general track foreman, and a master mechanic—who followed him to Fushun under contract with Mantetsu to train the workers there in novel mining methods and the operation of new extractive equipment. Handpicked not only for their individual skills but also for their "durability and steadiness under the very different working conditions in Manchuria," these men would, Davenport promised, "demonstrate the most skillful and efficient method of earth moving."[44]

The ability to move earth on the scale that Davenport would propose was a relatively recent development, enabled by deploying coal-fired steam technologies. This capability bred a certain arrogance intrinsic to carbon technocracy that was rooted in the presumption that human ingenuity might replicate the forces of nature. But if Davenport seemed confident before his return to Fushun, he was to find his time there vexing. Writing in the fall of 1922 to Inoue, who by then had been forced to resign from

41. Whitten and Whitten, *Handbook of American Business History*, 71. On the history of the Iron Range around this time, see Manuel, *Taconite Dreams*, 5–13.
42. "Bei gakusha dan."
43. "Tech Man Who Moves Towns."
44. Davenport to Inoue (November 29, 1921), in *ITM*, reel 105, no. 05310.

Mantetsu following a major company scandal, Davenport was exasperated: "I have accomplished practically nothing since I have been here."[45] Part of the problem, as he saw it, was that he was simply being ignored. In one episode he recounted, he had gone down into the open pit to show a drilling foreman and his crew how to blast "gopher holes." Miners typically drilled vertical holes on top of a bank, into which they inserted black powder that they detonated to structurally weaken the bank before stripping it by shovel. But gopher holes, made near the base of the bank and on a slight horizontal incline, were intended for when the ground was too dry and sandy for vertical holes to stay open and when the bank was too high to drill from the top.[46] After Davenport's demonstration, the workers decided that it was easier for them to keep drilling top holes and proceeded to disregard his directions.[47]

Frustrating as he found this indifference on the ground, Davenport's greatest annoyance stemmed from his ongoing disagreement with the superintendent of the open pit, Onuma Tokuei. The two men contended over plans for the open pit, specifically the important matter of how to transport extracted material out of the excavation. Davenport advocated a "railway approach" method that involved several locomotives ferrying loads from the workings—overburden, debris, shale, and coal—along two tracks between the open pit and the processing plants and waste dumps. As these trains ran their cycles, they occasionally switched between the two tracks to complete their routes. Davenport highlighted that this method was employed in many major mines, including those of the Chile Exploration Company in Chuquicamata, the world's largest copper mining complex, and that his plan was merely an adaptation to Fushun's conditions. Onuma, who had been involved in opening Guchengzi in 1914, was unimpressed.[48] According to Davenport, Onuma claimed that he "knows all about open cut mining and can learn nothing new from America." The Japanese engineer had dismissed the railway approach method as "difficult" and "dangerous," his main critique being that the constant switching of tracks could easily lead to collisions. Onuma proposed an alternative, an "inclined winding" method, which consisted of a more extensive system of double tracks between each piece of equipment and its corresponding

45. On the Mantetsu scandal and Inoue's resignation, see Iguchi, *Unfinished Business*, 12–13.

46. "Loading Black Powder"; Hustrulid, *Blasting Principles*, 16–18.

47. Davenport to Inoue (October 22, 1922), in *ITM*, reel 105, no. 05312.

48. Yamaoka, "Two Strip Pits," 945.

destination point, with the cars carrying the extracted material hoisted up the incline of the pit by cable en route.[49]

Davenport was quick to defend the safety and simplicity of his technological design. He argued that stationing a flagman at each switch to direct train traffic would prevent collisions, adding that this danger existed in Onuma's plan as well, since it was present whenever trains shared the same track. Davenport then turned around and critiqued Onuma's inclined winding method, which he found wanting in both economy and efficiency. This plan was more costly, he contended, for it failed to factor in interest and depreciation and hence underestimated expenses by more than 900,000 yen. It was also far more "complicated and cumbersome." Using the example of shale transport, Davenport pointed out that Onuma's plan involved nine steps and twenty-two new pieces of equipment, while his required just three steps and a mere seven new pieces of equipment. Another issue that critics frequently brought up against Davenport's plan was that the level of the technology involved did not match the quality of labor available to operate it. As the charge went, "Chinese coolies are not as intelligent as the common labor in the American mines." Davenport refuted this too, stating that in the iron mines of Minnesota, "less than 10% of the employees are English speaking," most of them being "Europeans, Montenegrins, Austrians, Italians etc. of the lowest class"—a defense of the superiority of his design that, in a manner consistent with technocratic tendencies, all too readily took for granted the inferiority of those who were to actually carry it out.[50]

The quarrel between the two engineers is not surprising—scientific and technical experts prescribe different solutions when faced with the same problem all the time. It does, however, remind us that while Japanese technocrats were generally enthusiastic about bringing the latest techniques and technologies from abroad for application to their industries, they were not uncritical about the process. In his study of architecture and seismology in post–Meiji Restoration Japan, historian of technology Gregory Clancey demonstrates how Japanese and foreign architects, engineers, and scientists came to question the inherent superiority of foreign building methods when the great Nōbi earthquake of 1891 leveled many of the new constructions of brick and iron while leaving older Japanese wooden structures intact—remarkable given the high pedestal Western knowledge

49. Davenport to Inoue (October 22, 1922), in *ITM*, reel 105, no. 05312.
50. Davenport to Inoue (October 22, 1922), in *ITM*, reel 105, no. 05312.

occupied in that era.[51] By the 1920s, Japanese experts like Onuma were self-confident and well positioned enough to regularly challenge assumptions that imported technologies were necessarily better or well suited to local conditions.

In order to mitigate the dispute between Onuma and Davenport and decide on a course of action for the open pit, Mantetsu appointed a special committee to review both proposals. By winter 1923, the committee drafted a temporary plan that combined aspects of each proposal. Because the railway approach method seemed to work better at the shallower portions of the pit and the inclined winding method at the deeper parts, railway approach would be used to transport overburden and shale closer to the surface and inclined winding to transport coal farther down. This hybrid plan laid the groundwork for the development of an excavation that would transform the economic and environmental landscape of the mining town.[52] The colliery imported more machines from the United States to enable this effort, adding to its existing array of American-made Bucyrus excavators and shovels.[53] By 1924, just a year later, coal output from the open-pit mine was 1.8 million tons, over a third of the colliery's total output. Daily production had reached 7,000 tons.[54] Considering that the sixteen coal mines owned by American industrialist Henry Ford averaged around 13,000 tons per day in the 1920s, the productivity of this one mine was remarkable.[55] As part of this whole process, the urban district that abutted the excavation was moved to a new location, Yongantai, even as the area it had occupied was consumed by the widening maw of the mined monstrosity.[56] In this way, then, did the open pit that so captivated Yosano Akiko emerge.

The product of transnational tensions and compromise in which machines and methods used to mine iron in Minnesota were adapted to the extraction of coal in Manchuria, Fushun's open-pit mine quickly became an icon of Japanese industrial modernity. As with Daniel Jackling's Bingham Pit copper mine near Salt Lake City, Utah, a great deal of headiness

---

51. Clancey, *Earthquake Nation*. For more on the frictions that underlay processes of "technology transfer" in Japan, see Gooday and Low, "Technology Transfer and Cultural Exchange."

52. "Rotenbori zenkōan no naiyō"; Minami Manshū tetsudō kabushiki gaisha, *Minami Manshū tetsudō kabushiki gaisha daisanji jūnen shi*, 1756.

53. Yamaoka, "Two Strip Pits," 946.

54. Davenport to Inoue (October 22, 1922), in *ITM*, reel 105, no. 05312.

55. Bryan, *Beyond the Model T*, 136.

56. Koshizawa, *Zhongguo Dongbei dushi*, 108–13.

THE SKIP-MACHINE OF THE OPEN-AIR COAL-MINE, FUSHUN

観壮のブッキス掘天露すか驚を眼の客訪（順　撫）

2.4. "The Skip-Machine of the Open-Air Coal-Mine, Fushun." Dump cars filled with coal, shale, and debris are pulled from the depths of the pit up the incline. The Japanese text reads, "(Fushun) The grand sight of the open-pit mine's skip that startles the eyes of visitors." The technological sublime that this excavation evoked was a product of both its scale of operations and its degree of mechanization, which the skip exemplified. (Postcard from the author's collection.)

and hubris surrounded this act of stripping the earth to plunder its riches.[57] By 1931, the workings had reached such depths that China-based American newsman J. B. Powell felt prompted to remark that it was a "gigantic excavation into the surface of the earth into which an entire city could be buried" and that "America's most modern sky-scraper could be stood on the bottom of the pit and the spire would not reach very far above the surface of the surrounding country-side."[58]

Apart from gracing the pages of Mantetsu publications and all manner of print media beyond, the open pit at Fushun was also a landmark for visitors like Yosano and many others before and after her. The appeal, it seemed, was in the awe. "Upon arriving at the observation platform at Guchengzi, you will be astounded by the full view of the grand open pit," one Japanese travel guidebook promised.[59] Among the most notable visitors to this excavation was the future South Korean dictator Park Chung Hee. In 1942, Park, who was then a trainee with the Manchurian Military Academy, had stopped by while on a weeklong class excursion to former battlegrounds and the towering heights of industry in the region. The open pit at Fushun was a site not to be missed.[60]

Rationalization through Mechanization

As much as Fushun's open-pit mine may have been famed for its magnitude, so, too, was it widely celebrated as the epitome of mechanization. Bulky steam excavators tore at the overburden, while hefty electric and steam shovels ripped up the shale and coal below. The materials, once removed from the earth, were loaded onto dump cars that traveled along tracks on a circuitous route up the winding terraces out of the pit. Those cars that carried coal would stop halfway and empty into skips—mining cars hoisted up slopes or shafts—that were pulled along an inclined railway by cables to the coal preparation plant above.[61] The extant pictures of these operations suggest that when everything was in motion, the scene resembled a clockwork apparatus of gears, chains, and other mechanisms rotating and revolving to the rhythm of industry. Coal powered this massive technological assemblage, made complicit, as it were, in its own exploita-

57. LeCain, *Mass Destruction*, 11–15, 129–68.

58. Powell, "The Marvellous Fushun Colliery," 374.

59. "Bujun shi yūran no shiori."

60. Eckert, *Park Chung Hee*, 232.

61. Minami Manshū tetsudō kabushiki gaisha, *Minami Manshū tetsudō kabushiki gaisha dainiji jūnen shi*, 588–93; Woodhead, *A Visit to Manchukuo*, 73.

tion. As the assemblage grew, so, too, did its demand for the black rock. Energy extraction itself was and remains an energy-intensive enterprise.

Mechanization of coal mining extended beyond the open pit. For one thing, it was central to coal dressing, in which coal was sorted according to size by human hands and motor-powered conveyors, screens, and jigs, removing rocks and other impurities in the process.[62] Mechanization also continued to transform the nature of underground mining, as it had done since Matsuda's first development plan.[63] The spectacle of the open-pit mine had caused Yosano to reconsider "the notion of what a coal mine was" by upending the idea that it should be "an excavation of horizontal and vertical tunnels dug deep into the ground."[64] Nevertheless, the subsurface workings that open-pit mining appeared to have rendered obsolete still persisted and flourished in Fushun. Coal production from the crisscrossing network of tunnels that ran beneath the earth consistently accounted for at least half—if not more—of Fushun's output.[65] And as with open-pit mining, the sustained expansion of these extractive efforts underfoot, too, owed a great deal to the deployment of all sorts of machines.

Fushun's mining engineers, like overseers of many other industries, regarded mechanization as an integral part of the "rationalization" (合理化) of work. Rationalization, as a concept, had gained much traction in business and bureaucratic circles in interwar Japan. Although there was never a solid consensus as to what exactly it entailed, the notion of rationalization was almost always associated with the application of science to industry— "'science' in this case connoting objectivity, the utmost efficiency, and the ascendance of 'fact' over tradition and the rule of thumb," as historian William Tsutsui has put it.[66] Rationalization coursed through the arteries of the wider industrial world. Its pulse could be felt in the productivist ideologies of Taylorism and Fordism and in the ubiquitous quest for more output at greater speeds and lower costs.[67]

At Fushun, the loudest proponent of rationalization through mechani-

62. Yu, *Fushun meikuang baogao,* 120.

63. Minami Manshū tetsudō kabushiki gaisha, *Minami Manshū tetsudō kabushiki gaisha dainiji jūnen shi,* 545–47.

64. Yosano and Yosano, "Man-Mō yūki," 140.

65. Kubo, "Bujun tankō no saitan jigyō," 28.

66. Tsutsui, *Manufacturing Ideology,* 63. This idea of replacing rule-of-thumb methods with standardized ones is a key principle in Tayloristic scientific management. See Taylor, *Principles of Scientific Management,* 10.

67. For a transnational account of how rationalization generally and Fordism specifically were taken up in regimes of varying ideological persuasions, see Link, *Forging Global Fordism.*

THE SELECTION PLACE OF THE OYAMA COAL MINE, FUSHUN.
日活動する探山大炭撰所炭場 (順 撫)

2.5. Inside the Ōyama mine's coal dressing plant. With the help of machinery, primarily the English-manufactured Marcus screen conveyor, workers clean and sort the raw coal by size into coal fragments, smaller pieces, and larger lumps. The Japanese text describes the plant as "operating day and night" (日夜活動する), and this facility did indeed run constantly, with three overlapping shifts of workers each laboring ten hours. (Image courtesy of the Harvard-Yenching Library, Harvard University.)

zation was the mining engineer Kubo Tōru 久保孚 (1887–1948).[68] Born in Kōchi prefecture in Shikoku, the smallest of Japan's four main islands, Kubo was the third child and eldest son in a family of seven children. His father, Gidō, was a local official, but one who was colloquially called a "well and fence" (井戸塀), as the family fortune was so exhausted paying for the affairs of office that only the "well and fence" were said to remain. It did not help that Gidō also invested in several business ventures that wound up failing. Despite the financial hardship at home, Kubo did well in school and secured external sponsorship for his studies at Tokyo Imperial University, where he majored in mining engineering.[69] Like all students in this field, Kubo needed to undertake a practicum at a working mine as part of his coursework. This practice had its origins at Japan's first modern engineering school, the Imperial College of Engineering, where Henry Dyer, the Scottish engineer who served as its first principal, incorporated

68. Kubo, "Saitan no gōrika."
69. Hara, *Harukanari Mantetsu*, 18–24.

the practicum into the curriculum to ensure that "the students were," in his words, "taught the relations between theory and practice, and trained in the habits of observation and original thought."[70] Book learning needed to be refined in the flame of application. It was this requirement that first brought Kubo to Fushun in the summer of 1912.

While most students completed their practicums in mines in the home islands, Kubo opted to do his at the Fushun colliery. Manchuria-bound students received higher stipends, and this appealed to Kubo since he was, at the time, shouldering his father's debts, as one biographer notes.[71] His experience in Fushun would set the course for the rest of his life. In his practicum report, Kubo expressed deep admiration both for Fushun's natural endowment and for the technological apparatus Japan was assembling on site: "Even as the magnitude and inexhaustibility of its coal seams below are known throughout the world, so too are the extensiveness and newness of the facilities above admired globally."[72] So taken was he by the coal mines that Kubo arranged to meet with colliery manager Yonekura toward the end of his practicum to discuss the possibility of working for the colliery after completing his degree. Graduating third in his class a year later, Kubo returned to Fushun, and he steadily rose through the ranks over the next few decades, serving at the helm as colliery manager between 1932 and 1937.[73] On his way to the top, he obtained an engineering doctorate from Tokyo Imperial University. His dissertation, based on his work in Fushun, was on hydraulic stowage.[74]

In 1934, not long after becoming colliery manager, Kubo penned a piece for an engineering journal on the importance of further mechanizing subsurface mining at Fushun: "Although the operations at the Open Pit Mines are perfectly mechanized, the mechanization of underground workings has not yet attained the degree of perfection." As he understood it, there were limits to "human power" (人力) that the colliery had already reached, and "in order to further leap forward in efficiency, there is no other way than to develop a new course and to aim for the mechanization of coal mining." Kubo noted Fushun's engineers had been making efforts toward those ends.[75] Through the 1920s and 1930s, these engineers

---

70. Dyer, *Dai Nippon*, 5. For more on the Imperial College of Engineering, see Duke, *History of Modern Japanese Education*, 172–81.

71. Hara, *Harukanari Mantetsu*, 22.

72. Kubo, *Bujun tankō*, 1.

73. Hara, *Harukanari Mantetsu*, 23–24.

74. Kubo, "Bujun tankō ni hattatsu seru shase jūten saikutsu hō."

75. Kubo, "Bujun tankō ni okeru saitanki," 264.

brought into the depths of the mines an assortment of tools to better facilitate coal extraction. There was, for instance, the coal cutter, a machine with
an arrangement of pick-like teeth on a chain, a bar, or a disk that miners
used to undercut a working face before employing other tools to collapse
the overhanging portion above. And there was the electric auger drill. With
this tool, miners could more easily make holes on the coal face into which
they inserted black powder to detonate so as to break the coal apart. Then
there was the coal pick hammer, a device designed for miners to work particularly hard rock or coal by pummeling it with percussive force. Mostly
imported from the United States and Germany, these mechanized tools
promised increased efficiency and, with that, greater productivity.[76]

It was, however, not so much extracting coal from the seam as moving it
out of the pit that seemed to have benefited most from mechanization. In
the past, miners broke the coal off the face, collected it into wicker baskets,
and loaded it onto two-wheel carts, which they then manually pushed to
the bottom of the haulage way. There, they emptied the coal into cars that
were hoisted to the surface. Alternatively, the coal could be transported out
of the tunnels through water channels. This proved problematic, though,
since coal dust often settled and accumulated in the water and choked the
system. Moreover, in winter months, the water frequently froze, rendering
these channels unusable. These challenges of muscle and materiality were
ultimately mitigated by installing conveyors that carried the excavated coal
and debris away from the working face by chain and sprocket.[77]

Other machines that had been central to the mining process since the
early days included water pumps and ventilation fans. While the Fushun
coal mines were naturally drier than those in the Japanese home islands,
the use of hydraulic stowage resulted in considerably more water needing
to be drawn to the surface in order to keep the mines operational. According to one estimate, for every ton of coal extracted, five tons of water had
to be pumped out.[78] As for ventilation fans, these were used not only to
ensure that fresh air flowed into the working areas but also to expel dangerous gases from those subterranean spaces.[79] In this way, then, pumps and
fans regulated the conditions beneath the surface so that this otherwise
hostile and unwelcoming environment could function as a safe and productive one. The energy regime of coal was grounded in the artifice of the

76. Kubo, 265–79.
77. You, *Mantetsu Bujun tankō no rōmu kanri*, 205.
78. Minami Manshū tetsudō kabushiki gaisha Bujun tankō, *Tankō dokuhon*, 472–73.
79. Minami Manshū tetsudō kabushiki gaisha Bujun tankō, 467–68.

mine, characterized by Lewis Mumford as "the first completely inorganic environment to be created and lived in by man."[80] There was, however, a certain precariousness to this environment, for it still remained dangerous and became immediately more so when those supporting apparatuses stopped functioning.[81]

The mechanization of mining above and below ground, while key to Fushun's expansion and transformation into a center of energy extraction, was itself enabled and maintained by harnessing considerable amounts of energy. The modern machine was, after all, inseparable from the power that drove it. In the beginning, coal-fired steam engines supplied much of the motive power for mine work, from haulage and winding to pumping and ventilation. Steam was supplemented by compressed air, which was sent underground at high pressure through pipes to power pneumatic devices. Steam and compressed air were gradually displaced—though they never did completely disappear—as Fushun underwent widespread electrification in the successive decades. Fushun's first power plant began operations in 1908 with just two coal-fueled 500 kilowatt generators. In 1915, the colliery brought on line another power plant, though this was driven not directly by coal but indirectly through Mond gas—named after German chemist Ludwig Mond, who discovered that heating coal with steam produced a fuel gas.[82] To these two plants, a third and then a fourth were added at Daguantun in western Fushun, both coal fired. By 1937, the Mond gas plant shut down because of threats of subsidence caused by mining, but the remaining thermal power facilities boasted a capacity of as much as 280,000 kilowatts. That year, subsurface mining consumed close to 200 million kilowatt hours, with over 60 percent of this going toward pumping facilities—an indication of how much energy it took to sustain the manufactured environment.[83]

80. Mumford, *Technics and Civilization*, 69.

81. As Liza Piper reminds us through her work on the industrialization of the large lake region in Canada's Northwest, industrial transformation often "did not weaken links between humans' work and nature's work but remade human relationships to nature." See Piper, *Industrial Transformation of Subarctic Canada*, 287. In an almost too literal way, the mechanization of the mines brought human workers into deeper contact with the subterranean environment and its many hazards, as we shall soon see.

82. One benefit of Mond gas was that in the process of its manufacture, ammonium sulfate could be recovered as a by-product. Given Fushun coal's high percentage of nitrogen—which together with hydrogen, sulfur, and oxygen make up ammonium sulfate—the colliery ended up producing much of this commercially valuable inorganic salt, which it sold to Japan on the side as fertilizer for the cultivation of crops such as rice, wheat, and barley. See Yamaoka, "Two Strip Pits," 947.

83. Oka, "Electric Power Development," 307; Minami Manshū tetsudō kabushiki gaisha Bujun tankō. *Tankō dokuhon*, 343–45.

The shift from steam to electricity as the primary form of inanimate power at the mines yielded several benefits for Fushun's operations. To begin with, electrification allowed the mines to make use of many of the latest technologies available from abroad, from coal cutters to conveyors, which increasingly ran on electricity.[84] Electricity drove local coal and passenger transport, including a high-voltage direct-current overhead trolley system that was "the first of its kind in the Far East."[85] At the open-pit mine, the switch to electricity also helped in overcoming challenges of clime. The company had initially faced problems with running steam-powered shovels in the notoriously cold Manchurian winter. There was, as one report noted, "great difficulty with water supply pipes and steam connections and fittings freezing up and breaking, causing considerable delay and expense."[86] As a corrective, Fushun began using more electric-powered shovels instead. Furthermore, in the underground workings, moving from steam to electricity meant that mining could be carried out deeper into the earth, as electrical transmission wires had a reach far beyond that of stationary steam engines.[87] In this way, electrification facilitated the extension of extraction at scales and depths hitherto unreached. This may have allowed for more coal to be taken out of the earth, but it also brought more miners into its far reaches, where, as we will discuss below, the dangers they were exposed to deepened with descent.

Fushun's operators may have pursued mechanization in the name of efficiency, economy, and the expansion of operations, but they were also interested in its promise to reduce the reliance on labor. World War I and its immediate aftermath saw a rise in labor activism in Japan, as the number of workers in factories almost doubled because of industrial expansion and many of these workers began to agitate for better working conditions and wages, especially in response to wartime inflation. Bureaucrats and businessmen alike found themselves troubled by what they called the "labor question" (勞働問題), which coincided with the specter of socialism following the 1917 Bolshevik revolution and the 1918 Japanese rice riots—a summer of unprecedented nationwide uprisings against the precipitous price increase of that staple grain. These elites regarded this as a problem for which they impatiently sought solutions.[88]

---

84. Devine, "Coal Mining."

85. Yamaoka, "Two Strip Pits," 946.

86. "Influence of the Coldness during the Winter upon Open Workings," in *ITM*, reel 109, no. 05954.

87. You, *Mantetsu Bujun tankō no rōmu kanri*, 137.

88. Garon, *State and Labor*, 40–42. On the rice riots as a watershed in the prewar Japanese labor movement, see Gordon, *Labor and Imperial Democracy*, 104–9.

Although Fushun did not experience labor uprisings in this period, tensions were nonetheless evident. "We have yet to see the disturbances of labor strikes here in Manchuria," Inoue Tadashirō observed as the rice riots were taking place, "but it does not mean that their undercurrents do not exist." In addition, he noted that the Fushun colliery had recently been experiencing a slowdown in production, which he attributed in part to difficulties in attracting as many "Shandongese coolies" as before because of rising wages elsewhere and the general mobility of labor.[89] Mantetsu's leadership regarded mechanization as one way to deal with this labor shortfall. For example, in the report he wrote outlining his proposed railway approach method for open-pit mining, Davenport stressed that one priority Mantetsu had laid out that his method would realize was for an increased output that could be "maintained in the face of possible labor shortage, strikes etc."[90] Yet for all its efforts at mobilizing the machine, the colliery never did seem to be able to truly diminish its dependence on labor. As its targeted production rose from year to year, this industrial apparatus regularly required tens of thousands of workers to keep its cogs turning.[91]

"The Driving Force in Coal Mining"

Essential as human energy was to the mines' productivity, Fushun's management became preoccupied with controlling the colliery's sizable labor force. By 1915, a few years into the second development plan, there were over twenty thousand workers in Fushun. This was more than a tenth of the total number of workers in the coal-mining industry in Japan that year.[92] While Fushun's majority Chinese workforce fluctuated with the vicissitudes of the coal market in the coming decades, it would not fall below that figure. The number of colliers almost doubled between 1916 and 1919, corresponding to production rising to meet increased demand for regional coal in the midst of World War I. Likewise, the workforce contracted in accordance with the economic downturn of the early 1920s and expanded again as the economy recovered later in the decade. Although the onset of the depression and the Japanese takeover of Manchuria in 1931 were

89. Inoue, "Manshū no tankō."

90. Davenport, *Open Cut Mining*, 1.

91. For an account that, through the case of the late nineteenth-century Assam tea industry, similarly elucidates the interdependence of capital- and labor-intensive industrialization, see Liu, *Tea War*, 115–51.

92. "Establishment and Persons Engaged in Mining."

accompanied by yet another dip in the number of laborers, Fushun's work-force steadily swelled in the years after. By 1937, on the cusp of the Second Sino-Japanese War, the colliery boasted as many as fifty thousand work-ers. The company adopted a paternalistic posture toward these workers. In one notice to the Chinese miners, the colliery manager began with the premise that as "our company generally has in mind the good fortune of you coolies, loving you, helping you," so should the "coolies" then "refrain from delaying your own duties" and "follow the rules."[93] At the same time, the company repeatedly referred to Fushun's workers in more objectifying terms, foregrounding the potential energy they embodied when it spoke of them as "the driving force in coal mining" (採炭の原動力).[94] Mantetsu expended much energy in trying to master this force.

Work for Fushun's miners, as for most miners elsewhere, was grueling. The hours were long, the environment unsettling, the labor hard. In prin-ciple, the colliery originally operated in two ten-hour shifts daily. In 1929, it moved to two or three eight-hour shifts a day, depending on the mine. In actuality, however, there was a lot of variance between mines, and miners at times found themselves working up to eleven- or twelve-hour shifts. This was, the company readily admitted, "an outcome of competition between various mines for the enhancement of efficiency"—of labor being squeezed as different units within the colliery strove to outperform one another, as they would in socialist production competitions years later (chapter 6).[95]

Those who labored in the underground workings typically reported for the beginning of their shift at the mouth of the coal pit. There, they pre-sented their coal vouchers—stamped in black ink if it was a day shift and red if it was a night one—and had their job titles and serial numbers re-corded. They then signed out safety lamps and, with these lighting their way, descended into the darkness, where, in the words of Lewis Mumford, "day has been abolished and the rhythm of nature broken" in an envi-ronment of "dogged, unremitting, concentrated work."[96] The miners' work revolved around two main efforts: excavating tunnels and extracting coal. More specifically, there was an assortment of individual tasks, including setting off explosives, hewing at the coal face, loading coal, removing de-bris, setting up pit props, and readying rooms for hydraulic stowage.[97] Tools

93. Bunkichi, *Manshū ni okeru kōzan rōdōsha*, 248–49.
94. *MT10*, 495.
95. Mantetsu rōmuka, *Minami Manshū kōzan rōdō jijō*, 84–91.
96. Yu, *Fushun meikuang baogao*, 172; Mumford, *Technics and Civilization*, 70.
97. Mantetsu rōmuka, *Minami Manshū kōzan rōdō jijō*, 11.

may have made work more efficient, but they did not necessarily make it any easier. The heft of equipment like the coal pick hammer, for instance, weighed heavily on the bodies of those who operated it.[98]

At the open-pit mine, too, workers were mainly engaged in manual labor, with tasks such as drilling bore holes for explosives, clearing rubble, and filling mining cars with coal.[99] A handful were involved in the more technical work of operating excavators and shovels. In regard to working such machinery, as in other areas, the Japanese management raised doubts about the competence of the Chinese miners on whom it nevertheless relied. A Sino-Japanese phrasebook for Japanese foremen at the open pit included scripted admonitions to use on their machine-operating Chinese subordinates, mostly for sins of omission that ranged from not being able to "hear sounds" that may indicate "problems" with "the machines that you yourself are responsible for" to not cleaning the insides of these machines or not greasing their gears. As much as the contrived phrases pointed to patronizing distrust, so did they also, in their nagging quality, suggest that Chinese miners occasionally found chances to circumvent the otherwise strict regimen of work—only doing what they were supposed to do when they were explicitly instructed.[100]

Chinese colliers found other ways to work the system, pushing against the strictures of carbon technocracy. Read against the grain, the regulations the company regularly issued give us a peek into such possibilities. If the work in the pits was exhausting, some miners sought out spaces to catch forty winks. One such space was where the mine cars were stored underground. Sleeping there, the company cautioned, was "extremely dangerous and may cost you your life." While a napping worker being run over by one of these cars in that dimly lit setting may have indeed been a danger, this warning was quite likely also an attempt to dissuade miners from snoozing on the clock. Many of the other infractions concerned the nature of the workers' output. Miners did not always fill up the mine cars entirely. Or if they did, the load may have included too many coal fragments of lesser value or pieces of rock mixed into it. There were, in addition, some miners who tried to pull a fast one on those who inspected the coal cars at the surface by "placing rocks on the inside and putting good coal on

---

98. In the example of the coal pick hammer, each weighed around sixteen or seventeen pounds, which may not seem too heavy, but given how miners had to hold it up in various positions for periods of time as the device struck the coal face, its use, too, was physically demanding. See Bulman, *The Working of Coal*, 206.

99. Mantetsu rōmuka, *Minami Manshū kōzan rōdō jijō*, 16.

100. *NMKY*, vol. 2, *Rotenbori hen* [Open-pit mine volume], 178–82.

the outside" to cover the worthless fillers beneath. When such instances were discovered, the colliery would dock the culpable workers' already meager pay.[101]

As it was at the very beginning, Mantetsu's enterprise at Fushun continued to rest heavily on cheap Chinese labor as inexpensive human energy. The wages Fushun's colliers received were determined by vocation, as was consistent with modern factory production and its division of labor. The organizational structure of this divided labor changed over the course of the four decades of Mantetsu's management, but the underlying pattern remained generally the same. In keeping with the tenor of technocracy, perceived technical expertise was a main factor in determining divisions. At the broadest level were employment categories. These had bearing on benefits enjoyed and whether one received a daily wage or piecemeal payment. In 1917, there were three main categories: "regular employee" (傭人), who was deemed to possess some measure of technical skill, enjoyed full benefits of a Mantetsu employee, and earned a daily wage; "full-time worker" (常役夫), who was regarded as slightly less technically proficient, enjoyed some benefits, and also earned a daily wage; and "coal-mining coolie" (採炭苦力), who did not receive any benefits and who was paid based on output. To supplement this labor, the colliery also hired "temporary workers" (臨時夫) as and when needed. Most of the Chinese colliers fell into the full-time worker or coal-mining coolie categories.[102]

Under this framework, colliers were sorted into different positions with disparate compensation based on whether they were Chinese or Japanese. Here, the logic of technocracy was further tempered by a dialectic of racial difference. In 1916, there were almost sixty positions.[103] Some positions, such as "coal inspector" (檢炭方), were more supervisory in nature and only employed Japanese. Others, such as "coal transport laborer" (運炭夫), which tended to be more physically exacting, were filled solely by Chinese. Chinese employees were consistently paid less than Japanese ones overall. This discrepancy was not solely because Chinese tended to be hired into lower-paying positions—a Japanese coal inspector was paid ninety-four sen a day, a Chinese coal transport laborer forty-five sen. The gap existed even when Chinese and Japanese occupied the same position. For example, Japanese "pitmen" (坑夫) earned on average ninety-two sen. By comparison, Chinese pitmen only received the equivalent of fifty sen.

101. Bunkichi, *Manshū ni okeru kōzan rōdōsha*, 250.
102. Bunkichi, 231–44; Mantetsu rōmuka, *Minami Manshū kōzan rōdō jijō*, 16.
103. Bunkichi, *Manshū ni okeru kōzan rōdōsha*, 128–32.

Across the board, Chinese workers' wages were around half to a third of their Japanese counterparts'.[104] This discriminatory disparity was further exacerbated by the fact that Mantetsu paid Japanese workers in the generally stable Japanese yen, the currency of choice in the railway zone, while paying Chinese workers in the more volatile local small silver coin, which was prone to drops in value.[105] So attractive was cheap Chinese labor to the company that it occasionally hired Chinese over Japanese, especially during economic downturns such as the slump of the early 1920s, which drew the ire of Manchuria-based Japanese blue-collar workers grappling with the very real prospect of joblessness.[106] Mantetsu was most likely only thinking about labor costs in that instance, but such practices further foreclosed possibilities of Chinese-Japanese worker solidarity against managerial action that racially determined asymmetries of position and pay already made exceedingly difficult.[107]

In managing Fushun's Chinese workforce, Mantetsu faced a number of persistent challenges. Two primary ones were labor mobility and labor volatility. Worker turnover had been an issue from early on. Mantetsu noted several reasons for this. As many of the migrant miners were from farming families, they often "came in the fall and left in the spring" (秋来リ春去ル), between harvest and planting seasons. Others who kept abreast with the economic situation across the region moved whenever better opportunities arose elsewhere. Still others, who had "extremely low standards of living" and were presumably content with little, returned home as soon as they had saved a small sum of twenty to thirty yen. Overall, one company report concluded, these "primitive farmers" were "obstinate and ignorant in character" (性質頑迷固陋), and, instead of "feeling fortunate" for "the variety of good things received" after arriving at the "big mine," they proved "unable to put up with" the "strict regulations" and "complicated life" therein and left soon after.[108] To get a sense of the turnover, in 1917, the colliery witnessed about thirty-five thousand arrivals and thirty-one thousand departures. From the colliery's perspective, labor mobility meant not only that a lot of resources were constantly put into recruitment but also that

104. Bunkichi, 129–30.
105. Teh, "Mining for Difference," 94–95; "Chinese Miners of Fushun Collieries."
106. O'Dwyer, "People's Empire," 143.
107. Labor activism by Japanese blue-collar workers in Fushun generally took place through the Mantetsu Employee Association's local branch, of which Chinese workers were not part. See O'Dwyer, *Significant Soil*, 218–19.
108. Bunkichi, *Manshū ni okeru kōzan rōdōsha*, 79.

the workforce was not accumulating the experience necessary for boosting productivity.[109]

In terms of labor volatility, Fushun's colliers may not have risen up during the 1918 rice riots, but, like their counterparts around the globe, they engaged in collective action against their employers at other times. Their importance to the energy system gave coal miners a kind of power that few other industrial workers enjoyed. Strikes in coal mines, as Timothy Mitchell argues, "became effective . . . because of the flows of carbon that connected chambers beneath the ground to every factory, office, home or means of transportation that depended on steam or electric power."[110] Historian Tim Wright maintains that coal miners in China were less likely to strike, deterred as they may have been by pseudokinship and secret society ties that undermined class solidarity or by the fact that mining remained a secondary occupation for many who used it to supplement their agricultural mainstay.[111] Still, in Fushun, disgruntled Chinese miners periodically gathered and took their demands to the Japanese management. As a 1927 *Manchuria Daily News* article put it, "the Chinese miners of Fushun Collieries are not certainly all docile like lambs."[112] Wages were the most frequent cause of strikes.[113] In one episode on December 16, 1923, 1,400 workers at the Yangbaipu mine went on strike to protest the lowering of wages amid concurrent food price hikes. By the end of the day, the company agreed to only reduce wages by half the intended amount and to leave food prices untouched.[114] Other motivators for strikes included the arrest of compatriots, the lengthening of working hours, disputes with Japanese foremen and other personnel, and, most notably, maltreatment by Chinese labor contractors.[115]

The contractor system through which Mantetsu recruited and employed many Chinese workers was rife with abuse. In the first few years, the colliery issued work to individual Chinese contractors, who then brought together the necessary labor to get the job done. The colliery paid the contractor for the work, and he, in turn, distributed wages to his workers from

109. Bunkichi, 80–81. For more on the problem of labor mobility at Fushun, see Teh, "Mining for Difference," 124–36.

110. Mitchell, *Carbon Democracy*, 21.

111. Wright, *Coal Mining in China's Economy and Society*, 182.

112. "Fushun Miners All Not Like Lambs."

113. Out of the twenty-seven strikes at the Fushun colliery involving more than one hundred participants between 1916 and 1930, twenty-one were related to wages. *FGY*, 7–33.

114. *FGY*, 13.

115. *FGY*, 7–33.

the sum he received. Later, as the enterprise expanded and the company got more directly involved in procuring labor, the contractor's main function became labor supervision. Under this system he received a cut of the wages of the workers he managed. He also collected bonuses if his workers stayed on for more than thirty days, an attempt to mitigate high labor turnover. Organizationally, there were "big contractors," who had more than a hundred workers in their charge, and "small contractors," who had around fifty. Big or small, contractors in Fushun, as elsewhere, were infamous for underpaying their workers, withholding or delaying their wages, and committing other misuses of authority. Some contractors issued loans to workers with high interest; some opened provision shops for miners with goods that may not have been fairly priced. "All with the intention of fattening their private coffers," one report contended.[116] As with Korean migrant workers in interwar Japan, whose contingent labor and daily struggles have been closely studied by historian Ken Kawashima, Fushun's Chinese workers were frequently subjected to the "intermediary exploitation" of their "countrymen" who signed them on and supervised their work.[117]

Much of the anger against contractors stemmed from the perception—and the reality—that they were enriching themselves at the expense of the miners under them. A burglary of a contractor's home in 1917 gives an indication of the extent of wealth these contractors could amass. The perpetrators made off with 270 yen in cash—around two full years' earnings for a regular Chinese pitman.[118] In January 1922, more than a thousand workers from the Laohutai mine participated in a strike against contractors for cutting their wages.[119] Although Mantetsu would retain contractors as "liaison agents between the management and the working force" at Fushun over the decades it owned and operated the coal mines, it did introduce some changes to curb abuse. In 1927, it changed how wages were disbursed so that workers were paid directly by the company and not through contractors. In 1930, it forbade contractors from issuing loans to workers and began fixing prices at the contractor-run provision shops.[120] Still, contractors remained much hated figures, firmly entrenched in local historical memory as exploiters of Chinese labor alongside their Japanese

116. "Labor Management at the Fushun Coal Mines," 378.
117. Kawashima, The Proletarian Gamble, 74–91.
118. "Burgulary at Fushun."
119. FGY, 12.
120. "Labor Management at the Fushun Coal Mines," 378; Murakushi, Transfer of Coal-Mining Technology, 76.

overlords.[121] As one local ditty puts it, "The [Japanese] devils eat our flesh; the contractors gnaw on our bones."[122]

Strikes by Fushun's Chinese miners, be these prompted by contractor abuse or other grievances, occasionally met with success. A conspicuous feature of these instances was that only very rarely did they arise simultaneously across multiple mines or spill over from one mine to another. This may have been testament to just how separate the units within the colliery were. But it was also possibly an outcome of the strikes being mostly reactive, taking place in response to specific changes or events within individual units, and the company responding quickly whenever they occurred. Regardless of whether Mantetsu conceded to worker demands, the majority of strikes lasted at most three days. In terms of the workers getting the concessions they sought, there was, perhaps unsurprisingly, strength in numbers. Five hundred colliers seemed to have been a critical mass that sufficiently threatened production to the point of moving the company toward conciliation. Confronted with worker pressure, Mantetsu now and again raised wages, shortened working hours, fixed prices in the face of inflation, and reduced scheduled pay cuts, as in the case at Yangbaipu mentioned earlier.[123]

Meanwhile, the colliery put in place "welfare facilities" (福利施設) for Chinese workers that may have slightly dampened any volatility. In 1925, the company set aside a large piece of land to the west of the newly constructed town area and developed it into an amusement district. There, workers could take strolls in the public park, offer prayers to an assortment of popular Chinese gods at the "magnificent green-tiled, red-painted temple," or pay for pleasures at theaters, bathhouses, restaurants, bars, and brothels. The company also brought entertainment directly to workers, screening motion pictures for them across the various mines.[124] Local chroniclers of Fushun's history, writing more than half a century later, would condemn these efforts as nothing more than attempts by Mantetsu to "trick and intoxicate [欺骗麻醉] the Chinese people."[125]

---

121. Teh, "Labor Control and Mobility," 95–96. For more on labor contractors in Fushun, see Murakushi, *Transfer of Coal-Mining Technology*, 64–79; You, *Mantetsu Bujun tankō no rōmu kanri*, 81–88; and Teh, "Mining for Difference," 143–88. On the contract labor system in the prewar Chinese coal mining industry more generally, see Wright, "'A Method of Evading Management.'"

122. Xiao and Jin, *Wangshi jiu ying*, 87.

123. *FGY*, 7–34.

124. Mantetsu rōmuka, *Minami Manshū kōzan rōdō jijō*, 182–83.

125. *FGY*, 15.

In the company's own account at the time, it had introduced these welfare measures with the understanding that should it overtax its Chinese
workers without allowing for sufficient recovery and recreation, there would
eventually be a "degeneration of work efficiency" (作業能率の退化).[126] The
point, it claimed, was purely productivity. It is possible that Mantetsu did
not regard such practices as strategies for curbing labor volatility, as American corporations that similarly expanded worker provisions in exactly the
same period had—historians of the United States would later frame such
efforts as expressions of "welfare capitalism."[127] Regardless, these welfare
facilities quite likely helped, however marginally, in reducing worker an-
tagonism toward the company by giving the latter a veneer of benevolence,
even if it had (as it itself admitted) established them for its own benefit.
With the setting up of the amusement district dividing the decade into
two roughly equal halves, the first half, by one count, witnessed sixteen
recorded strikes at the mines and the second half eleven.[128] Still, perhaps
Mantetsu was far too aware of how easily its Chinese employees could be-
grudge the inequalities of power and privilege through which the enterprise
otherwise operated to not remain troubled by the prospect of labor unrest.

As the colliery's management and workers negotiated persistent tensions between the former's desire for productivity and profit and the latter's demands for fair wages and working conditions, Japanese technocrats
constantly looked for ways to mitigate problems with labor mobility and
labor volatility. Mechanization presented a compelling solution in that it
promised to take the central factor—labor—out of the equation or at least
diminish its function. This point was not lost on some miners. One visitor
to Fushun in 1933 recounted how, at the open pit, he "saw one gang of
about thirty men who acted as though the devil was driving them. Dumping their baskets of coal in the freight cars on the siding, they would run to
fill their baskets again." In his assessment, they were so moved by a "fear
of being robbed of their livelihood by a machine." A distance from where
he was standing, "workmen . . . were putting finishing touches on a giant
hopper, which, in loading freight cars, will probably do the work of at least
thirty men."[129]

126. Mantetsu rōmuka, *Minami Manshū kōzan rōdō jijō*, 181.

127. On American welfare capitalism in the 1920s, see Cohen, *Making a New Deal*, 159–
211. For an example of welfare capitalism in the Japanese home islands, see Kinzley, "Japan in
the World of Welfare Capitalism."

128. This is based on my tabulation from *FGY*, 10–26.

129. Goforth, "Battle of the Coal."

2.6. Bucyrus electric shovel at the open-pit mine filling a dump car with excavated material. Near the upper right-hand corner of this image, one can make out a group of workers drilling vertical holes into the bench so that explosives can be inserted to break it apart later. (Image courtesy of Special Collections and College Archives, Skillman Library, Lafayette College, and the East Asia Image Collection, https://dss.lafayette.edu/collections/east-asia-image -collection/.)

Furthermore, although mechanization did not end up delivering on its labor-saving promise, it reordered the working environment in a way that made it harder for sparks of dissent to ignite. This was particularly so at the exposed work environment of the open-pit mine. If subsurface miners enjoyed a certain kind of autonomy due to the nature of underground mine work—in smaller teams away from direct supervision—and if this autonomy was a major factor in why miners often became militant against the incursions of capital, then it is perhaps not surprising that open-pit miners, whose work was more visible and easily observed because of the structure of the work environment, were less likely to mobilize.[130] This was evidently the case with open-pit mines in other parts of the world, as historian Brian Leech has shown in the case of copper mining in Butte, Montana.[131] Out of the twenty-six major strikes (strikes with more than one hundred participants) that broke out in Fushun between 1914 and 1930, only three

130. Goodrich, *The Miner's Freedom*.
131. Leech, *City That Ate Itself*, 157–59.

took place in an open-pit mine, and none after operations expanded in the early 1920s.[132]

Technologies of Control

In its management of Chinese labor, the Fushun colliery also turned to new technologies. Most prominent among these was fingerprinting, which Mantetsu introduced to the coal mines in August 1924. The primary aim was, in the words of Kuribayashi Kurata, who headed the colliery's Chinese labor bureau, the "elimination of undesirable elements" (不良分子排除)— a catchall category that included those who used opium, failed to repay their debts, were in trouble with the law, ran away from another work site, or had participated in strikes.[133] The modern practice of fingerprinting had arrived in Japan in 1908 through Ōba Shigema, an English-educated attorney who had studied the technique in Germany and who sought to apply it not just to criminal justice matters—as it was used elsewhere—but to the management of individual citizens under the national family registration system (戸籍).[134] Its employment at Fushun, however, marked the first time that fingerprinting was used for labor management on a large scale.[135] The fingerprinting revolution that had its origins in British India did not diverge too far from its roots as a tool of colonial control.[136]

At Fushun, the concern with "undesirable elements" was partly a product of the enterprise's size, which made keeping track of individual laborers difficult. With multiple mines and factories under the colliery coupled with the large labor force, a worker dismissed or denied employment from one unit for whatever reason could go and try to get hired in another—the "roaming and roving between pits" that the colliery had long found deeply troublesome.[137] Fingerprinting provided the means to track the waves of wandering workers flowing in and out of the mines, making legible through arches, loops, and whorls the figure of the individual worker and the patterns of his movement. In so doing, it also helped meet several overlapping objectives. It served as the starting point for the company to keep more de-

132. This is based on my tabulation from *FGY*, 6–34.

133. Kuribayashi, "Konwakai ni shimon jisshi," 51; Teh, "Labor Control and Mobility," 112; Minami Manshū tetsudō kabushiki gaisha, *Minami Manshū tetsudō kabushiki gaisha dainiji jūnen shi*, 572.

134. Takano, *Shimon to kindai*, 51–52, 54–55.

135. Tanaka, "Shimon ōnatsu no genten."

136. Sengoopta, *Imprint of the Raj*, 37–52.

137. Minami Manshū tetsudō kabushiki gaisha, *Minami Manshū tetsudō kabushiki gaisha dainiji jūnen shi*, 572.

tailed accounts of its Chinese workforce, as was consistent with its embrace of scientific colonialism. The identification cards on which the prints were made also contained employment information, such as department, vocation, and pay, and personal information, such as name, age, hometown, residence, and next of kin. If a worker's status changed because of death, desertion, dismissal, or resignation, an explanation was added on the copy of the card kept at the general affairs office.[138] On the basis of these records, the company could more effectively dole out rewards and punishments to workers and also get a better handle on the causes of labor mobility.

Fingerprinting proved an efficient way to vet workers and weed out "undesirable elements." When each worker applied for employment, he would have two sets of fingerprints taken. One set went to his hiring department and the other to the Chinese labor bureau. The bureau then compared these fingerprints to the ones it had on file, and if they matched someone the company took to be a troublemaker, the worker was dismissed immediately. A year after the colliery introduced fingerprinting, it denied employment to 1,640 workers, a small but still significant number considering that the total employed in 1925 was just under 35,000.[139] The success of fingerprinting in Fushun prompted Mantetsu to apply this practice to all its enterprises, and other Japanese-owned, Manchuria-based companies soon followed suit. After Japan established Manchukuo in 1932, the government of the new client state put fingerprinting into widespread use, incorporating it into the identification cards that it issued to both migrant workers and subject citizens.[140] In this way, this technology of control that was first implemented for managing Chinese workers in Fushun ended up as a tool for governing colonized subjects across the wider region.

The technoscientific means by which Mantetsu attempted to exert control over the bodies of workers extended beyond fingerprinting. These materializations of scientific colonialism took forms that ranged from the medical examinations and housing arrangements discussed in the previous chapter to the electric fences and watchtowers I will touch on in a subsequent one. Interestingly, they also included subjecting to scrutiny workers' digestive tracts. If the company regarded Chinese labor as something approaching machines ("the motive power for coal mining"), then it also took much interest in inputs that fueled their functions. The colliery moni-

138. "Fushun meikuang zhiwen guanli," 316.
139. Teh, "Labor Control and Mobility," 111.
140. For more on the history of fingerprinting in Manchuria, see Takano, *Shimon to kindai*; Ogasawara, "Bodies as Risky Resources"; and Tan, "Science and Empire."

tored what workers ate, carefully noting the protein, fat, and carbohydrate content of their staple foods of sorghum, maize, and beans. The imperative here was that these workers "maintain their health" and "possess great physical strength" for the toil ahead of them.[141] By the 1940s, Fushun was counting workers' caloric intake, reflecting a concern with quantifying the energetic basis of human labor that seemed consistent with the further dehumanization of Chinese workers during the war years.[142] Under a wartime system marked by squeeze (chapter 4), this appeared to have been less about how much it needed to feed these workers and more about how little it could.

The company's use of comparatively subtler technologies of control were matched by its exercise of overt constabulary force. In particular, the Japanese police stationed in Fushun seemed effective in crushing the sprouts of Chinese Communist activity that now and again took root among the mines' workers as challenges to the high hand of carbon technocracy. Since Fushun was one of the industrial bastions of Manchuria, the fledgling Chinese Communist Party, which had yet to more fully shift its focus from the proletariat to the peasantry, targeted it as a site for labor organization.[143] When the Communists began their activities in Fushun in 1927, they were entering an environment with a combustibility akin to that of the firedamp-filled spaces in subterranean mines. The dismal living conditions, the difficulty and danger of the work, and the discrimination that Chinese workers faced served as fertile soil for sowing seeds of discontent and revolution. The Communists, however, struggled to sink their revolutionary roots in the Coal Capital.

There are several plausible explanations as to why the Chinese Communists met with limited success in Fushun from the late 1920s into the 1930s. The strength of the Japanese policing and surveillance apparatus was certainly a leading factor. Fushun's Japanese police proved adept at detecting and crippling underground Communist activity. In August 1929, they carried out their first major roundup and arrested over a dozen leading Communist operatives. Most of those who were not caught fled, and the local party branch and the red unions into which it had been recruiting workers folded as a result. Through this operation, the police also seized several documents that gave them important insights into how the Communists organized themselves and what their plans for Fushun and

141. Bunkichi, *Manshū ni okeru kōzan rōdōsha*, 206.
142. Abiko and Shimeno, *Manjin rōdōsha no eiyō*.
143. Fushun shi shehui kexueyuan, *Zhonggong Fushun difang shi*, 25.

the wider region were. This knowledge no doubt helped them stay a step ahead of the Communists, who, not deterred by this setback, tried again and again to mobilize Fushun's miners. Moreover, in the roundup of 1929 and those that followed, the Japanese police greatly benefited from spies they had placed among the Communists, most notably one by the name of Fan Qing, who had been leaking information to the Japanese authorities since his arrival in Fushun in 1927. Fan's long-standing betrayal, which did not come to light for years, would severely undermine the Communists' repeated efforts to establish themselves in the colliery town.[144]

And yet, leaving aside the (admittedly nontrivial) issue of Japanese policing, the Chinese Communist operatives dispatched to Fushun faced difficulties in winning over the masses. This seems to have been largely a matter of approach. These operatives did not manage to replicate the earlier successes of their predecessors at the Anyuan coal mine, "a cradle of the [Communist] labor movement." There, as Elizabeth Perry has shown, Mao Zedong and his comrades Li Lisan 李立三 (1899–1967) and Liu Shaoqi 劉少奇 (1898–1969) utilized what she calls "cultural positioning"—"the strategic deployment of a range of symbolic resources (religion, ritual, dress, drama, arts, and so on) for purposes of political persuasion"—to gain access to the mining community, organize its workers for a victorious strike in 1922, and set up on site China's "Little Moscow."[145]

In contrast, the Communist operatives in Fushun appeared to have lacked the familiarity with or sensitivity to local conditions necessary for cultural positioning. They were, consequently, ineffective in their messaging. This was evident, for instance, in the posters and pamphlets the Communists put up and scattered around the colliery, which Japanese authorities found right before the August 1929 roundup. On the one hand, these materials urged workers to protest imperialist war by observing "red international day" on August 1.[146] On the other hand, they also defended the Soviet Union in the recent Chinese Eastern Railway Incident, which, to many Chinese, had underscored the imperialistic character of that socialist state.[147]

The Soviet Union, which inherited the Chinese Eastern Railway from its tsarist predecessor, maintained control over it through an ostensibly joint Sino-Soviet administration. This was in spite of repeated Chinese requests

---

144. *FGY*, 19, 25, 32–33; "Bujun ni okeru Chūgokujin kyōsan undōsha."
145. Perry, *Anyuan*, 4, 46–123, quotation on 4.
146. On "red international day," see "Long Live May Day."
147. Lee, *Revolutionary Struggle in Manchuria*, 104.

that this concession be relinquished. In July 1929, the Manchurian warlord Zhang Xueliang 張學良 (1901–2001), with the backing of the Nationalist government in Nanjing, took the railway by force, and the Soviets began preparing for a military intervention in response. Although the Chinese Communists were initially divided over what position to take, they soon came down in support of the Soviets. Chen Duxiu 陳獨秀 (1879–1942), one of the founders of the Chinese Communist Party, opposed the decision. For that, he was expelled from the party as a Trotskyite. By endorsing Soviet actions in Manchuria, the Communists opened themselves up to critique. Their Nationalist foes gleefully labeled them traitors.[148] Even the Japanese police in Fushun, perusing the confiscated propaganda defending the Soviet Union, questioned its efficacy in light of widespread anti-Soviet sentiments.[149]

The Communists continued to try and find their footing in Fushun. A year after the first roundup, the reconstituted local party branch received orders from the regional committee stating that "Fushun's central task is to organize a political alliance for strikes, prepare for an armed insurrection, take hold of power, and establish in Fushun a local soviet and a red army."[150] While the Communists managed to get some workers behind that vision, it seemed almost too abstract, too far removed from the realities of everyday life and labor in the mines, to attract a larger following. Their presence in Fushun, routinely buffeted by the Japanese policing machinery and waves of arrests, remained relatively weak throughout the period before 1945. The Chinese Communists would only really succeed in seizing Fushun in 1948, by which point they were no longer interested in toppling carbon technocracy but in rebuilding it (chapter 6).

Managing Disaster

If the Fushun colliery met with modest success in disciplining labor, it was less effective in exerting control over the risks that accompanied mining. Common perils such as roof falls, fires, flooding, and the release of asphyxiant, toxic, and explosive gases constantly threatened the mines' operations and the lives of those who labored within. These were grim and murderous reminders that there were limits to humans' ability to mediate the natu-

---

148. On the Chinese Eastern Railway Incident and the Sino-Soviet conflict of 1929, see Elleman, *Moscow and the Emergence*, 192–205; and Patrikeeff, "Railway as Political Catalyst."

149. Lee, *Revolutionary Struggle in Manchuria*, 104.

150. "Manzhou shengwei yi Fushun."

ral forces of the subterranean environment. Although hydraulic stowage helped reduce the risk of roof collapse from weak or compromised support, cave-ins remained an ever-present danger. Within the enclosed space of the underground pit, the opposing elements of fire and water were also fearsome. In fires, there was the horror of burning to death or suffocating from smoke. Fires could start for a number of reasons, from the spontaneous combustion of exposed coal seams to gas or coal dust explosions. Once they did, they were difficult to extinguish, fueled as they often were by the surrounding coal. In floods, most often a product of insufficient or ineffective pumping, there was the terror of a watery tomb. Furthermore, gases, built up over time in the cracks and pores within the coal seams and released through mine work, could kill by asphyxiation (with carbon dioxide and nitrogen) or poisoning (with carbon monoxide). Deadliest of all were explosions, caused when flammable gases containing high levels of methane, coal dust, or a combination of the two came into contact with a source of heat, even a single spark. At Fushun, such blasts periodically killed tens or hundreds of miners. There is no record of whether the colliers here, like their counterparts in the coal mines of Colorado, formed symbiotic relationships with rodents underground, feeding them and taking them as pets in return for their company and unwitting service as detectors of danger—these animals fainted or died when they inhaled carbon monoxide and could sense tremors that foreshadowed tunnel collapses.[151] Still, when a fire broke out at the Laohutai mine in 1915, it was noted that the rats in the pits had all fled.[152]

Fushun's biggest disaster took place at the Ōyama mine in 1917. Just over an hour before midnight on the evening of January 11, an explosion, "like a thunderbolt," rocked Ōyama's underground workings with such force that it instantly shattered the windows and blew off the roof of the ventilation fan room above. So strong was this force that it even triggered a blackout in the nearby town. Because the equipment for regulating airflow in the mine had been destroyed in the blast, there was no way to immediately shut off the air circulation, and air was sucked into the mine through the up shaft (going in the opposite direction from which it normally would), which fed the fire that had started near the base of the down shaft. Half an hour after the explosion, the surface opening of the down

151. Andrews, *Killing for Coal*, 129–30. On mice in coal mines, see also Long, *Where the Sun Never Shines*, 34–36. Animal behavior had long been taken to be predictors of environmental events in China as in many other parts of the world. This was so into the socialist period as well. See Fan, "Can Animals Predict Earthquakes?"

152. "Laohutai Fire Scented by Rats."

shaft "spewed out black smoke, which was soon followed by flames swirling up into air, reaching more than seventy feet."[153]

Before the fire spread, there was an attempt to evacuate the trapped miners using the hoist in the up shaft, but the cage got stuck, and only a few Chinese workers managed to get out this way. At the same time, two relief teams with "men noted for pluck, clad in fire-proof attire and carrying oxygen," descended a separate incline for hydraulic stowage. When they got to the bottom, the men were "speedily attacked by the suffocating smoke." They called out for survivors until there was purportedly "no more to be heard from" before being "compelled to beat a retreat." To contain the fire by cutting off the air supply, Fushun's management ordered that all openings from the surface be sealed in succession. Decades later, one survivor would contend that "for the sake of coal and without regard for the lives of the Chinese [miners], the small [Japanese] devils forced the people at the top of the pit to close off the opening with mud, and, because of that, the workers below had no way to escape."[154] In total, 170 miners were rescued and 917 perished—17 Japanese and 900 Chinese.[155]

To Shinozaki Hikoji, a Kyushu Imperial University mining engineering student who did his practicum at the Ōyama mine several years later, the disaster of 1917 was a lingering tragedy. "With a single boom, a thousand precious human lives vanished in a cloud of smoke," he wrote, "In the minds of those signing labor contracts, they thought that they would eventually come into good fortune after months and years of hardship. Who would have known that what awaited them was such a hell of a miserable and tragic death?"[156] Few were quite as sympathetic.

Soon after the incident, colliery manager Yonekura issued a statement with theories as to the cause of the disaster. Given several factors, including that the section of the coal seam where the explosion took place contained very little gas to begin with, he deduced that the explosion originated not from gas but from coal dust. As to how the coal dust ignited, he ruled out suggestions that it was because of blasting work or the spontaneous combustion of coal—the former because blasting was seldom conducted in Ōyama, and "whenever such operation is undertaken a trustworthy Japanese is charged with the task"; the latter because spontaneous combustion

153. "Colliery Disaster"; Yu, *Fushun meikuang baogao*, 212.

154. "Lao gongren Sui Xingshan 1959 nian huiyi cailiao" [Old worker Sui Xingshan's recollections from 1959], Fushun Archives, as cited in Fu, "'Jiu-yi-ba' qian de Fushun kuanggong," 81.

155. "Colliery Disaster"; Yu, *Fushun meikuang baogao*, 212.

156. Shinozaki, *Bujun tankō Ōyama saitanjo*, 1.

is a slow process, with preceding signs of "an offensive smell . . . followed by a veil of thin grey smoke," which would have, he argued, alerted the watchmen. The explanation he thought most plausible was that a miner working at a point where small concentrations of gas and coal dust had been noticed before "smoked a cigarette, or blundered in some way in the handling of one of the safety lamps, setting the gas afire, which ignition inflamed the coal dust in the neighborhood."[157] This assumption was certainly consistent with Mantetsu's position, reflected in a 1918 report, which took for granted the "deep ignorance" (無知蒙昧) of Chinese workers in accounting for mining accidents.[158] It seemed to be a common enough belief, echoing presumptions about Chinese ineptitude. "Because Chinese laborers are generally low in intellect and deficient in basic mining knowledge," Seta Hidetoshi, another Japanese student on practicum at Fushun, reasoned, "they have a complete lack of sense concerning matters like mine safety."[159]

Following "numerous unforeseen disasters" in its first few years of operation, the Fushun colliery set up a system of relief squads, which it outfitted with German-manufactured Dräger mine safety equipment—breathing apparatuses, oxygen tanks, pulmotors, and other devices. Under Fushun's relief squad system, each mine selected several "able-bodied" workers for training. After a weeklong course in which they learned how to use the safety equipment and practiced rescue drills, these men returned to their regular jobs, ready to be mobilized in the event of a disaster. The first recruits were halfway through their training when they were dispatched to provide aid after a fire broke out at the Laohutai mine. As more were trained in subsequent decades, the system of squads become quite extensive. Over the years, the colliery upgraded the safety equipment, most notably with the complete adoption of Proto breathing apparatuses from England, which, with their flexible breathing bag, fewer tubing connections and joints, and comparatively lighter weight, promised greater safety, comfort, and ease of movement.[160] The relief squads, in providing rescue and remediation, complemented the colliery's efforts at preventing disaster. The reduction of such risk was, after all, one main reason behind the adoption of hydraulic stowage. It was in this vein, along with the continued distrust of Chinese workers, that the mines began making Chinese miners strip

157. "Colliery Explosion,"
158. Bunkichi, *Manshū ni okeru kōzan rōdōsha*, 175.
159. Seta, *Bujun tankō Rōkudai kō*, 2.
160. Manshikai, *Manshū kaihatsu yonjūnen shi*, 91–92; "The 'Proto' Breathing Apparatus."

completely naked and put on outfits issued by the colliery at the beginning of their shifts so as to forestall "undesirable and dangerous practises such as carrying matches and tobacco on one's self into the coal mine."[161]

The Fushun colliery's interest in matters of safety conformed to the economic rationale and productivist logic of this energy enterprise. The company had committed to offering compensation for the loss of life and limb accompanying mining disasters and other accidents. According to a 1916 stipulation, it was supposed to provide 25 qian a month for medical fees to Chinese "coolies" who had been injured on the job and up to 10 yuan to cover funeral expenses for those who died.[162] After the 1917 Ōyama explosion, the company, which organized an elaborate ceremony that included a ritual to gather up the wandering spirits of the deceased "by imitating the crowing of cocks," paid out 50,000 yen to the bereaved families—almost one-seventh of the total estimated damages from the disaster. More importantly, from the company's perspective, disasters also halted operations. The Ōyama mine was closed for more than half a year following the explosion, and the colliery's output and profits suffered accordingly.[163]

For all the precautionary and relief measures the colliery undertook, it was, however, unable to prevent the regular occurrence of mining disasters at Fushun or to fully mitigate the human toll of such calamities. Although it had attempted to improve safety conditions upon the understanding that disasters meant losses in productivity, the colliery's same obsession with output tempted it to take risks that frequently resulted in catastrophe. According to one count, almost seven thousand Chinese workers perished in accidents big and small over the decades of Mantetsu's management. Countless more were injured or maimed through their work in this extractive enterprise.[164]

Conclusion

The mechanization of production at Fushun underlay its transformation into the Coal Capital. By applying the latest mining technologies, Mantetsu excavated the open pit and expanded other operations above and below ground, generating the level of output that made Fushun the largest coal mine in Asia. But even as it supplied energy to masses of consumers

161. "The Ryuho Coal Mine," 64–65.

162. Bunkichi, Manshū ni okeru kōzan rōdōsha, 175.

163. "Fushun Collieries"; Minami Manshū tetsudō kabushiki gaisha, Minami Manshū tetsudō kabushiki gaisha dainiji jūnen shi, 604–5.

164. "1907–1945 nian Fushun meikuang shigu siwang."

across the empire and beyond, so, too, did Fushun itself become a massive consumer of energy. The wide array of mining machines, from dragline excavators to coal cutters, were powered by coal directly through steam engines or indirectly through electricity generated in one of the colliery's power plants. In this manner, Fushun's development came to capture carbon technocracy's vision of progress predicated on engines and energy.

Yet Fushun's mechanization, in enabling operations to reach unprecedented depths and scale, created a work environment that placed more colliers in harm's way. Historian Brett Walker, in writing about a devastating explosion at the Hōjō colliery in Japan's Chikuhō region, frames the disaster as a product of "hybrid contexts and causations" in which "carefully engineered subterranean environments conducive to extracting thousands of tons of coal interfaced with naturally occurring methane-gas deposits."[165] This was all too true in Fushun as well. What made accidents here even more destructive, I would add, was that the engineered environments in which they occurred drew more labor as operations expanded. The result was potentially more casualties when calamity struck. While few, if any, could predict the onset of such disasters, racialized inequalities between Chinese and Japanese often determined who would get out alive.

In 1928, a flood broke out in a pit at the Ōyama mine. The adjacent abandoned pit had been filled with water as a precaution against spontaneous coal fires—presumably, with incidents like the 1917 explosion in mind—but the wall dividing the two pits could not withstand the pressure and gave way. The Japanese foremen "were each equipped with an electric pocket lamp enabling them to quickly escape from the crushing waters"—there were no Japanese casualties—but the Chinese miners, "having only oil lamps which were all extinguished by the strong gust of wind caused by the pressure of the water, were all left helplessly groping in the darkness." In total, 470 Chinese workers died that day.[166] In expanding its extractive endeavors, the company inadvertently generated new risks for the miners in its employ for which little could be accounted in advance save the unequal distribution of unpredictable costs.

Philosopher André Gorz once contended that the "total domination of nature inevitably entails a domination of people by the techniques of domination."[167] In Japan's empire, Fushun's open-pit mine was so celebrated precisely because it represented a particular domination of nature

165. Walker, *Toxic Archipelago*, 177.
166. "Explanation Given of Disaster at the Fushun Mine."
167. Gorz, *Ecology as Politics*, 20.

in which industrial technologies were used to turn this landscape of extensive resource potential into a site of intensive resource production on such a scale as to replicate the hand of nature itself. This act of domination also entailed the domination of tens of thousands of Chinese workers. Historian Timothy LeCain has rightfully decried the scarring effects of open-pit mining on the planet, characterizing this extractive act as one of "mass destruction."[168] At Fushun, the open-pit mine that survives to the present has marred the surrounding environment. Aside from its historically situated connections to the intensive extraction and consumption of carbon resources, the gaping hole in the ground continues to pollute local water sources with its drainage. According to one environmental impact assessment, this "adversely affects farmlands and water wells, growth of plants, and the health of residents."[169] Even so, the deep damage that this technological enterprise has wrought was rooted in another act of mass destruction: the exploitation of labor that left many bodies broken and lives lost. Perhaps Yosano Akiko was not wrong, after all, when she first laid eyes on Fushun's open pit and saw not a marvel but a monster.

168. LeCain, *Mass Destruction*, 11–15, 108–71.
169. Gilpin, *Environmental Impact Assessment*, 152–53.

THREE

Fueling Anxieties

Early one summer afternoon in 1925, Gotō Shinpei delivered a lecture before an expectant crowd at the Imperial Railway Society in Tokyo's central Marunouchi district. "Taller than the average of his countrymen," with white hair and beard cropped short and pince-nez perched atop his "delicately lined aquiline nose," Gotō was, at least as one admirer described him, "a man of commanding mien, whom once to see is never to forget."[1] Years of high office, from which he had recently retired, may have added to his aura of authority. After his short but significant stint as Mantetsu's first president, Gotō went on to serve in a succession of cabinets as communications minister, home minister, and foreign minister. From late 1920 to a few months before the 1923 great Kantō earthquake, he was mayor of Tokyo. In the wake of that catastrophe, he again occupied, for several months, the position of home minister, busying himself with ambitious plans for the capital's reconstruction.[2] A spirited statesman whom critics would dismiss as a "big talker" (大風呂敷) incapable of following through on his wild schemes, Gotō brimmed with ideas and opinions about almost everything, including energy.[3]

His lecture that day was titled "Human Life and the Fuel Question" (人生と燃料問題). The audience, numbering over three hundred, consisted of coal and oil industry representatives, scientists, engineers, government officials, and military men. It was the third anniversary of the Fuel Society of Japan, and Gotō's talk was part of a commemorative symposium

1. Hayakawa, "New Foreign Minister," 189.
2. On Gotō's career, see Hayase, "Career of Gotō Shinpei." On Gotō's largely unrealized plans for Tokyo's reconstruction in the aftermath of the great Kantō earthquake, see Schencking, *Great Kantō Earthquake*, 153–225.
3. Hayase, "Career of Gotō Shinpei," 254.

recognizing the society's ongoing work to tackle the so-called fuel question (燃料問題) or fuel and power question (燃料動力問題)—an articulation of rising anxieties about Japan's fuel future and its implications for "national defense, the national economy, and the daily lives of citizens."[4]

Founded in 1922, the Fuel Society brought together individuals from both public and private spheres who either conducted research on fuels or were otherwise involved or invested in the energy sector. Its activities, aimed at "widening and improving the arts and sciences relating to fuels and power," included organizing lectures, seminars, and roundtables; hosting an essay contest on the fuel question; drafting proposals and petitions calling for a national fuel policy; and publishing the monthly *Journal of the Fuel Society of Japan* (燃料協會誌).[5] One contributor to an early issue of that journal spoke of the society's work in terms of the ethical burdens of expertise: "When a moving train is about to cross over a broken bridge, it is incumbent upon those passing by, as people near to the train, to endeavor to save it. As those who are close to our country's fuel question, we should alert our compatriots to the current danger and work to mitigate the fuel situation for the sake of our national livelihood."[6] Operating with this deep sense of urgency, the society inadvertently further fueled such anxieties in the process.

In his talk, Gotō similarly framed the fuel question as one of life and death: "If fuels were exhausted, human life would be a dead thing [死物] . . . their oxidation [酸化] completes our existence." This was, he insisted, especially so in the post–industrial revolution present, as the consumption of fuels such as coal, oil, and firewood greatly intensified in tandem with the growth of mass production. Like many other commentators at the time, Gotō questioned whether Japan had the resource endowments to survive in such an age. He noted that according to current estimates, Japan had 8 to 10 billion tons of coal, while the United States and Britain had close to 4 trillion tons and more than 200 billion tons, respectively.[7] Although it was several more years before Japan was widely spoken of in public discourse as a "have-not country" (持たざる國), concerns over its relative poverty of material assets—which such country comparisons foregrounded—already ran deep.[8] In terms of fuels specifically, many Japa-

---

4. "Nenryō kyōkai sōritsu," 724.

5. Takenob, *Japan Year Book, 1923*, xliv.

6. Yoshimura, "Nenryō kyōkai jūnen shi," 1196, 1200–1210, quotation on 1196.

7. Gotō, "Jinsei to nenryō mondai," 29–31.

8. On the discourse of Japan as a "have-not country," which took hold of the public imagination in the 1930s, see Dinmore, "Small Island Nation Poor in Resources," 59–107; and Satō, '*Motazaru kuni' no shigen ron*, 62–68, 138–54.

nese regarded limited domestic reserves as threatening the endurance of Japan's expanding industrial modern order, powered as that order was by carbon-energy resources.

The answer to the fuel question, Gotō suggested, lay in the "power of science" (サイアンスの力). It was, he contended, the "surprising ignorance" of politicians and law bureaucrats concerning matters of "contemporary science" that had resulted in their hitherto inadequate response to the problem. This was a common critique for men of science like Gotō to make. While he, as a physician, had advanced to the state's topmost echelons, most other scientific and technical experts in public service could not reach the upper rungs of the government career ladder, as those positions were overwhelmingly occupied by men holding law degrees. Dissatisfied with their exclusion, many scientists and engineers agitated for more access to power by arguing that their expertise, which those who only had legal training lacked, was essential for good governance.[9] Within this context, Gotō stressed the importance of the Fuel Society and its research, encouraging members to "prepare for comprehensive, scientific work" in the service of national interests.[10] The fuel crisis, in this way, presented an opportunity for realizing technocratic governance.

Gotō brought up Fushun twice in his talk. First, he spoke to long-standing concerns among Japanese coal producers that Fushun coal would flood the markets in the home islands and wreck their coal-mining businesses. Gotō began by saying that he understood the calls to protect metropolitan mines by restricting colonial imports. He nevertheless urged his audience to take "a broader scientific view" regarding "the connections between demand and supply." The implication here was that if the Japanese empire's fuel needs were considered as a totality, more coal would indubitably be a good thing, including coal from Japan's overseas possessions like Fushun. Second, Gotō mentioned ongoing efforts at setting up Japan's first shale oil industry in Fushun.[11] Oil distilled from the shale resting on top of Fushun's coal seams promised a solution to Japan's poverty in petroleum. Mantetsu and the Imperial Japanese Navy would put much time and money toward its development here in this laboratory of empire. Against

9. The preponderance of law graduates in the higher bureaucracy was a function of the civil service examinations, which were designed to test legal training and not specializations in other areas. On attempts by scientists and engineers to gain ground in government vis-à-vis law bureaucrats, see Mizuno, *Science for the Empire*, 22–25, 40–41.

10. Gotō, "Jinsei to nenryō mondai," 29–30, 34.

11. Gotō, 33–34.

the backdrop of the fuel question, Fushun appeared central to the Japanese empire's energy future.

Just as carbon technocracy had come to define Mantetsu's extractive economy in Fushun, so, too, did it, in the aftermath of World War I, transform the wider Japanese empire's relationship to energy. A galvanizing factor was the rise of autarkic thinking and, with that, the fuel question that Gotō and others aspired to address. At the heart of the fuel question lay, after all, a fear of present and projected energy scarcity that fundamentally challenged ascendant ideals of national self-sufficiency.[12] In this, the fuel question pointed to one of the central paradoxes in the energy regime of carbon technocracy. As much as it may have been inspired by a confidence in the imagined inexhaustibility of energy, extensive mining was equally spurred by abject anxieties over insufficient supply and limited access.

This chapter takes us from Fushun and Manchuria back to the Japanese home islands, where planners, scientists, and engineers wrestled with the fuel question, seeking solutions in public education, national policy, and technoscientific fixes like enhanced efficiency or alternative fuels. Members of the Fuel Society were at the forefront of such efforts. At the same time, as those measures started falling short of expectations, rising numbers of fuel experts came to postulate that the energy crisis could only truly be resolved through imperial extraction, whether of Manchurian coal, Fushun shale oil, or fuel resources in foreign territories on which empire builders harbored increasingly expansionist designs.

A Growing Appetite for Coal

Underlying the fuel question was a deepening dependence on carbon energy, mainly in the form of coal. Coal, which fueled Japan's industrial and military takeoff in the Meiji era, continued to power the island nation's socioeconomic transformation in the period that followed. Coal extracted in mounting quantities from seams across the Japanese archipelago and in Fushun and elsewhere on the continent fed engines, boilers, and furnaces in Japan's growing urban centers and beyond. In 1909, Japanese coal con-

---

12. In regard to scarcity—and certainly energy scarcity specifically—one central issue is, as Frederick Albritton Jonsson, John Brewer, Neil Fromer, and Frank Trentmann have pointed out, whether it arises from natural limits or from social factors. I am sympathetic to their suggestion that we appreciate the "hybrid character of scarcity," for "the problem of biophysical limits cannot be separated from questions of power, distribution and value." This perspective has informed my treatment of energy scarcity in this book. See Jonsson et al., "Introduction," 2.

3.1. Women, children, and men loading coal by relay onto a ship in the port of Nagasaki. These workers typically approached ships entering the harbor in coal barges and, when close enough, set up temporary structures from which they could get the coal onto the ships. A striking visual that highlights how it took considerable amounts of human energy not just to get coal out of the ground but to move it from sites of extraction to sites of consumption too. This photograph dates from the 1910s. (Image courtesy of MeijiShowa/Alamy Stock Photo.)

sumption per year was just over ten million tons. By 1918, this more than doubled to twenty-three million tons.[13]

The bulk of this coal went toward industry. The Japanese manufacturing sector, which progressed at a modest pace from the late nineteenth to the early twentieth century, made significant strides during World War I. Not only was Japan producing munitions and other wares of war for the Entente powers, but the inability of many Western firms to continue supplying Asian markets amid the fighting created openings across the economic landscape that Japanese manufacturers moved in to fill. From 1914 to 1919, Japan's manufacturing output rose by 72 percent and real gross national product by 40 percent.[14] Industries that prospered included spinning, weaving, ceramics, paper making, chemicals, brewing, and machine building. Coal provided much of the power for this expanding production.

13. Inouye, "Japan's Position," 53.
14. Crawcour, "Industrialization and Technological Change," 436, 439.

Factories that ran primarily on inanimate energy increased from 10,300 in 1914 to 17,700 in 1919, while those driven mainly by manual power decreased from 6,700 to 6,200 during that half decade. Notably, this development was accompanied by an expansion of the industrial workforce from around 850,000 to 1.6 million workers—another testament to how the intensification of carbon-energy use in this period went hand in hand with an increase rather than a decrease in demand for human labor.[15]

The iron and steel industry, another mainstay of Japan's industrial economy, consumed much coal as it grew over this period too. A ton of coke was typically needed to produce a ton of iron; three or four tons for a ton of steel. By 1918, smelting iron and manufacturing steel accounted for over a tenth of the coal used in industry.[16] Moreover, coal fueled the transport sector so essential to the larger industrial assemblage in moving material inputs like iron and steel and finished products around the home islands and to markets afar. Japanese shipping and railways each burned about a million more tons of coal a year by the end of the war than at its beginning. Overall, between 1914 and 1918, annual industrial coal consumption went up from about 8.4 million to 14.2 million tons.[17]

Beyond its direct application as fuel, coal was used to produce electricity and gas, other forms of energy that similarly served a range of functions throughout the industrializing archipelago. Japan's first coal-fired power plant went online in 1887. Owned and operated by the Tokyo Electric Light Company, this modest outfit with a 25 kilowatt homemade generator had been set up to power incandescent lamps in the nation's capital. The industry expanded rapidly in capacity and use thereafter. By the early twentieth century, Japan extensively employed electricity not only in lighting but also for driving electric tramways that plied city streets, powering electric motors in factories, and completing electrochemical and metallurgical processes in industries such as copper production, in which the metal was refined electrolytically.[18] As time passed, Japan generated its electricity less by combusting coal than by harnessing "white coal," or hydropower,

---

15. Takenob, *Japan Year Book, 1920–21*, 593; *Japan Year Book, 1921–22*, 469. One caveat here is that these official figures are for factories employing ten or more workers.

16. Berglund, "Iron and Steel Industry," 629, 635, 637, 640.

17. Inouye, "Japan's Position," 53.

18. Rutter, *Electrical Industry of Japan*, 5–6, 8–9. The use of electricity in factories here and elsewhere was significant in that its transmission was steady as compared to steam power, which was conveyed from engines to machines by gears, shafts, and belts that moved more slowly the farther they were from the power source. Switching to electricity allowed for different factory layouts, including ones amenable to the assembly line. See Nye, *America's Assembly Line*, 17–18, 25–27.

the first facility for this having been established in 1891 on the then re-cently cut canal between Lake Biwa and the old imperial capital of Kyoto. The amount of power produced by thermal electric plants as compared to hydroelectric ones was 76,000 to 38,000 kilowatt hours in 1907, 229,000 to 233,000 kilowatt hours in 1912, and 422,000 to 711,000 kilowatt hours in 1919.[19] This trend persisted in the years that followed, the latter dramati-cally overtaking the former. Nevertheless, electricity generation continued to rely in good measure on coal. This was partly seasonal. In winter, before snowmelt replenished the rivers, and occasionally in summer, water levels might drop to a point where hydroelectricity could not be sufficiently pro-duced and had to be supplemented by thermally derived power.[20] It was also partly geographic. Some areas depended more on coal-fired power plants because of their distance from riverine systems or their proximity to coalfields. In 1943, imperial Japan's peak year for electricity genera-tion, thermal power still accounted for more than a fifth of production at 8.6 million kilowatt hours.[21]

Coal gas, manufactured from heating coal in the absence of oxygen (as in the production of coke), predated electricity as an industry in Japan as it did in most parts of the industrial world.[22] The country's first gasworks, lo-cated in the port city of Yokohama, was designed by French engineer Henri Pelegrin in 1870 and began operations in 1872. This facility sat along a wa-terway on which barges floated down to deliver coal that workers bearing baskets unloaded, weighed, and stored, to be later shoveled by stokers la-boring twelve-hour shifts into retorts to produce gas that was then piped to the surrounding municipality.[23] Gasworks soon appeared in other cities as well. Originally intended for lighting lamps on the street and in domiciles, gas quickly faced competition from and gradually lost out to electricity in the provision of illumination both public and private.[24] At its most trou-bled times, the gas industry was "a barely paying business," and producers relied more on the sale of by-products such as coke and coal tar—used in the manufacturing of dyes—in order to stay afloat.[25]

19. Inhara, *Japan Year Book, 1933*, 611.

20. "Railway Electrification in Japan," 516.

21. United States Strategic Bombing Survey, Electric Power Division, *Electric Power Industry of Japan*, 12, 15.

22. Almost all the manufactured utility gas in prewar Japan was made from coal. A very small percentage was acetylene produced from calcium carbide.

23. Graham, "Round the World," 820–21.

24. "Gas Companies in Japan," 425.

25. Takenob, *Japan Year Book, 1919–20*, 611. The coal gas industry was "the sole domestic source of coal-tar materials" that went toward the making of dyes. Coal in Japan, as elsewhere,

Gas nevertheless gained ground over the years as a fuel for cooking and heating in urban homes. In order to create this domestic demand, gas companies designed and developed household appliances such as rice cookers and stoves that ran on gas. Using gas was promoted as convenient, in that such appliances could be lit, as one advertisement boasted, "with just one match."[26] Some opposed these new contraptions, finding the rice cooked with gas to have "an unpleasant flavor." Nevertheless, as one commentator claimed, many "realize that taste and convenience cannot always agree and prefer convenience."[27] Gas was also touted as clean, since it did not leave behind soot and ash after burning as coal, charcoal, or firewood did.[28] That it took the combustion of "dirty" coal to produce this "clean" gas seemed to escape the notice of boosters and the consumers to whom they pushed these wares. The carbon-energy regime, marked in part by a widening distance between sites of production and sites of consumption, encouraged such inattention—a problem that still plagues our present.

Japan's coal use, be this directly as fuel or indirectly through electricity and gas, even rose over the course of the economically troubled 1920s. The boom of World War I had been followed by a bust. Throughout the 1920s, the Japanese economy was assailed by crisis upon crisis, and the coal-mining industry was not spared.[29] Between 1924 and 1926, coal output, which had been steadily climbing over the years, declined. This resulted from and was reflected in the high price of coal, "about double that in America." Behind this high price lay several factors, especially high production costs and the relative inefficiency of Japanese miners, "chiefly due," according to one assessment, "to the poverty of the Japanese seams and the greater difficulty experienced in using mechanical contrivances."[30] In 1927, total coal output per worker in Japan was about 130 tons. In contrast, American miners each produced on average 872 tons in bituminous mines or 485 tons in anthracite ones.[31]

Still, in spite of the drop in domestic production and the slowdown in

---

was important for the chemical industry beyond serving as a source of energy. See Delahanty and Concannon, *Chemical Trade of Japan*, 15. On the larger significance of coal tar dye for the scientific enterprise, see Beer, "Coal Tar Dye Manufacture."

26. Sand, *House and Home*, 78.

27. Fukushima, "Japanese Gas Works," 490.

28. Sand, *House and Home*, 78.

29. On the economic crises of the 1920s, see Nakamura, "Depression, Recovery, and War," 451–67.

30. Takenobu, *Japan Year Book, 1927*, 505.

31. Hoar, *Coal Industry of the World*, 45 and 225.

economic activity, Japanese coal usage expanded from twenty-six million tons in 1921 to thirty-two million in 1930.[32] In 1923, Japan became, for the first time, a net importer of coal. Although Japan had long brought in coal from China and Indochina, these imports were, until then, consistently less than the volumes that Japanese producers exported to China, Hong Kong, the British Straits Settlements, the Russian Far East, the Philippines, and beyond. Japan would largely remain a net importer in the years that followed.[33] Speaking to this apparent contradiction of rising coal consumption amid industrial recession, coal expert and Fuel Society member Naitō Yū suggested that it was an issue of thermal inefficiency. As the economy started to stagnate, factories that used to run day and night began limiting their operations to daylight hours. The result, he explained, was that steam boilers on site, left to rest at night, lost most of their heat and so needed more coal to fire them up the next day.[34] As Naitō and other fuel researchers fought a seemingly ceaseless war against such waste, coal consumption continued to swell.

Fushun fed Japan's carbon cravings. A good proportion of the archipelago's coal imports was coking coal for blast furnaces and steel mills since domestic deposits held little coal of this sort. Fushun coal may have been ill suited for making coke when used just by itself, but when the coal from the eastern part of the coalfield was mixed with metallurgical coal from the nearby Benxihu colliery—at a ratio of 70 percent Fushun coal to 30 percent Benxihu coal—it produced a strong foundry coke for iron smelting and steel manufacturing.[35] Moreover, the coalfield's central and western parts contained a highly volatile bituminous coal that worked well in driving steam engines and generating gas, as early Mantetsu tests had determined.[36] In 1928, over half of the coal that Fushun exported went to Japan, accounting for as much as two-thirds of Japanese coal imports that year.[37] We will see later in this chapter how Fushun's outsized presence in the Japanese coal sector—which, as Gotō noted, had been a constant point of concern for local producers—eventually became a target of metropolitan miner protests.

32. Takenobu, *Japan Year Book, 1927*, 505; Inahara, *Japan Year Book, 1933*, 542.

33. Takenobu, *Japan Year Book, 1927*, 505; Bradley and Smith, *Fuel and Power in Japan*, 7.

34. Naitō, "Kōjō keizai," 336–37.

35. Benxihu was a coal-mining and iron- and steel-producing enterprise that had been developed from 1905 as a Sino-Japanese joint venture and that was largely under the control of the Ōkura zaibatsu.

36. Quackenbush and Singewald, *Fushun Coal Field*, 9–10.

37. *MSZ*, vol. 4, no. 1, 243.

The Emergence of the Fuel Question

Anxieties over fuel, wrapped up as they were in Japan's growing reliance on energy, coalesced around the fuel question in the period after World War I. There were, of course, antecedents, the worries about shortages of suitable naval coal before the Russo-Japanese War being but one such instance (chapter 1). Still, the interwar fuel question distinguished itself not so much by pinpointing a problem specific to a situation (such as securing adequate fuel supplies before an imminent war) as by naming a condition with which Japan was presumed afflicted. "Japan is, roughly speaking, far behind European countries and America, as regards the development of various forms of fuel in their essential aspects," grumbled Fuel Society secretary Okunaka Kōzō in 1927. This deficient condition called for correction. "The question," Okunaka urged, "is really one of vital importance in that its satisfactory solution is closely interwoven with all issues of national strength, defence, and industrial development, in short the whole system of national well-being."[38]

The fuel question took its place alongside other questions that arose in late nineteenth- and early twentieth-century Japan like the "women's question" and the "social question"—questions that represented efforts to grapple with both changes in Japanese society and Japan's place in a changing world.[39] This trend of posing questions had a global dimension. Historian Holly Case has pronounced the long nineteenth century "the age of questions." She argues that this period witnessed the "question" emerging in Europe and the United States as "an instrument of thought with special potency, structuring ideas about society, politics, and states, and influencing the range of actions considered possible and desirable." Those who wielded this instrument used it to etch images of envisioned futures onto the present.[40] So, too, was it with the fuel question in Japan. Its interlocutors raised it not merely to express dismay over yet another measure by which Japan might be deemed inferior to the Western powers; they also did so to give weight to the issue of fuel security and lend credence to their accompanying prescriptions, which, in addressing that problem, promised to ensure the empire's continued advancement.

The fuel question first began taking shape through concerns over petro-

38. Okunaka, "Fuel Problem in Japan," 30.
39. On the women's question, see Koyama, *Ryōsai Kenbo*, 76–82. On the social question, see Garon, *State and Labor*, 23–29.
40. Case, *The Age of Questions*, xv, 211–12.

leum after World War I. The preceding conflict had made dreadfully clear how indispensable this liquid fuel now was to warfare. Over land, on and under water, and in the air, vehicles driven by petroleum-powered internal combustion engines granted combatants an unprecedented range of speed and mobility. In so doing, they redefined the terms of battle. "Petroleum supply now dictates victory or defeat," Matsuzawa Dentarō of the dominant Japanese petroleum producer Nippon Oil declared.[41] Fellow promoters of petroleum's strategic significance would insist that Germany's loss had largely resulted from shortages of this vital resource, often pairing this claim with French statesman Georges Clemenceau's memorable and much-quoted assertion that "a drop of oil is worth a drop of blood."[42]

The Imperial Japanese Navy took that lesson to heart. Having first fully outfitted a destroyer with oil-fired boilers in 1915, it subsequently moved to convert its entire fleet to run on oil.[43] As a naval fuel, oil held several key advantages over coal. Nakazato Shigeji, a vice admiral who later headed the petroleum concession Japan secured in Soviet Sakhalin, would contend that oil's higher calorific value allowed for greater speed, that its loading process by pipes rather than by workers saved considerable time and labor, and that its cleaner combustion gave off less tactically compromising black smoke.[44] The navy's shift to oil coincided with (and may have, in fact, been hastened by) restrictions on its fleet expansion. In 1922, the world's three leading naval powers, the United States, Britain, and Japan, agreed by treaty to contain the size of their fleets so as to forestall the arms race that had been brewing among them. Under the final terms, Japan was allowed 300,000 tons, 60 percent of the American and British allotments.[45] At the time, this put an end to a long-standing dream Japanese naval planners had for an enlarged "eight-eight fleet," so called for the eight dreadnoughts and eight armored cruisers that would have served as its backbone. Writing just as the treaty was being finalized, naval fuel specialist Mizutani Kōtarō 水谷光太郎 (1876–1962) cautioned against thinking that the pending limits on fleet growth would lessen demand for petroleum. He predicted, instead, a steady rise in both military and civilian consumption and proposed that Japan should prepare accordingly.[46] The navy, for one, endeav-

41. Matsuzawa, *Kokubōjō oyobi sangyōjō*, 1.

42. See, for example, "Waga sekiyu mondai."

43. Evans and Peattie, *Kaigun*, 184.

44. Nakazato, "Kaigun to sekiyu."

45. On the Washington conference at which these limits were decided, see Kajima, *Diplomacy of Japan*, 465–501.

46. Mizutani, "Gunbi seigen to sekiyu mondai."

ored to work around the imposed constraints and to continue expanding its operational capacities. Exclusively installing oil-burning boilers across its vessels was a means to that end. By 1929, the Japanese fleet was almost completely powered by oil.[47]

Mizutani was right. As he had anticipated, civilian demand also grew, at least in industrial applications. Petroleum straight out of the ground (termed in Japanese, as in English, "crude oil" [原油]) is of little human use. Distilled in refineries into different liquid and solid derivatives, it acquires a variety of potentialities, fueling engines being the main but not the only one. Between 1914 and 1921, Japan's consumption of liquid petroleum products (not factoring in the fuel oil the navy burned, the figures for which were not publicly circulated) increased slightly from 2.4 million to 2.5 million barrels a year. But this marginal rise was in spite of kerosene consumption falling by half, as this light oil, which had previously constituted the bulk of liquid petroleum products in Japan, lost out to electricity as an illuminant. All other oils saw remarkable gains. Aside from lubricating oil, which greased the gears of industrial Japan's motors and machines, the rest of these oils provided energy to engines: gasoline for airplanes, automobiles, and an assortment of machinery; and neutral and fuel oils mostly for marine vessels.[48] Considering oil's importance for national security and its mounting use as an industrial fuel, the dearth of domestic production and the reliance on imports became sources of deep anxiety.

The modern Japanese petroleum industry owed its origins to foreign assistance but consistently chafed under competition from abroad thereafter.[49] In parts of Japan, petroleum had been used for centuries. Where it leaked from the earth, collectors used straw to gather what they could of the viscous liquid, which they then burned in lamps for lighting.[50] Because of the stench that accompanied its combustion, petroleum was often called "stinky water" (臭水). Only the poorest illuminated their dark days with this fetid fuel. Following the Meiji Restoration, Japan began importing refined petroleum products, primarily from the United States, and local entrepreneurs, attempting to profit from the rising demand for these commodities, started sinking wells on home ground—initially meeting

47. Bōeichō bōei kenshūjo senshishitsu, *Kaigun gunsenbi*, 702.

48. Kurita, "Oil Consumption Index," 43–45.

49. For more details on the history of the prewar Japanese petroleum industry, see Samuels, *Business of the Japanese State*, 168–77; and Hein, *Fueling Growth*, 46–49.

50. This process bears resemblance to how crude oil was collected in eighteenth- and nineteenth-century Galicia. See Frank, *Oil Empire*, 50–51.

with limited success.[51] In 1876, American mineral expert Benjamin Lyman Smith—who had been serving as the Meiji government's chief geologist and mining engineer since 1873 and who had recently led an extensive survey of Hokkaido that uncovered the wealth of the region's coal riches—was tasked with studying Japan's oilfields. His findings confirmed the potential of select sites, most notably in Niigata, where Nippon Oil was established and commenced drilling just over a decade later.[52]

In the years that followed, Japan's petroleum production climbed, albeit slowly. Output jumped in 1914 after the introduction of the American rotary drill, which outpaced older cable tools (also imported from the United States) in allowing producers to reach greater depths. Domestic extraction peaked in 1916 at around 3 million barrels. Production subsequently declined, undermined by war-induced shortages of American drilling equipment, rising operational costs, and, most notably, cheap imports of both refined products and crude oil.[53] Japanese producers faced fierce competition from foreign firms, particularly Standard Oil of New York and Rising Sun Petroleum, an import and distribution company under the Royal Dutch Shell group.[54] Despite tariffs, domestic crude consistently cost more than imported equivalents, at times as much as double.[55] By the early 1920s, local producers, confronted with this price onslaught and the reality of Japan's limited reserves, began redirecting their energy from extraction to refining, importing foreign crude for this purpose. Between 1920 and 1922, domestic crude production fell from 2.2 million to 2 million barrels and domestic output of refined liquid products from 1.7 million to 1.4 million barrels, while imported crude rose from 99,000 to 563,000 barrels and imported refined liquid products from 1.2 million to 1.3 million barrels.[56] This trend only continued. In 1931 Japan imported 10.1 million barrels of refined liquid petroleum products. That year, it produced 5 million barrels of the same, but less than half, or 1.9 million, were from domestic crude while the rest were from imports.[57] The fear of an overdependence on foreign petroleum that Japanese planners had begun to harbor a decade or so earlier proved to be warranted.

51. Scott, "History of Oil in Japan," 483.
52. Smith, *Geological Survey of the Oil Lands*, 1–2, 45–46.
53. Coumbe, *Petroleum in Japan*, 2.
54. Coumbe, 19–20.
55. Samuels, *Business of the Japanese State*, 174.
56. Coumbe, *Petroleum in Japan*, 2, 14, 17.
57. Inhara, *Japan Year Book, 1935*, 552.

That Japan should be self-sufficient in petroleum or coal or any other key resource was an idea that fit within a wider commitment to autarky, which had also taken root through World War I. Recent prior wars in Asia and elsewhere had been relatively short. For Japanese military strategists, then, the main objective was the rapid mobilization and deployment of men and munitions. As historian Michael Barnhart has traced, the protracted fighting in the Great War, typified above all by the trench, challenged Japanese officers to rethink the conduct of war and appreciate the centrality of self-sufficiency.[58] This turned out to be a lesson well learned by most countries, whether or not they were directly involved in the fighting, particularly as blockades and the wartime commandeering of shipping put many of them under some degree of de facto autarky.[59] In Japan, the Ministry of War appointed army colonel Koiso Kuniaki to study the nation's strategic considerations and capacity under the new mode of warfare. Koiso drew two conclusions from his research. First, Japan, lacking sufficient resources for modern war, needed to establish control of territories that held these in abundance, such as China. Second, the domestic economy should be organized to allow for a quick mustering of necessary goods in the event of war. Koiso's recommendations garnered much interest among civilian and military leaders alike, and the Ministry of War backed the drafting of a comprehensive plan in line with his second conclusion that eventually became the Munitions Industries Mobilization Law in 1918.[60] This law and subsequent ones that similarly availed industries to military takeover presupposed and proceeded from a desire for autarky that began as presumed military necessity but was soon broadly understood as imperative for the survival of the modern state.

While recognizing its important military implications, Yoshimura Manji, the Fuel Society's inaugural president, insisted that what made the fuel question truly pressing was how it impinged upon social order and civilizational progress. Yoshimura began with the observation that labor movements were springing up across the post–World War I world, from miner and railroad worker strikes in Britain to trade union activities in the United States. "A trait of the times," he remarked, "was the awakening of the masses." As he saw it, the masses, no longer content with just satisfying their basic needs, sought higher-order things for which a robust material

58. Barnhart, *Japan Prepares for Total War*, 22–23.

59. Carr, *Twenty Years' Crisis*, 123.

60. Barnhart, *Japan Prepares for Total War*, 23. This was, incidentally, one of the laws that Inoue Tadashirō, as a member of a committee within the House of Peers, helped to shape.

foundation was critical. Two factors stood in the way of laying this material foundation. The first was rapid population growth. "In the past decade, our country has grown by 137 people for every 1,000. We can expect our population to double in size in about fifty years," Yoshimura wrote. This would, he added, echoing Malthusians in Japan and elsewhere, put immense pressure on national resources.[61] Such, in turn, was compounded by a second obstacle: rising material expectations. As the range of commodities available to everyday consumers expanded, many of these items became regarded as absolutely essential, spurring further increases in the mass production that had made them widely accessible in the first place. To Yoshimura, here was the source of the fuel question. The expansion of industrial production and the transportation system that circulated its finished goods demanded ever-escalating volumes of fuel. "[Coal] consumption grows in step with the progress of civilization," he declared. Labor unrest, he intriguingly suggested, was an unintended outcome of this progress, incited not by workers recognizing the conditions of their exploitation (that had enabled such progress) but by them coming to hold elevated expectations amid fossil-fueled material abundance. Following this reasoning, the awakened masses might be pacified and social order maintained if there were sufficient supplies of coal to sustain a continuous scaling up of mass production for their consumption.[62]

Equally interesting was Yoshimura's singling out of coal. Writing in the first issue of the society's journal in 1922, he proclaimed that "the solid fuel age is passing, and the world is entering the liquid fuel age." Yet he concurrently contended in the very same article that "coal will be the primary fuel of the future," citing recent remarks by British mining engineer John Cadman (who, incidentally, went on to become deeply immersed in the petroleum industry as chairman of the Anglo-Persian Oil Company). In Yoshimura's estimation, this was a matter of output and reserves. The global annual production of petroleum then was around 600 million barrels, which amounted to, he pointed out, about 100 million tons of coal. Coal output exceeded that by more than tenfold at roughly 1.3 to 1.4 billion tons per year. Furthermore, he asserted that while calculations of deposits might differ, there was little doubt that the volume of coal reserves worldwide and hence their longevity were "by far greater" than those of petroleum. In addition, there was the prospect of coal directly substitut-

61. On overpopulation anxieties in Japan, see Lu, *Making of Japanese Settler Colonialism,* 183–86.

62. Yoshimura, "Waga kuni ni okeru nenryō mondai," 1–3.

ing for petroleum. With limited access to the petroleum that powered the many machines now running on liquid fuels, countries like Japan should develop alternatives, Yoshimura advised, with one promising possibility being liquefying coal through low-temperature carbonization.[63] The persistence of "older" sources of power across energy transitions may be a given, but the way that these sources were sometimes repurposed in new forms is worth noting.

At the same time, Yoshimura, having read William Stanley Jevons's 1865 *Coal Question*, seemed to share the British economist's reservations about the long-term sustainability of coal-fired growth. He accepted the book's premise, namely, that Britain's strength was predicated on cheap coal, which had allowed it to produce cheap iron, then cheap machines, and finally cheap commodities for export. "This one book discusses in detail how the fate of nations—their waxing and waning—is deeply connected to coal," Yoshimura wrote.[64] But he also accepted its central concern. "The question," as Jevons had put it, "is, not how long our coal will endure before absolute exhaustion is effected, but how long will those particular coal-seams last which yield coal of quality and at a price to enable this country to maintain her present supremacy in manufacturing industry."[65] That is, once all the easily mined deposits were dug out, even if coal still remained in the ground, it would become increasingly more costly to extract, as was already the case in Japan. To aggravate matters, coal was, Yoshimura stressed, "a thing without hope of regeneration" (再生の望まないもの) with a rate of consumption that was quickly going up. Hence, even though the availability of this resource was not a problem at present, it soon would be. Given global estimates, which, for example, placed Japan's coal reserves at 0.2 percent those of the United States and 1 percent those of China, "it is not difficult to imagine that our country will exhaust [its coal] much faster than the United States, China, and other countries," he warned.[66] If coal were to be the primary fuel of the future, then Japan's future looked rather bleak indeed.

By the end of the 1920s, the fuel question became an issue of widespread interest across Japan. Often also spoken of in terms of its constituent parts with attendant particularities—mainly the "coal question" or the

---

63. Yoshimura, 3–4, 10.
64. Yoshimura, 6–7. On *The Coal Question*, see Seow, "*The Coal Question.*"
65. Jevons, *The Coal Question*, 35.
66. Yoshimura, "Waga kuni ni okeru nenryō mondai," 5.

"petroleum question," although the "charcoal question," for instance, was not altogether unheard of—it reflected a broad set of concerns surrounding ongoing and anticipated shortages in domestic sources of energy. It was, as a 1929 *Japan Advertiser* article acknowledged, "a momentous issue to solve."[67] While different commentators often understood the fuel question differently in terms of what it entailed and how it should be addressed, the suggestion that Japan faced an energy crisis went basically unchallenged. Its accompanying anxieties would fuel the expansion of carbon technocracy.

The Fuel Society of Japan

The establishment of the Fuel Society of Japan in 1922 represented one response to the fuel question. This was, however, not the first organization formed to confront the crisis. In 1921, the Imperial Japanese Navy expanded its briquette factory in the port of Tokuyama, turning it into a fuel depot that soon led naval research on the liquid fuels with which it expected to power the entirety of its future fleet. A year earlier, the Ministry of Agriculture and Commerce set up in Saitama the Fuel Research Institute (燃料研究所) to study coal and its efficient usage, prompted by worries over dwindling deposits.[68] Extant organizations, from corporations like Nippon Oil to academic institutions like Hokkaido Imperial University, also redoubled their efforts on or redirected their attention toward research aimed at tackling the fuel question. In general, scientists and engineers across these organizations focused on two sets of solutions: improving the efficiency of fuel extraction and use and developing new fuel substitutes— the effort in Fushun to turn shale oil into an economically feasible petroleum alternative being an example of the latter.

The Fuel Society was distinct in that it itself did not conduct research into fuels. Rather, it brought together researchers and other interested parties from industry, academia, government, and the military (many of whom worked for or were affiliated with one of the other organizations mentioned here) in order to, in Yoshimura Manji's words, "further the exchange of knowledge about fuel and power among [ourselves] and inform the general public [of our country's fuel question]."[69] This being so, the

67. "Basic Fuel Problem."

68. Kashima, "Japan Letter."

69. Okunaka, "Fuel Problem in Japan," 34–36; Yoshimura, "Nenryō kyōkai jūnen shi," 1198.

society functioned as what historian Tessa Morris-Suzuki has described as a "social network of innovation," through which loci of research and production were connected and by which, she contends, post–Meiji Restoration Japanese technological development was propelled.[70] But while the exchanges between society members may have helped advance fuel research, they also heightened the sense of urgency around that work. In relentlessly bringing up the fuel question, particularly as justification for their research, the Fuel Society's crosscutting coalition of stakeholders and experts both reinforced the importance of acting expediently to address the problem and established their authority as the ones to do the job.

The Fuel Society began as the Fuel Discussion Society, founded in 1921 not long after the Fuel Research Institute was created. In this earliest iteration, it was a loosely organized gathering of fuel researchers who met regularly to discuss such topics as the low-temperature carbonization of coal or coal pulverization in fuel use. Within a year, the twenty-three members of this group, seeking a more formalized arrangement and a broader mandate, reconstituted themselves as the core of the new Fuel Society.[71] The organization expanded quickly thereafter. By the end of 1925, the year Gotō gave his lecture on "Human Life and the Fuel Question," its membership reached 1,500. Two years later, it was registered as a corporate entity, "in order to further strengthen social trust [toward it] and solidify its [institutional] foundations," its board reasoned.[72] As it expanded, the society appointed regional directors to oversee affairs across the home islands and in the colonies. The Manchurian director was Kido Chūtarō, who still headed Mantetsu's Geological Research Institute almost two decades after leading its first study, the survey of the Fushun coalfield.[73] Headquartered in Saitama at the government's Fuel Research Institute, the Fuel Society was invested in encouraging the development of energy resources all the way to the edges of Japan's empire.

As founding president, Yoshimura Manji, who at the time also led the Fuel Research Institute, was instrumental in determining the society's direction. A mining engineer by trade, Yoshimura, like many other prominent mining engineers—including, as we have seen, Fushun's colliery managers—graduated from Tokyo Imperial University, finishing in 1906. He spent the years between 1913 and 1915 first in Europe and then in the

70. Morris-Suzuki, *Technological Transformation of Japan*, 7–9.
71. Yoshimura, "Nenryō kyōkai jūnen shi," 1197.
72. Yoshimura, 1199.
73. Yoshimura, 1198.

United States, where he completed graduate work in chemistry at Stanford University.[74] While abroad, Yoshimura was struck by how the British and Americans had, to his eye, internalized the idea of "preserving and efficiently using natural resources" in spite of the material wealth that their "many colonies" and "expansive territories" had afforded them. The Japanese, who were poorer on these fronts, would do well to cultivate a similar consciousness and a general appreciation for the gravity of the fuel question, he decided.[75] Yoshimura thus ensured that the Fuel Society adopted a strong focus on public outreach from the start.

To meet this primary purpose of exchanging and disseminating fuel-related knowledge, the society organized numerous events. It held monthly and special public lectures ("not resting," Yoshimura boasted, "even in the sweltering heat of the summer"). These talks, delivered by "experts of all stripes," covered a range of topics, from the latest fuel research to the meaning of the fuel question in Japan and elsewhere. They drew sizable crowds. In 1925, for example, over seven hundred attendees in total came to the monthly lectures. Periodically, the society hosted roundtables among specialists on select issues such as fuel considerations in urban reconstruction following the great Kantō earthquake. These, too, were open to the public and had wide appeal. Every spring, the society put together a large conference consisting of lectures both popular and academic and field trips to nearby industrial sites and research institutes. The location of this conference changed from year to year. The main stipulation, though, was that the places picked be "deeply connected to the production or consumption of fuel." The first year, it was held in Osaka, the "Manchester of the Orient" with its "forest of smokestacks." The next year, the organizers got more ambitious, holding the conference in Korea and Manchuria, with participants traveling to several sites across the two. One of these sites was, unsurprisingly, the Coal Capital, Fushun.[76]

Of all the events the society put on, the biggest was a "fuel exhibition" (燃料展覧會) in Tokyo in 1930. Held at the Chamber of Commerce and Industry in the center of the capital, this exhibition showcased a host of products, from novel fuels such as "artificial charcoal" that burned without ash and "synthetic petroleum" made from fish oil to new appliances such as coal-saving boilers, stoves, and kettles. Through an assortment of line

74. Takenobu, *Japan Year Book, 1927*, 123; Leland Stanford Junior University, *Graduate Study*, 58.

75. Yoshimura, "Nenryō kyōkai jūnen shi," 1194.

76. Yoshimura, 1201–4.

graphs, bar charts, and drawings, detailed information and statistics about fuel were also on display—presentations that drew, one observer approvingly noted, much interest, "especially from women and children." Attendees, of whom there were close to forty thousand in the ten days the event ran, could also sit in on lectures about the fuel question, pick up pamphlets bearing such titles as *On the Use of Fuels* and *A Fuel Reader*, and even watch short films about fuel. Its organizers would declare the exhibition "unimaginably successful."[77] Whether one accepts their self-congratulation, it is hard to deny that the Fuel Society did more than any other entity to make the fuel question a public concern, as it worked to condition the Japanese populace to crave fossil fuels and fear energy shortages.

The society further circulated knowledge about fuels in print, with its monthly journal as the main platform. Transcripts of many of the lectures and discussions that the society organized were published there after the event, ensuring an audience far beyond those who had attended. The journal also contained editorials on the state of the fuel and power industries; articles and reports on all manner of issues pertaining to fuel from the highly technical to the easily accessible; summaries of research findings and industry surveys abstracted from foreign publications; statistics on the production, import, and export of coal and petroleum; and news about the society itself and the larger energy economy.[78]

In 1925, the society, in the spirit of fostering public engagement, held an essay contest in conjunction with its third anniversary. The prompt was to write a piece "to inform the general public of important issues concerning the fuel question." The judges' pick, which was published in the society's journal the following year, was one they deemed "outstanding . . . in content, structure, and style," distinguishing itself from the forty other submissions by attending not only to the expected economic, national security, and social policy aspects of the fuel question but to its "ethical" (倫理) dimensions as well. Its author, army officer Nagai Mosaburō, had contended that since "natural resources are gifts bestowed upon humankind by heaven," access to them should be open to all countries under principles of "co-existence and co-prosperity" within "international community." He argued that "while it may be unjust for resource-poor countries to invade countries with resource abundance, it is also unjust for resource-rich countries to deny other countries access to their resources." By this logic, "forcing oneself in" (割込) to secure fuel rights may be de-

---

77. Yoshimura, 1208–10; "Mokutan ya sekiyu"; Mitsuki, "Jūshūnen no omoide."
78. Yoshimura, "Nenryō kyōkai jūnen shi," 1194.

fensible on ethical grounds, he concluded. The representative case in this, as he saw it, was the United States, who "having arrived late to the struggle for oil" (in the Middle East) was now "crying for equal opportunities and the open door."[79] The Japanese empire, it seemed, had something to learn from the American one.[80]

Committed as it was to raising public awareness of the fuel question, the Fuel Society also tried to address the problem more directly. For instance, its journal may have sought to reach a general readership interested in "the arts and sciences relating to the technology of Fuels," but it also served as an important forum for fuel experts across sectors to discuss among themselves how they might solve the fuel question.[81] In addition to ongoing pursuits of fuel efficiency and fuel alternatives (which the next two sections will cover in greater detail), interlocutors agreed that a national fuel policy was imperative, and the society would actively agitate for one.

While opinions differed on the finer points of such a policy, what was clear was that many fuel experts saw a prominent role for the state in settling the fuel question. "The Government should lose no time in framing a definite fundamental policy, so that by their combined efforts the present limited resources and precarious supplies may be worked to the best advantage and the most economic uses may be made of them," the society's secretary, Okunaka Kōzō, wrote.[82] In 1924, Yoshimura and several society leaders drafted a proposal that they submitted first to the Kenseikai, the dominant political party of the day, and then to the cabinet recommending that a "fuel and power investigation commission" be established to lay the groundwork for a fuel policy.[83] Two years later, the Ministry of Commerce and Industry, upon further pressure from the navy, set up the Fuel Investigation Committee (燃料調査委員會) to do just that.[84]

To proponents, the fuel policy they envisioned was a means of secur-

79. "Nenryō mondai no jūyo," 736–37, 746. On the United States' efforts to enter the struggle for oil in the Middle East after World War I, see DeNovo, "Movement for an Aggressive American Oil Policy"; and Yergin, The Prize, 194–206.

80. On the United States as an empire, see Maier, Among Empires; and Immerwahr, How to Hide an Empire. For a treatment specific to energy resources, see Shulman, Coal and Empire.

81. Takenob, Japan Year Book, 1923, xliv. The trade journals Coal Times (石炭時報) and Petroleum Times (石油時報) also served similar functions, with contributors ranging from company presidents to naval vice admirals, though neither quite had the same breadth of coverage as the Journal of the Fuel Society of Japan in terms of multiple fuel types, focused as they were on their respective industries.

82. Okunaka, "Fuel Problem in Japan," 30.

83. Yoshimura, "Nenryō kyōkai jūnen shi," 1206–7.

84. Sohn, Japanese Industrial Governance, 44.

ing state involvement and, perhaps more crucially, funding on some or all of the following fronts: discovering and opening up coal and oil fields in the home islands, regulating prices and distribution among domestic producers, developing fuel substitutes, and acquiring mining concessions abroad. It would take the committee several years to help work out a formal fuel policy, and even then this 1934 law was specific to the petroleum industry and consisted mainly of protectionist stipulations that guarded Japanese refiners from foreign competition.[85] Still, although there was not and would not be an overarching fuel policy, the state ended up advancing the aforementioned objectives in a piecemeal manner. For example, in 1927, the Fuel Investigation Committee convinced the Ministry of Commerce and Industry to factor into its budget 2.5 million yen of subventions over five years for machinery purchases and boring tests to encourage domestic explorations for oil resources.[86] In regard to obtaining extractive rights on foreign soil that was not already a colony or, in the case of Manchuria, part of the leased railway zone, the Japanese government, after a failed attempt to acquire oil concessions in Mexico, concluded an agreement with the Soviet government that allowed it to establish in northern Sakhalin a state-run enterprise that drilled for petroleum there from 1925 to 1944.[87]

Behind the various provisions that fuel policy advocates put forward was the conviction that Japan needed access to increasing amounts of fossil fuels. Without doubt there were some, namely local producers, who sought to protect their profitability with restrictions on output or imports, as we will see in the tension between Japanese and Fushun coal. But most of those who called for a fuel policy—certainly those from the Fuel Society—wanted more carbon resources out of the ground and in Japanese hands, including carbon resources that lay beneath foreign soil. One might take this dynamic of expansion for extraction to be the natural interplay between empire and energy until we consider that to other imperial powers like the United States and Britain, which were engaged in the contest for Middle Eastern oil during this period, controlling petroleum was actually less about producing escalating quantities than about limiting output (and thus maximizing profits for national firms) amid the worldwide glut of the

85. On the 1934 Petroleum Industry Law, see Sohn, *Japanese Industrial Governance*, 49–56; and Samuels, *Business of the Japanese State*, 177–79.

86. Okunaka, "Fuel Problem in Japan,"

87. On Japan's advance into northern Sakhalin and its securing of oil concessions on that half of the island, see Hara, "Japan Moves North."

late 1920s.[88] All too aware of its comparative poverty in energy resources as in other endowments, Japan, in contrast, just could not get enough of that stuff.

Conserving Fuel for the Nation

For many experts, one solution to the fuel problem was to use less of it. They were, however, not talking about scaling back the industrial and other activities that consumed coal, petroleum, and other sources of energy. Rather, speaking in terms such as reducing "the sacrifice of fuels," they promoted methods and mechanisms that would allow "small amounts of fuel to burn completely and to give off great heat" to support existing and expanding applications—in other words, increased efficiency along with increased use.[89] This interest in fuel conservation was in response not only to anxieties about present and projected scarcity at the heart of the fuel question but also to worries about rising costs of coal and petroleum.[90] Some framed it as a moral imperative. To Naitō Yū, Japan needed to recover a lost "sense of respect for material things" (物資尊重の念) that it purportedly possessed as a "closed country" (鎖國) in the premodern era.[91] "Since the Meiji Restoration and the importation of Euro-American civilization, great increases in production and exchanges with foreign countries have made commodities more easily available, and, as such, we have stopped abiding by our traditional admonitions [to be mindful of how we use resources]," he lamented.[92] A founding member of the Fuel Society with a degree in applied chemistry from Tokyo Imperial University, Naitō worked for Tokyo Gas for several years before setting up his own laboratory for fuel research. He would be one of several fuel researchers who concentrated their efforts on conserving fuel through maximizing efficiency, whether through improving methods of fuel selection and handling or designing engines and other devices that retained more heat and thus saved more fuel.[93]

The first major attempt at applying fuel conservation to industry took place in Osaka in the late 1920s. As historian Kobori Satoru contends in his study of transwar Japan's "energy revolution," this was a key episode in

88. Ross, *Ecology and Power*, 219–20; Mitchell, *Carbon Democracy*, 43–45, 96–97.
89. Naitō, "Nenryō kenkyū no igi," 6.
90. "Japan Industries Retarded"; Kobori, *Nihon no enerugii kakumei*, 34–35.
91. For resource efficiency in early modern Japan, see Hanley, *Everyday Things*, 51–76.
92. Naitō, "Nenryō kenkyū no igi," 4.
93. Naitō, 6.

"the birth of energy conservation policy."[94] Behind this effort was the Fuel Combustion Guidance Bureau (燃焼指導部) under the Osaka Industrial Efficiency Research Institute (大阪府立産業能率研究所). Formed by local officials and business leaders in 1925, the institute sought to bring about the "rationalization of production" (産業合理化) in this most industrial of cities.[95] In consultation with Fuel Society member and mining engineer Tsujimoto Kennosuke, it created the bureau in 1929 with the main goal of increasing industrial efficiency by promoting better fuel burning. The bureau dispatched its experts to factories to give guidance on matters of fuel combustion, particularly in regard to selecting suitable fuel, certifying the proper installation of equipment, and ensuring correct operations, from the way that the coal was fed into boilers to ventilation control. Without investing in additional facilities, factories that followed the recommendations they received were able to reduce their fuel consumption by 16 percent on average.[96] In pursuit of similar ends, the bureau also conducted extensive training courses for boilermen, who, from the early 1930s, needed to be licensed (in the proper handling of fuel) before they could be hired by factories.[97]

It was, however, not directly in the name of industrial efficiency but of smoke abatement that fuel conservation became policy in Osaka. In 1932, the Osaka prefectural government passed regulations on smoke abatement, of which the boilermen licensing requirement was one important component. These regulations, Kobori claims, were "the first regulations on air pollution in Japan."[98] They had been long in the making. Concerns about the adverse effects of smoke emitted from coal-burning factories had been present in industrial Japan since the late nineteenth century.[99] Although many saw in smoke a sign of progress—as was the case with the ironworks city of Yahata, whose municipal anthem included a line celebrating the fact that "billows of smoke overspread the sky"—some had their misgivings.[100] Writing in 1888, Gotō Shinpei, then a young doctor in the Home Ministry's Bureau of Health, observed that "some people seem to rejoice idly the increase in the number of smoke-stacks, showing no resistance to the in-

94. Kobori, *Nihon no enerugii kakumei*, 40–51. My account of this case draws from Kobori's work.

95. Tsutsui, *Manufacturing Ideology*, 79.

96. Kobori, *Nihon no enerugii kakumei*, 48.

97. Kobori, 50–51.

98. Kobori, "Development of Energy-Conservation Technology," 223.

99. Britain served as a point of reference for many Japanese critics. On smoke pollution and smoke abatement in Britain, see Stephen Mosley, *Chimney of the World*.

100. Shigeto, *Political Economy of the Environment*, 28.

3.2. Osaka, "Manchester of the Orient" and "Capital of Smoke." In this photograph from the 1880s, one can already spot multiple factory chimneys in the background, the beginnings of the "forest of smokestacks" that would later come to define this industrial city and inspire Japan's first successful smoke abatement campaign, which prevailed when advocates managed to link smoke prevention to industrial efficiency. (Image courtesy of MeijiShowa/Alamy Stock Photo.)

creasing installation of them in the heart of our cities, believing apparently that the spectacle is thereby enhanced." He demurred: "Such a situation evokes grave apprehension for the country's future."[101]

In Osaka, also nicknamed the "Capital of Smoke" (煙の都), there were several failed attempts to regulate chimney exhaust in the decades that followed. One representative example was a smoke abatement ordinance put forward by a circle of influential citizens in 1913 that included clauses mandating the installation of emission-reducing devices. When the Osaka Chamber of Commerce was asked to comment on the proposal, it delayed its response for over a year, before concluding that the cost of such devices was too high for operators to bear: "If the proposed ordinance were to be enforced, an inevitable consequence would be the closing down of many of the factories which would strike a deathblow to the vital sinews of the

101. Cited in Shigeto, *Political Economy of the Environment*, 30.

city of Osaka."[102] It proceeded to veto the proposal. Here and elsewhere, smoke pollution worsened, even as its harmfulness to health became more widely recognized. In a Fuel Society lecture, nutrition expert Arimoto Kunitarō warned both of how extended smoke inhalation could lead to "city lung" (black lung disease) and of how the decreased exposure to sunlight due to smog-darkened skies could cause deficiencies in vitamin D, which was essential for "the prevention and cure of rickets" and "the normal growth and development of bones and teeth."[103]

That the Osaka prefectural government finally managed to introduce smoke abatement regulations in 1932 was a result of its collaboration with Fuel Society representatives like Tsujimoto. Together, they made a case that strongly equated smoke prevention with increased fuel efficiency. A more complete combustion of coal, after all, used less fuel and released less particulate matter into the atmosphere. On that basis, they convinced local businessmen that the necessary measures for smoke abatement would actually reduce their costs rather than add to them.[104] The regulations put in place in Osaka were soon adopted by other urban industrial centers, including the Manchurian port city of Dairen.[105] This might have not been the first and certainly was not the last time that environmental protections saw introduction only after corporate interests were assured.

Beyond industry, fuel conservation also made its way into the household. As Ōshima Yoshikiyo 大島義清 (1882–1957), then Fuel Society president, remarked in 1926, "Domestic fuel consumption may seem small for individual households, but were we to regard the cumulative total for the whole country, it is not something to sneeze at."[106] As part of its ongoing efforts in public education, the society committed to pushing for more prudent fuel use in the home. A society meeting held in conjunction with the 1930 Tokyo fuel exhibition featured a lecture on "Cooking and Fuel" by Japan Women's University professor Wakahara Tomiko. In similes that recalled Clemenceau's likening of oil to blood, she declared: "A piece of coal is like a slice of bread; a drop of oil like a cup of drinking water." Fuel, much like these staples, sustained the social body and hence should not be wasted, she seemed to imply. Like many other urban reformers, Wakahara promoted coal gas as the household fuel of choice. This was not only

102. Cited in Shigeto, 30; Adachi, "Gaien bōshi undō," 1468.
103. Arimoto, "Toshi kūki," 984–87.
104. Kobori, Nihon no enerugii kakumei, 46.
105. Kobori, 53–59.
106. Ōshima, "Katei nenryō," 642.

because it was convenient and hygienic, as boosters liked to advertise, but also because it was, in her assessment, more cost and heat efficient.[107]

But fuel conservation in the home was more than just a matter of fuel choice. It also involved how the fuel was used. Wakahara's talk included several practical measures on how to minimize fuel consumed in cooking based on tests that she herself had conducted. For example, when making rice, one should, she advised, first maximize the heat applied to bring the water to boiling point, then turn the heat down to the lowest level possible to keep the water boiling, and, in the last stage, adjust the heat carefully to avoid burning the rice. To illustrate this process, she provided a graph of temperature against time, with the three phases clearly marked out.[108] In cooking, as in other fuel applications, the search for efficiency was buttressed by the trappings of science. For middle-class women who labored in the home, the new domestic sciences and technologies, in their exacting demands, saved fuel while often adding time and toil, resulting in what historian of technology Ruth Schwartz Cowan, in her examination of household tools and gendered work in industrializing America, has identified as "more work for mother."[109]

To Wakahara and other advocates of household fuel conservation, saving fuel at home served national needs. In order to solve the fuel question, Japan, no doubt, had to "discover new fuel sources" and "develop overseas resources," Wakahara contended. And yet, she continued, it was also essential that its fuel use be "effectively rationalized," including by "women as consumers of kitchen fuel."[110] Frugality with fuel had long been championed by proponents of household austerity.[111] With the rise of the fuel question, though, such economizing practices became explicitly less about the virtue of thrift as about the interests of the state and its fuel future. In the context of Japan's "material shortages," which extended to fuel, "there is nothing more urgent than for ordinary households to cut back massively on consumption," wrote Tokyo-based teacher Kunisawa Kiyoko in a handbook she authored on the home economics of fuel.[112]

In both factory and domicile, then, fuel conservation was touted as a

---

107. Wakahara, "Ryōri to nenryō," 1114–16, quotation on 1114.
108. Wakahara, 1116–20.
109. Cowan, *More Work for Mother*.
110. Wakahara, "Ryōri to nenryō," 1115.
111. On household austerity in early twentieth-century Japan, see Garon, "Fashioning a Culture."
112. Kunisawa, *Katei ni okeru nenryō*, 3.

way to get more out of the energy resources available to Japan. What differentiated the two was that when fuel truly became scarce, it was the home and not the mill that was treated as the primary site of sacrifice, where women and men were asked not just to burn fuel efficiently but, at times, to forgo burning it altogether. "State policies" should come before "personal interests," Wakahara insisted.[113] As will be apparent in the next chapter, this became painfully so in the squeeze of the wartime era.

A Pipe Dream

In tandem with efforts at conserving fuel were attempts by Japanese researchers to develop fuel substitutes. If Japan lacked the energy resources that made states great, then a solution many of these scientists envisioned involved deriving similar types and amounts of power from alternative sources. For instance, chemist Kobayashi Kyūhei made a small splash in 1921 when he extracted "a volatile oil" with "a green fluorescence and a petroleum-like odor" out of a mixture of herring oil and fuller's earth.[114] Similar experiments with fish oil and other substances followed. One particularly infamous episode came at the tail end of the Pacific War. In a bid to secure sorely needed aviation fuel, naval scientists pushed for the widespread distillation of pine root oil, an immensely labor-intense initiative that resulted in the decimation of much of the archipelago's dense pine forests while producing little fuel that could or would be put to actual military use.[115] Another notable alchemic endeavor (which the next chapter will take up) was the coal liquefaction that Yoshimura had mentioned. While Japanese scientists worked on operationalizing this process through the 1920s and 1930s, it was only during World War II that plants were set up in Fushun and elsewhere to produce fuel by liquefying coal. Alongside these efforts were attempts by engineers and fellow tinkerers to design engines that ran on nonstandard fuels, the foremost application of this technology being the charcoal-powered automobile that traversed roads across metropole and colony with increasing frequency as wartime petroleum shortages intensified.[116] For the most part, however, less attention was paid to fashioning such devices, which never seemed to catch up in capability with the engines they were meant to replace. In regard to technoscientific

113. Wakahara, "Ryōri to nenryō," 1115.
114. Kobayashi, "Gyoyu yori sekiyu seizō"; "Artificial Oil Made in Japan"; Colby, *New International Year Book*, 131.
115. Tsutsui, "Landscapes in the Dark Valley," 300; Yergin, *The Prize*, 363–64.
116. Shirato, "Mokutan gasu jidōsha."

fixes, fuel substitutes remained the primary preoccupation, and of these none were quite as significant as Fushun shale oil.[117]

Shale oil is a liquid hydrocarbon distilled from the insoluble organic matter contained within the finely textured sedimentary rock called oil shale. To the delight of many Japanese imperialists, a huge deposit of this rock sat directly atop Fushun's thick coal seam. According to Akabane Katsumi, a Mitsui man turned Mantetsu executive who played a pivotal part in establishing Japan's first shale oil industry in Fushun, this was nothing less than "good news from heaven" (天来の福音), for the estimated 5.5 billion tons of Fushun oil shale might yield over 2 billion barrels of oil, which was, he noted, about a fifth of the reserves of the petroleum-rich United States and enough to meet Japanese demand for 300 years. Moreover, "in supplying the empire's various military, industrial, and transport activities," this fuel "would contribute to the furthering of the country's fortunes," Akabane wrote in 1924, making a case for the value of shale oil as interested parties explored the possibility of exploiting it.[118]

Local lore has it that the oil-bearing properties of the shale overlaying Fushun's storehouse of coal were discovered by an old Chinese worker in 1912. One supposedly scorching summer afternoon, this worker was taking a smoke break by some exposed shale on the edge of the Qianjinzhai mine. After finishing his last drag and emptying out his pipe, he noticed that the shale right by him was oozing a strange liquid that caught fire when it came into contact with the disposed ashes. A Japanese foreman passing by saw this curious sight and reported it to his superiors. Thereafter, as the story goes, the presence of oil in the rock was confirmed and quickly became widely known.[119]

Japanese accounts make no mention of a Chinese worker and date the discovery to 1909. They do, however, talk about a "flammable rock" (燃える石) found in Fushun that was taken to Mantetsu's Central Laboratory in Dairen, where it was analyzed by company chemists.[120] The shale sample initially subjected to tests came from a part of the formation closest to the coal, where the quality was lowest. The oil it yielded was only 2 percent of its mass.[121] Although further tests were carried out over the next

117. On the rise of Fushun's shale oil industry, see Iiduka, "Mantetsu Bujun oirushēru jigyō"; and Yamamoto, "Manshū oirushēru jigyō." My account of the establishment of Fushun's shale oil industry draws in part from both of these articles.

118. Akabane, *Nihon no sekiyu mondai*, 9–10.

119. *FKZ*, 290.

120. Mizutani, "Manshū ni okeru ekitai nenryō," 822.

121. Iiduka, "Mantetsu Bujun oirushēru jigyō," 3.

few years, the prospect of turning this into a worthwhile enterprise seemed unlikely.[122]

And yet, as the fuel question surfaced after World War I, Fushun's oil shale began receiving unprecedented attention. The navy had taken some interest in this resource early into the war, commissioning studies of existing shale oil operations in Europe and the United States. Before long, it reached out to Mantetsu to explore the viability of developing a shale oil industry in Fushun. To that end, the navy ran tests on Fushun shale at its Tokuyama briquette factory, sent tons of the rock to Scotland for industrial-scale trials, and dispatched personnel to the colliery to conduct experiments on site.[123] For its part, Mantetsu intensified its own investigations into shale oil. Like the navy, it shipped huge quantities of shale halfway around the world to undergo trials in not only Scottish but also Estonian, German, and Swedish retorts.[124] In-house, its scientists and engineers carried out further research on the shale's properties. Overall, the various results left the navy slightly hopeful and Mantetsu much less so as to whether shale oil could be produced in Fushun in a cost-effective manner.[125]

Two factors did, however, come into play to make this enterprise seem workable. As Fuel Society founding member and navy officer Mizutani Kōtarō explained over a decade later, one was that through boring tests into the shale stratum and a thorough examination of the core samples retrieved, Mantetsu researcher Kimura Tadao determined that the portion of the shale deposit farthest away from the coal was of good quality and could yield more than 10 percent of its mass in oil. Kimura, who had, under company orders, spent several months observing operations at an oil mill outside of Edinburgh, concluded that the closer one got to the coal, the lower the quality of the shale.[126] This being so, if the company were to leave out the more coal-adjacent shale, it might be able to get an oil-recovery rate of 7 to 8 percent.[127] The other, perhaps more important, factor was that with the advent of extensive open-pit mining (chapter 2), colliers had to remove the shale to get to the coal beneath anyway. As such, the unearthing of this rock to process fuel, another Mantetsu researcher

122. *ITM*, reel 107, nos. 05612–05615.

123. Yamamoto, "Manshū oirushēru jigyō," 181–82.

124. See, for instance, "Oil-Shale-Tests in Sweden, Sample of Oil Obtained" (June 9, 1922), *ITM*, reel 107, no. 05616.

125. Yamamoto, "Manshū oirushēru jigyō," 182–84.

126. Mizutani, "Manshū ni okeru ekitai nenryō," 824; "New Fushun Shale Oil Plant," 58.

127. Iiduka, "Mantetsu Bujun oirushēru jigyō," 5.

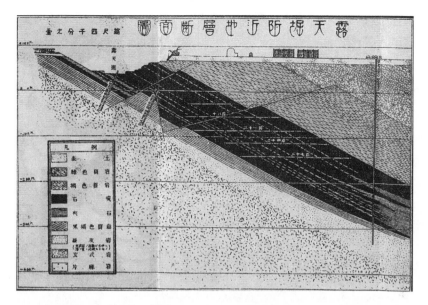

3.3. Cross-sectional diagram of the open-pit mine. This diagram comes from a pamphlet Mantetsu published in 1929 titled *Bujun tankō gaiyō* [An overview of the Fushun colliery]. The black band marks the coal seam, while the tier right above it is the layer of oil shale. (Image courtesy of the Harvard-Yenching Library, Harvard University.)

contended, "may be regarded to have been free of cost."[128] The prior expansion of extraction here hence unexpectedly opened up possibilities for exploiting another carbon resource. But while there was indeed no additional expense for getting the shale out of the ground, once one factored in the capital required for retorts and other processing facilities—or, in hindsight, the environmental impact of producing and consuming this unconventional oil—the costs were in fact ample.[129]

Costs arose as a major consideration in deliberations over launching Fushun's shale oil industry. To discuss steps toward establishing such an enterprise, Mantetsu hosted a weeklong meeting with the navy and the army at its main office in Dairen in May 1925. That its president, Yasuhiro Banichirō 安廣伴一郎 (1859–1951), himself chaired this meeting suggested how seriously the company took the prospective endeavor. Among

128. Mizutani, "Manshū ni okeru ekitai nenryō," 824; Ishibashi, "Present Status," 183.

129. And there was, of course, the continued cost of maintaining this site of extraction. As Aaron Jakes puts it in his examination of Egypt's cotton economy, "nature's 'free gifts' were never exactly free. . . . [Their recovery] entailed diverse arrangements for suppressing the real costs of these profound ecological transformations." See Jakes, "Booms, Bugs, Bust," 1043.

its other representatives were Mantetsu Technical Advisory Board chief and Engineers' Association of Manchuria president Kaise Kingo, Fushun colliery manager Umeno Minoru, and the colliery's industrial affairs section chief Okamura Kinzō, who had just returned from observing shale oil operations across Europe and the United States.[130] The navy and army personnel in attendance included officers who oversaw supplies and armaments as well as engineering professors Ōshima Yoshikiyo, Kurihara Kanshi, and Tanaka Yoshio who were on military secondment (all of them were, incidentally, also active Fuel Society members).[131]

The main issue on the table was how the shale oil was to be recovered. The Scottish distillation method had immediate appeal. While several other countries also had shale oil facilities, Scotland was "the only one with large-scale and industrial [capacity] at present," Kaise stressed.[132] From the point of view of the "navy's fuel oil needs" and the principle of "petroleum self-sufficiency," Ōshima and other naval delegates contended that the tried and tested Scottish method was "the most indispensable."[133] However, as Okamura, referencing what he had learned from his recent trip, pointed out, the Scottish method, in which retorts were almost always externally heated, was "bad in terms of heat economy" and would incur heavy operational costs on top of the high initial capital outlay. He instead advocated for developing a more efficient internally heated method, such as what he witnessed in Germany, in which fuel and feedstock reacted directly with one another within the retort. Most of his Mantetsu colleagues, while not directly endorsing this idea, took the position that trials on both of these methods should be carried out to determine the optimal—that is, most cost-effective—option.[134]

Over the next few years, Mantetsu garnered much public attention as it slowly ushered the shale oil industry into existence. To begin with, it secured a commitment from the navy to purchase the entirety of the projected yield, guarding itself against a situation in which there was no market for this costly petroleum substitute.[135] Still, the question of how to produce the shale oil remained. It had quickly become apparent that the Scottish

130. Iiduka, "Mantetsu Bujun oirushēru jigyō," 6.

131. Minami Manshū tetsudō kabushiki gaisha, *Bujun yuboketsugan jigyō,* 1–2.

132. Minami Manshū tetsudō kabushiki gaisha, 252.

133. Minami Manshū tetsudō kabushiki gaisha, 248.

134. Minami Manshū tetsudō kabushiki gaisha, 232–37; Yamamoto, "Manshū oirushēru jigyō," 187–88. On the Scottish method of shale oil distillation, of which there were several variants, see Gavin, *Oil Shale,* 62–74.

135. "More on Fushun Oil Shale Question"; "Fushun Shale Oil Industry Decided."

method would be economically untenable: it was capital intensive relative to output, and there were limits as to how much the government would allow the navy to pay. According to Ōkochi Masatoshi—the president of the Institute of Physical and Chemical Research, Japan's premier scientific research center—who visited Fushun in the summer of 1925, only in the most optimistic of circumstances might the price of petroleum be high enough to justify producing shale oil by the Scottish method. "However, in launching a new undertaking, the worst contingency in the line of that particular business should form the basis of calculation, so as to minimize the attendant risk," the enterprising engineer reasoned.[136] Mantetsu's executives seemed to have operated by that very logic.

Ōkochi noted that "the Fushun experts are taxing their intellect and effort for inventing a new process more advantageous to them than the Scottish process."[137] By September 1925, their experiments started yielding encouraging preliminary results. Fushun shale oil proved to have "a far lower melting point" than shale from Scotland, Germany, and Estonia. When superheated by steam to 1,100 degrees Fahrenheit in the oven the experts had contrived, it melted completely. But when the heat was lowered to 1,000 degrees Fahrenheit, "heavy oil of superior quality was obtained."[138] Based on this outcome, the experts—Okamura and two of his subordinates at the colliery's industrial affairs section, engineers Hasegawa Kiyoji and Ōhashi Raizō—went on to further improve upon the process. They scaled up the workings from the initial ten-ton pilot furnace to a full forty-ton (later, fifty-ton) industrial retort. Along the way, they redesigned the vessel to allow the surplus gas generated during distillation to be fed back into the producer for heat so that auxiliary fuel was no longer needed. Finally, through measures such as adjusting the rate and temperature of distillation, they made it possible to recover not only profitable volumes of ammonium sulfate and coke (standard by-products in shale-oil retorting) but also large quantities of paraffin wax. On account of these valuable incidental substances, Fushun shale oil became "worth tapping."[139] At the beginning of 1928, Mantetsu's directorate, headed by new president Yamamoto Jōtaro 山本条太郎 (1867–1936), finalized a plan to establish the Fushun shale oil plant on a scale doubling that which had been previously proposed, so confident were they of this industry's forecast. The plant would consist

136. "Anshan Pig-Iron."
137. "Anshan Pig-Iron."
138. "New Oil Extracting Contrivance."
139. "New Fushun Shale Oil Plant," 60; "Fushun Shale Oil Experiments."

3.4. Two images related to Fushun's shale oil enterprise. *Top*, the pilot plant that Mantetsu engineers Okamura Kinzō, Hasegawa Kiyoji, and Ōhashi Raizō had set up as they worked on a more economically viable distillation process for Fushun oil shale. *Bottom*, the finished shale oil facility in operation. (*Top*: Image courtesy of Special Collections and College Archives, Skillman Library, Lafayette College, and the East Asia Image Collection, https://dss.lafayette.edu/collections/east-asia-image-collection/. *Bottom*: Image courtesy of the Harvard-Yenching Library, Harvard University.)

of eighty fifty-ton retorts that could handle 1.25 million tons of shale annually to produce over 360,000 barrels of crude. Construction costs were estimated at close to ten million yen, raised, upon government approval, by supplementary estimate.[140] While Mantetsu did purchase a million yen worth of distillation and refining equipment from Czechoslovakia, most of the facilities were Japanese produced—the retorts, for example, were fashioned by the Kawasaki Dockyard Company in Kobe. By the end of 1929, the assemblage was finished, and the first fire lit in its grates.[141]

The Fushun method of shale oil production may have been, as its proponents insisted, less complicated than the Scottish one—internal heating at the very least meant fewer fittings—but the process was still involved. Beginning at the open pit, where laborers drilled into the working face and then set off charges to break it apart, the loosened shale rock was loaded by shovels and excavators onto dump cars and carried along rails to the nearby plant. There, the shale was unloaded into a storage bin that directly fed two sets of crushers working like front and back teeth: the first, Fairbanks single-roll crushers, chomped on shale of up to eight feet and turned it into five-square-inch pieces; the second, Symons cone crushers, ground the smaller pieces into finer ones of two square inches most suitable for distillation. The crushed shale was then passed through a vibrating screen, with those pieces that were too small conveyed off to a separate storage bin, from which they were later sent to the subsurface mines to be used as packing material in hydraulic stowage. The appropriately sized pieces, on the other hand, were lifted by bucket elevator to the top of the retorts and emptied into an overhead conveyor that distributed and discharged them into individual devices. The crude oil produced through gasification and condensation within was then piped to the distillation plant, where it would be refined and its by-products recovered. The spent shale was released from the bottom of the retorts and, like the pieces of shale too small to be distilled, transported off to the underground mines for use in hydraulic stowage. (Between the spent and small shale, the colliery would obtain enough packing material to stop dredging sand for this mining measure.)[142] By one estimate, each ton of shale could yield 1.5 gallons of gasoline, 4.2 gallons of kerosene, 5.6 gallons of diesel, 1.5 gallons of lubricating oil, 20 pounds of paraffin, and 40 pounds of ammonium sulfate.[143] The process

140. Minami Manshū tetsudō kabushiki gaisha, *Dainijūhachi-kai*, 9; "Fushun Shale Oil Industry: Almost Ready."

141. "New Fushun Shale Oil Plant," 60.

142. "New Fushun Shale Oil Plant," 60; Okamura, "Bujun yuboketsugan jigyō."

143. "New Fushun Shale Oil Plant," 63.

of extracting shale oil is an example of what chemists refer to as "destructive distillation"—in which the large molecules in the input materials are cracked under high heat. It was evidently also rather productive.

Heralded by one booster as a "god-send to Japan to head off fuel famine" and by another as the harbinger of "a new age in the world of fuels," Fushun's shale oil industry generated much excitement at home and some abroad.[144] The apparent value of Fushun shale oil was such that the Chinese government protested its extraction, arguing that the existing Sino-Japanese agreement extended only so far as the mining of coal.[145] This matter would not be resolved until the formal Japanese occupation of Manchuria, at which point the question of rights became essentially moot. The plant at Fushun was, at its establishment, the world's largest shale oil facility.[146] Mantetsu continued to invest in its expansion in the following years, deepening its strategic significance to the Japanese empire. By 1940, Fushun shale oil accounted for about a quarter of the oil produced in Japan and its colonies.[147] Fushun's shale oil industry also left an important legacy for the post-1949 Chinese Communist state. The refinery may have been gutted by Soviet forces and have fallen into further disrepair under the brief Chinese Nationalist occupation of Fushun, but it was restored during the socialist industrializing efforts of the 1950s (chapters 5 and 6). By the beginning of the new millennium, almost half a billion tons of oil shale were mined and processed into liquid fuel at this site, which today remains the country's biggest producer of shale oil. Although modifications have been made over the decades, the Fushun method of dry distillation that Okamura, Hasegawa, and Ōhashi devised years before has by and large served as the basis of the shale-oil production process here.[148]

However, if the prime objective of launching Fushun's shale oil industry had been Japan's energy self-sufficiency, then the outcome came up short. For even when Fushun shale oil made up a fourth of the Japanese empire's petroleum production in 1940, more than 60 percent of the petroleum Japan consumed that year was imported.[149] "It was a story with a long chain of

144. "God-Send to Japan"; "Ketsuganyu no jitsuyōka."
145. *MSZ*, vol. 4, no. 3, 829–36.
146. "World's Largest Shale-Oil Plant."
147. Department of Justice, War Division, Economic Warfare Section, "Report on Fushun, Part 1," 1.
148. He, "Mining and Utilization."
149. Department of Justice, War Division, Economic Warfare Section, "Report on Fushun, Part 1," 1.

difficulty, and at the best not a very lucrative venture, either," wrote Mizutani Kōtarō, who, while remaining in naval service, was attached to Mantetsu for years as an adviser on fuel issues.[150] That Fushun's shale oil enterprise would serve as anything approaching what one hopeful commentator termed the "axle of Japan's fuel supply" turned out to be but a pipe dream.[151]

In a way, though, Fushun shale oil, as a technoscientific artifact, encapsulated the character of the post–World War I Japanese empire. Its development had been prompted by the fuel question and related anxieties around establishing autarky. Its prospect as a potential energy resource was facilitated by earlier efforts imperial Japan took to realize carbon technocracy in the concession it carved out in Fushun, particularly through the excavation of the open-pit mine. Its production and transformation into an industrial object involved collaboration between Japan's largest colonial enterprise and the military, both readily mobilizing Japanese fuel experts who drew on transnational science and technology even as they carried out their own experiments within their laboratory of empire in Manchuria.[152] In his work on plant and animal breeding in Italy, Portugal, and Germany before and during World War II, historian of science Tiago Saraiva demonstrates how the cultivation of organisms from wheat to pigs by geneticists and the regimes they served gave form to fascist ontologies.[153] Although the organic life in oil shale perished millions of years earlier, its extraction and conversion into a liquid fuel substitute, in a similar fashion, not only reflected but also helped produce Japanese imperialism. The highly publicized progress toward the creation of Fushun's shale oil industry made concrete the unremitting arguments about the importance of energy to national security and strength even as the process of assembling the industry itself demonstrated the lengths to which the Japanese state would go to obtain petroleum resources, heavily subsidizing this enterprise in which liquid fuel was essentially squeezed out of a rock. Perhaps most importantly, by presenting shale oil as a viable and, for many, even vaunted solution to the fuel question, boosters reinforced expectations that answers to Japan's problems might be best found in empire.

150. Mizutani, "Liquid Fuel."

151. "New Oil Shale Plant."

152. For a few other examples in which the Japanese military worked together with industry to develop technologies that had potential wartime application, see Morris-Suzuki, *Technological Transformation of Japan*, 124–26. Japan's technoscientific militarization had parallels around the world. For the case of the United States, see Lindee, *Rational Fog*.

153. Saraiva, *Fascist Pigs*.

The Threat of Continental Coal

While Fushun, with its bounty of coal and oil shale, was widely regarded as key to Japan's fuel future, where many saw a treasure house, some spied a threat. The idea, extrapolated from Japanese geological surveys, that Fushun coal was "limitless" and "inexhaustible" contributed to and benefited from a larger notion that, as "a certain influential financier" put it in 1910, "the abundance of [Manchuria's] natural resources is truly that of a bottomless vault."[154] This wealth held much promise, proponents averred, for advancing Japan's economic interests and, with that, its position in the world. To Ishiwara Kanji 石原莞爾 (1889–1949), one of the architects of the Mukden Incident that led to the Japanese invasion of Manchuria in 1931, the resources of the region were "sufficient to build up our present heavy industrial base . . . to relieve our present plight [of economic depression] and to build a foundation for a great leap forward."[155] Yet not everyone took this plenty as positive. To Japanese coal producers, the Fushun colliery was an enduring menace that undermined their prospects in markets at home and abroad.

If coal as a resource had been pivotal to the rise of Japan's modern industrial economy, then coal mining as a sector was a pillar of that economic structure. As such, its concerns warranted attention. Following the restoration, the Meiji government initially forced the privatization of coal mines across the archipelago as these were sources of income for the daimyo, the feudal lords of domains whose continued existence ran counter to the new state's authority. Soon after, though, the government changed its course, first centralizing the coal-mining licensing system, then asserting the emperor's rights over all subsurface resources, and finally enacting a coal-mining industry law that permitted the state to own and operate mines. With these provisions in place, it took over the country's most promising collieries and invested heavily in their development. By the late 1880s, however, the government, strapped for cash from its panoply of modernizing projects, sold off its coal mines to emerging conglomerates. Through these purchases, the conglomerates came to manage the country's biggest coal-mining operations that were equipped with the latest in extractive technologies. For these conglomerates, the mines were highly lucrative. One year, for example, Mitsubishi derived four-fifths of its earnings from its mining division alone. While other business divisions would later take up

154. "Mantetsu no keizaiteki ichi."
155. de Bary, Gluck, and Tiedemann, *Sources of Japanese Tradition*, 295.

a more substantial share of conglomerate profits, many of these businesses were, in fact, first financed through coal revenue.[156] Beyond the big mines, which produced both for domestic sale and for export and which would come under ten large firms by the interwar period, there were hundreds of smaller mines that served local markets, the smallest of these referred to as "badger-dog burrows" (狸彫り).[157] Coal mining was also a huge employer. In 1930, there were over 170,000 workers in the industry, about 16 percent of whom were women, including those who hewed and hauled coal in the mines (unlike in Fushun, China, and, indeed, most parts of the world, where women tended not to be engaged in this kind of labor).[158]

Not long after Japan took control of Fushun in 1905, coal mine operators in Kyushu and Hokkaido began worrying that this colliery, once developed, might "deal a debilitating blow" to their businesses.[159] Concerns over continental coal had predated this moment, stretching back to the 1870s, when the discovery of large deposits in China prompted at least one Japanese industrialist to agitate for a price war to stymie the Chinese coal industry's growth.[160] Fushun, with its massive potential, posed a similar challenge, albeit one that the Japanese state was in a position to do something directly about. Mantetsu tried to assuage concerns. As historian Yoshihisa Tak Matsusaka has argued, it was partly in an attempt to safeguard the interests of Japanese coal producers that the colonial company introduced stipulations on the use of Fushun coal that relegated exports to the fourth and last category of allocation. Taking precedence were first, company use, which included powering its railway, shipping, factory, and mining operations as well as providing fuel for ancillary facilities like office buildings and power plants; second, sale within Manchuria, where Fushun consistently supplied much of the region's household and industrial needs; and third, sale as steamship fuel for vessels coming through the Manchurian ports of Dairen, Yingkou, Lüshun, and Andong. Only after meeting these priorities was the "surplus coal" (殘炭) exported to the Japanese home islands and elsewhere.[161]

Mantetsu did, however, grow its exports over the years, eventually openly targeting Japanese markets. In 1908, when the colliery first started selling

156. Hein, *Fueling Growth*, 32.

157. Samuels, *Business of the Japanese State*, 69–73, 81.

158. Inhara, *Japan Year Book, 1933*, 543. On women workers in Japanese mines, see Burton, *Coal-Mining Women in Japan*.

159. Kantō totokufu, *Manshū sangyō chōsa shiryō (shōgyō seizōgyō)*, 219.

160. Matsusaka, *Making of Japanese Manchuria*, 137.

161. Matsusaka, 137–38; *MT10*, 605–27.

coal abroad, exports made up 5 percent of output at 21,000 tons. Five years later, on the eve of World War I, exports comprised almost 50 percent at 1.2 million tons. From 1922 onward, exports to Japan exceeded those to other destinations, which included Taiwan, Korea, China, and Southeast Asia. That year, Japan imported 622,000 tons of Fushun coal. By the end of the decade, the figure was more than three times as much.[162] In 1923, Mantetsu established a subsidiary company with exclusive rights to market Fushun coal, allegedly as part of its larger plan "to increase the output of Fushun coal, and to increase the exports to Japan"—exactly what Japanese coal producers had been worried about from the very beginning.[163]

The tensions between Mantetsu's Fushun colliery and Japanese mines came to a head following the Mukden Incident. In 1930, the Japanese coal industry was already suffering from the worldwide depression that co-incided with the government's decision to return to the gold standard, a move that made the yen appreciate and Japanese goods less competitive in domestic and foreign markets just as global demand dipped.[164] Shrinking demand, falling prices, and the resultant inability to meet operating costs led to the closure of hundreds of small- and medium-sized mines. Unemployment among miners spiked as mines that remained in business dismissed almost 40 percent of their workers.[165] After the outbreak of hostilities in 1931, Chinese consumers, in an act of resistance, began boycotting Japanese goods, including Fushun coal (we will look at this side of the story in chapter 5). Mantetsu, now unwelcome in Chinese markets, unloaded much of its China-bound coal on Japan instead. In April 1932, domestically produced coal sold in Japan for 9.4 yen a ton on average, while Fushun coal went for 6 to 7 yen a ton.[166] Japanese coal mine operators quickly decried this as dumping.[167]

Two months later, hundreds of Japanese miners took to the streets of Fukuoka, at times "still in work clothes that were covered in coal dust and stained with grease."[168] Mobilized by the Coal Mining Mutual Aid Society (石炭鑛業互助會), they were part of a movement demanding restrictions on Fushun coal imports. The mutual aid society had been founded in Sep-

162. Manshikai, *Manshū kaihatsu yonjūnen shi*, 67.

163. Manshikai, 75; "Dark Future for Coal."

164. On the gold standard and the Japanese yen at that crucial point in 1930–1931, see Metzler, *Lever of Empire*, 217–39.

165. Nakamura, "Depression, Recovery, and War," 464.

166. Zhu, "Re-examining the Conflict," 238.

167. Samuels, *Business of the Japanese State*, 80–81; Tōa keizai chōsakyoku, *Honpō o chūshin*, 289–302.

168. "Kōfu sū hyaku mei."

tember 1930 by small- and medium-sized mines seeking to safeguard their interests, which, given the times, were increasingly embattled. On June 12, 1932, amid worries over Mantetsu's dumping of Fushun coal, the society passed a resolution stating, "To prevent the crushing loss of 500 million yen in investments and to deliver 700,000 workers from the brink of starvation and death, we pledge to petition the government to completely ban the import of Fushun coal."[169] The movement that it started in Fukuoka soon spread to the rest of the country.[170]

Within weeks, the mutual aid society managed to initiate negotiations over import limitations with Mantetsu and the Coal Mining Alliance (石炭鑛業聯合會), a production and sales cartel the large Japanese coal firms had formed in 1921 to maintain market share among themselves. When these negotiations stalled, the government, represented by the Ministry of Commerce and Industry and the Ministry of Colonial Affairs, came in to broker an agreement.[171] Mantetsu would reduce its annual exports to Japan by 200,000 tons. But the large coal firms under the alliance were to correspondingly cut their yearly output by 800,000 tons. If the protest against Fushun coal had been in effect about improving the lot of small- and medium-sized mines, then it theoretically did not matter that attempts to meet that objective (or at least to appear to be trying to do so) involved less the curtailing of Fushun coal imports and more the curbing of production (and hence profits) of large domestic mines.[172]

When considered from the perspective of the relationship between metropole and colony, it was significant that this compromise was reached. As one newspaper analyst pointed out, this was "the first economic frontal clash between Japan and Manchukuo," and its resolution had bearing on the settlement of similar future cases and, as such, the viability of "the proposed Japan-Manchukuo economic bloc"—an autarkic sphere in which Manchuria functioned as a "lifeline" pumping essential resources like coal to sustain the metropolitan heart of the imperial body.[173] Although the Japanese state clearly recognized how colonial coal had hurt smaller do-

169. Tōa keizai chōsakyoku, *Honpō o chūshin*, 294. Note the discrepancy here between the 700,000 workers mentioned and the over 170,000 recorded in Inhara, *Japan Year Book, 1933*, 543, as referenced earlier in this chapter. As the latter source is more reliable, the 700,000 was probably either an exaggeration, a typo in the transcription, or a figure that tried to factor in workers' family members and dependents as well.

170. Zhu, "Re-examining the Conflict," 230.

171. "S.M.R. against Plan."

172. "Fushun Coal Fight." For one analysis of the rationale behind the Japanese government's decision, see Zhu, "Re-examining the Conflict," 247–54.

173. "Fushun Coal Fight." On Manchuria as "lifeline," see Young, *Japan's Total Empire*, 88.

mestic producers, and although it acted to appease the latter in this instance, it evidently took for granted the necessity of continuing energy imports from the colonies and was not ready to so casually sever the link with Japan's frontier of extraction.

Prominent Fuel Society member and former director of the Meiji Mining Company Ishiwata Nobutarō, addressing residual dissatisfaction about the outcome of this episode a year later, would write, "In the home islands, one might regard the annual increase in the output of Fushun coal in an oversensitive [神經過敏] manner, but I think there really is no reason to worry." Aside from being well positioned in the coal-mining industry, Ishiwata may have felt particularly qualified to address this matter, having conducted one of the first geological surveys in Fushun a quarter of a century back.[174] He contended that, barring any exceptional circumstances, the demand for coal in Japan would gradually increase, and Fushun coal could "merely supplement" any shortfalls "should domestic productive capacity lose its elasticity." Estimating Fushun's yearly output to peak around seven million tons, he suggested, though, that if that point were to be reached, Japan and Manchuria might begin to experience "coal shortages" (石炭不足) as their fuel demands continue to rise. Given that possibility, Ishiwata recommended, as "a coal provision plan in preparation for the least unreasonable future," that more coalfields across Manchuria be opened up to fulfill the region's needs while Fushun be slated to cater solely to Japan's demands.[175] Several years earlier, at the society's first spring conference in Osaka in 1924, Ishiwata had actually brought up similar points in a lengthy talk he delivered on the coal question. There, he called for Mantetsu to survey and develop coalfields that seemed promising in the railway zone over which it held jurisdiction. This was in anticipation of coal shortages that he expected Japan to face ten years hence, for which Fushun alone would not be able to supply.[176] A decade later, Japan seemed to be struck with not a lack of coal but, as illustrated in the dumping incident, an abundance of it—an abundance that seemed, at least to Japanese coal producers, an aberration. Soon, however, demand would catch up with supply as Japan began its military-industrial buildup in anticipation of war. In Manchuria, the empire, whose reach now extended far outside the boundaries of the railway zone, had more ready access to the region's depths and was well poised to further exploit them.

174. Ishiwata, "Bujun tankō."
175. Ishiwata, "Nenryō jūōdan," 1286–87.
176. Ishiwata, "Waga kuni shōrai," 310.

While probably none of them had foreseen the exact sequence of events that led to the invasion and occupation of Manchuria, the members of the Fuel Society, like much of the wider population, came to accept that development.[177] Some even celebrated it. In his 1932 New Year's address, the society's president, Vice Admiral Sakamoto Toshiatsu, raised a figurative glass of "spiced liquor" to "wish good health to the expeditionary forces in Manchuria, Mongolia, and North China," whom he likened (in a somewhat overburdened simile) to "weeds that were fighting against both bitter cold and lacking conditions and that, when the moment arises, kill [their] foes with basic instincts like a bolt out of the blue." Turning then to his members, he urged them to "be vigilant about the fuel question before getting drunk," presumably from start-of-the-year libations or from drinks downed to commemorate Japan's campaign on the continent.[178] In the seizure of the region and its resources lay a promise that the fuel question might be resolved, a promise that proved more intoxicating than any spirit. Held in its stupor, Japan's empire builders staggered along an expansionist path from which they would find it increasingly impossible to backtrack.

Conclusion

In his 1925 lecture to the Fuel Society, Gotō Shinpei recognized the recent progress that had been made in applying science to the "utilization of calories" (カロリー應用). But there were limits, he noted, to these attempts at maximizing energy efficiency, for even in the best of circumstances, as much as 80 percent of calories were still wasted in combustion. This being so, economizing practices alone were insufficient to solve the country's fuel question. What was needed instead, he suggested, was for Japan to leverage its "advancement in science" to acquire and use coal and petroleum in countries where such "treasures" lay abundant but largely untapped by "cooperating with one another under co-existence and co-prosperity" (互に相協力して共存共榮の下). As he saw it, Japan had the science but not the energy resources, while these countries had the energy resources but not the science. Thus, Japan should supply the science and these countries the resources to their mutual benefit. Such was "the mission of the Yamato race" (大和民族の使命), he concluded.[179]

177. On widespread support across Japanese society for the idea of Manchukuo, see Young, *Japan's Total Empire*.

178. Sakamoto, "Nentō no kotoba," 2.

179. Gotō, "Jinsei to nenryō mondai," 31–32.

In light of Japan's turn to hypermilitarism in the 1930s, it may be easy for us to read Gotō's charge as part of an overture to later expansionist acts, especially when we consider how the "co-prosperity" he evoked would be mobilized by the Japanese wartime state as an ideal to justify its seizure of territories across Asia.[180] To do so, though, may be overdetermining that outcome. Not every word and deed from the 1920s needs to be taken as foreshadowing the decade that followed.[181] And yet there was something unmistakably imperialistic about how Gotō discussed harnessing science to gain control of resources abroad. This, however, did not necessarily stem from the particularities of the Japanese experience. In the post–World War I era, all major imperial powers would, in fact, be invested in securing foreign petroleum reserves. This included the United States, which was greatly admired by Japanese fuel experts. Many of them envied the American abundance of coal and petroleum. As we have seen, the United States repeatedly served as a point of contrast to Japan and its comparative dearth in fossil fuels. Some of these experts, such as Fuel Society essay contest winner Nagai Mosaburō, extolled the American audacity at muscling into the competition among empires for petroleum overseas as a latecomer to the global scramble for oil.[182] In her study of the United States Department of the Interior and its global reach, historian Megan Black has demonstrated how this ostensibly insular department was deeply implicated in the American acquisition of raw materials through settler colonialism on the continent and the projection of power offshore.[183] The work of Black and others reminds us that the United States had been and remains an empire. To Japanese observers in the 1920s, there was little doubt that it was one. To suggest that Japan may have been taking notes on how the United States and other Western imperial powers made efforts to lay claim to energy resources beyond their borders does not, however, absolve Japanese empire builders from attempting to do the same. Nor should it imply that Japanese imperialism was entirely derivative or inexorable. Nevertheless, it does allow us to situate Japan among the multiple empires jostling for mastery of energy in the carbon age.

180. Louise Young contends that hypermilitarism was one of the four elements in what she characterizes as Japan's "fascist imperialism" from 1931 to 1945. The three others are Asianism, red peril, and radical statism. See Young, "When Fascism Met Empire."

181. For an account of 1920s Japan that attends more to progressive political developments in both domestic and international spheres than to what many in hindsight have regarded as signs of the coming plunge into the "dark valley" of the 1930s and 1940s, see Dickinson, *World War I.*

182. Ross, *Ecology and Power,* 218–19.

183. Black, *The Global Interior.*

Which brings us to the question of the relationship between energy and empire. As necessary as coal was to Japan's emergence as an imperial power, it was not for the control of carbon, at least initially, that the Meiji state extended itself beyond the archipelago.[184] Japan had, after all, arrived in Fushun first, and only upon determining the site's wealth of coal did it start to set up the immense social and technological assemblage with which it plundered this resource, as earlier chapters have established. But as Japan consolidated its empire and continued its industrial expansion in the years that followed, it became ever more dependent on fossil fuels in the form of coal and, later, petroleum too.[185] At that juncture, the existing empire served as a source not only of greater volumes of coal resources than before but, over time, also of the alternative liquid fuels that were shale oil and, subsequently, liquefied coal. When these were deemed insufficient to feed its growing appetite for energy, Japan would look to accessing resources in other areas too.

It may have been that the Japanese who were increasingly interested in acquiring foreign mining concessions in the 1920s amid the anxieties of the fuel question had indeed assumed that consent for their extractive endeavors might be secured from locals without the application of force, as Gotō had suggested.[186] Yet given the turn to "a new celebration of military action and the aesthetics of violence," which historian Louise Young regards as central to the "fascist imperialism that took root in Japan after the Mukden Incident," most of them probably recognized by the 1930s that such concessions required coercion to capture.[187] When Yoshimura Manji wrote in 1936 that Japan should take into its hands "China's several hundred million [tons] of coal resources" that, "against the spirit of the modern world, is completely unused at present and hoarded away in vain," for instance, it is difficult to believe he would have really imagined that happening without the Chinese first staring down the barrel of a Japanese gun.[188]

Toward the end of his lecture, Gotō, to reiterate his point about the im-

184. While slightly dated, the contributions by Bonnie Oh, Ann Harrington, Hilary Conroy, and Okamoto Shumpei in Harry Wray and Hilary Conroy's edited collection of historiographical essays on modern Japanese history provide a good introduction to the debates surrounding the origins of Meiji imperialism. See Wray and Conroy, *Japan Examined*, 121–48.

185. Peter Shulman advances a similar argument in regard to the United States and the rise of its island empire in the late nineteenth century. See Shulman, *Coal and Empire*, 9, 125–63.

186. It should be noted, though, that Japan's establishment of an oil concession in Soviet Sakhalin in 1925 followed the Japanese navy's occupation of that half of the island. See Hara, "Japan Moves North."

187. Young, "When Fascism Met Empire," 275.

188. Yoshimura, "Shunjitsu nenryō dansō," 91.

portance of adopting a scientific approach to the fuel question, recounted an episode from his time as mayor of Tokyo. Having taken an interest in the issue of household fuel consumption, he had ordered an investigation of usage patterns across the metropolis. The findings suggested that considerable savings could be made if coal gas were to replace firewood as the city's primary domestic fuel. Although this involved constructing four additional gas plants and although gas was pricier than firewood per unit, because gas was also more efficient than firewood, it would consequently save Tokyo residents fifteen million yen a year. Gotō's proposal to effect such a change was, however, dismissed by the municipal council because its members, he claimed, "lacked scientific knowledge." It would be ideal, he contended, if politicians and the broader public learned to think not merely in terms of base cost but in terms of calories, which was the basis of modern life.[189] The calorie, in both its quantifiability and its ability to collapse qualitative differences between different forms of animate and inanimate energy, proved central to the calculative considerations of carbon technocracy. In the years after, Japanese leaders became increasingly preoccupied with securing calories from fossil fuels. And as we will see in the next chapter, this was something they would pursue with less and less regard to cost. Amid these developments, concerns over carbon persisted. For although Japan eventually settled on imperial extraction as its preferred response to the fuel question, this ultimately failed to provide it with the decisive solution that it sought.

189. Gotō, "Jinsei to nenryō mondai," 35–37.

FOUR

Imperial Extraction

They looked "overbearing and murderous," Mo Desheng recalled decades later. "Wearing steel helmets on their heads and holding rifles affixed with gleaming bayonets in their hands," the soldiers of the Kwantung Army showed up on the morning of September 16, 1932. Mo, then eight years old, had spotted them riding in a train of vehicles that snaked its way up the dirt road to his village of Pingdingshan in southeastern Fushun. As soon as they arrived, the soldiers went around searching for spearheads and other supplies that might suggest "collusion with rebels." Around noon, they ordered the villagers to gather by the hillside. Mo's father, a miner home after the night shift, grabbed hold of the boy's hand; his mother picked up his three-year-old sister. Along with his visiting grandparents, the family stepped out of their house to join the growing crowd.[1]

As they got into place, Mo took stock of his surroundings. The soldiers had positioned themselves around the assembled villagers. To one side there was "a device that looked like a camera, covered with a black cloth." Before long, an officer stepped forward and began addressing the villagers through a translator. He assured them that the army had merely come to protect them and their property from the "red rebels" in the area. "We are going to fight a battle here at Pingdingshan, but you need not be afraid, for once we drive away the 'red rebels,' we will let you return to your homes," he promised. He then told the villages that they would be photographed and instructed them to face the covered device. What followed, Mo remembered, was the "sound of bullets falling like rain onto people's bodies." Amid that, a deluge of screams. After the firing stopped, the soldiers walked among the fallen, driving bayonets into those still alive. Not even children

1. "Mo Desheng huiyi," 228–29.

"crying for their mothers" were spared. Mo, who had thrown himself to the ground as soon the shooting started, shut his eyes and played dead.[2]

When the soldiers finally left, Mo clambered to his feet and started looking for his family. He found the lifeless bodies of his mother and sister under a bloodied blanket. His grandparents, too, had been slain. Mo then spotted his father lying nearby. Hoping that he had merely passed out, the boy went over and attempted to rouse him by biting down hard on his arm. There was no response. It was only then that he noticed that his father's neck was gushing blood. Out of his family of six, Mo was the only one who survived.[3]

The massacre at Pingdingshan claimed as many as three thousand lives.[4] Retaliation had purportedly prompted this savage act. After the establishment of the Manchukuo state, Chinese resistance fighters continued to oppose the new order, carrying out attacks on centers of Japanese control. The Fushun colliery was one such site. The night before the Pingdingshan massacre, the coal mines came under assault from rebels who allegedly bore connections to the Red Spears—a secret society typically focused on local self-defense that got its name from the red-tasseled spears its members were said to carry.[5] Attacking under the cover of darkness, these rebels torched various mining facilities with firebrands fashioned from lumps of coal and other flammable items they purportedly obtained from workers at Yangbaipu. The night was windy, and the fire quickly spread.[6]

When Yamaguchi Yoshiko, then twelve years old, peered out of her home in Fushun's Japanese quarter that night, the "roofs of buildings, along with the poplar trees lining the curb, emerged before my eyes like a dark silhouette, their background engulfed in a ferocious sea of red, with tongues of flame blazing wildly into the distant night sky." In the chaos that ensued, the rebels killed seven or eight Japanese employees from the Yangbaipu and Tōgō mines. With some effort, the colliery's understaffed security forces routed these rebels—most of the Kwantung Army's Fushun garrison had been out patrolling the surrounding territory. As the rebels retreated, they passed through the southeast in the vicinity of Pingding-shan. The next morning, Yamaguchi witnessed the military police interro-

---

2. "Mo Desheng huiyi," 229–30.

3. "Mo Desheng huiyi," 230.

4. The number of victims in the Pingdingshan massacre is disputed, ranging from four hundred to three thousand.

5. On the history and historiography of the Red Spears, see Perry, "The Red Spears Reconsidered."

6. Yamaguchi and Sakuya, *Ri Kōran*, 22–23.

gating a Chinese man bound to a pine tree in the courtyard outside her house. When the man refused to speak, a policeman smashed his forehead in with the butt of his rifle, killing him in a single blow. It was years before she would piece together the different parts of the story: the fiery attack by night, the interrogation and murder of a man suspected of assisting those who carried it out, and the massacre of a nearby village under similar charges of collusion.[7]

The Pingdingshan massacre was among the most notorious of atrocities committed by Japanese militarists after the founding of Japan's client state of Manchukuo. As with imperial projects elsewhere, unbounded violence became a tool all too easily employed to discipline (and, indeed, decimate) populations deemed challenging to control. Often, as in the case here, those who suffered most for resistance were those least able to offer any of it. The massacre cast a long shadow over Fushun and Japanese Manchuria more broadly. Although the Japanese burned and buried the bodies and razed the village itself, news of the atrocity broke soon after thanks to an American journalist, Edward Hunter, who had gone undercover to Fushun to investigate rumors of the massacre's occurrence. Japan received international condemnation for this and other infractions surrounding its invasion of Manchuria, prompting it to withdraw from the League of Nations in February 1933.[8]

While this episode seems exceptional in its inhumanity, it is in many ways paradoxically connected and in contrast to the systematic violence of both the imperial project and the energy regime of carbon technocracy. We have seen earlier in episodes like the 1917 mining disaster how the dangers of the engineered environment exacted a heavy human toll. Such episodes would continue into this period. As an act of unstanched brutality, the Pingdingshan massacre further sullied sanitized narratives of industrial development that Manchukuo's Japanese boosters were fond of advancing. The client state was unmistakably bloodstained from the start.

This chapter explores how the pursuit of Manchurian industrial development and the accompanying intensification of coal extraction centered on the Fushun colliery were closely aligned with imperial Japan's war mobilization in the 1930s and 1940s. It traces consistent efforts by Japanese empire builders in the region to dominate coal production under the auspices of the "managed economy" (統制経済), even as their plans for Manchukuo shifted from it being a self-sufficient industrial base to it becoming

---

7. Yamaguchi and Sakuya, 17–24.
8. Mitter, *The Manchurian Myth*, 112–14; Meng, "More Evidence."

but a supplier of raw materials to the Japanese metropole with the war's outbreak and progress. Throughout, the Fushun colliery resisted the client state's attempts at monopolistic industrial management, pointing to fissures in the workings of carbon technocracy. The chapter then zooms out to look at Manchurian coal mining alongside the extractive endeavors across Japan's wider empire of energy. It shows how imperial expansion, motivated in part by acquiring more coal and other resources, produced what I call a "warscape of intensification," which ended up requiring increasing amounts of energy to sustain. Under mounting productivist pressures, sites of extraction ultimately could not keep up with growing demands. Energy in empire tended toward overextension and eventual exhaustion. At Fushun, this was evident not only in the failure to meet planned production goals but also by the use of Chinese forced labor, euphemistically called "special workers" (特殊工人) and controlled through violence, in a bid to increase output that was as ineffectual as it was inhumane.

"Releasing the Suppressed Energies for Industrial Expansion"

The Manchurian Incident of 1931 signalized the opening up of an unprecedented industrial era in Manchuria by releasing the suppressed energies for industrial expansion. The overthrow of the military dictatorship and the consequent establishment of Manchukuo meant a complete shift from the famed squeeze system and irrational exploitation of natural resources to the controlled economy and rational development of those resources.

—South Manchuria Railway Company, *Fifth Report on Progress in Manchuria to 1936*

To the architects of the Manchukuo state, framing the current order as having overcome the previous mismanagement of Manchuria's natural resources was foundational to legitimizing their new regime. In a similarly sympathetic contemporary account, the claim that "the Chinese did not attempt to develop this country" was contrasted with the assertion that "Japan's investment in the economic development of Manchuria has been the basis of the creation of wealth in that region."[9] As with Mantetsu's portrayal of the Fushun coalfield as lying largely unmined in the Qing era

---

9. Japanese Chamber of Commerce of New York, *Manchukuo*, 31. To highlight the problematic way in which Japanese empire builders in Manchukuo framed the preceding period is not to discount the fact that there were indeed deep Japanese economic interests in Manchuria. For a helpful account of this, see Schiltz, *Money Doctors from Japan*, 159–85.

because of superstitions over "dragon veins," depicting the region as previously subjected to "the famed squeeze system and irrational exploitation of natural resources" was a means by which Japanese imperialists justified colonial appropriation. Coal was a focal point in this regard. Boosters complained that while several coal mines—most of all those in Fushun—were already being exploited, there were more untapped deposits across the region that deserved directed development.[10] This goal was of paramount importance to the project of Manchukuo: under the energy regime of carbon technocracy, the act of ushering in Manchuria's "unprecedented industrial era" quite literally depended on drawing forth "suppressed energies" from the region's subterranean storehouses of fossil fuel resources.

Coal had long sustained Japan's industrial and infrastructural projects in Manchuria. Raising high the banner of development, Mantetsu and other Japanese capitalists pursued the expansion of industrial production centered on processing raw materials or fashioning partly finished goods, such as bean oil, distilled spirits, flour, sugar, tobacco, cotton, silk, and ceramics. In 1909, there were 152 factories producing 6.1 million yen of goods in the railway zone and other areas under Japanese jurisdiction. In 1927, there were 750 factories with an output valued at 140.4 million yen.[11] Coal supplied much of the energy for these economic endeavors. It also fueled the railway transporting people and goods around Japanese Manchuria. In Mantetsu's first two decades, the number of locomotives doubled to 441 and freight cars more than tripled to 7,260. Increasing volumes of coal drove this enlarged fleet.[12] Remarkable as this growth may have been, Manchukuo's planners proposed to further accelerate Manchuria's industrial transformation, intensifying coal extraction to power their developmentalist ambitions.

Official rhetoric notwithstanding, Manchuria's coal-fired industrial development was not an end unto itself. It was a means of meeting Japan's strategic and economic needs, which had motivated the Japanese annexation of the region to begin with. Strategically, Manchuria presented a bulwark for colonial Korea and Japan's imperial metropole against the Soviet Union, as it did against tsarist Russia at the beginning of the century. In the late 1920s and early 1930s, Japanese imperialists understood the threat in terms both military and ideological. It was not only about responding to troop buildup in the Soviet Far East but also about containing the spread

10. Imura, "Reconstruction of Manchuria."
11. Minami Manshū tetsudō kabushiki gaisha, *Report on Progress*, 131–38.
12. Minami Manshū tetsudō kabushiki gaisha, 77.

of communism and its perceived danger to social and political order. Economically, Japanese proponents of continental expansion believed that resource-rich Manchuria held the key to Japan's "deadlocked" economy amid the global depression. By the latter half of the 1920s, ideas of Japan's "special interests" in Manchuria ran up against a surge in Chinese nationalism that (not incorrectly) took Japanese activities in the region as expressions of encroachment and exploitation. To several officers within the Kwantung Army, invasion seemed so inevitable that they felt prompted to engineer it.[13]

There was a false start in 1928. The "Old Marshal" of Manchuria, Zhang Zuolin, with whom Japan had been cooperating as it pursued its interests in the region, proved himself, by the late 1920s, an unreliable ally. Zhang became increasingly attentive to growing expressions of anti-Japanese nationalism playing out in the streets of China's urban centers from the middle of the decade and began to harden his stance in negotiations with Japan regarding the future of its concessions in Manchuria. Furthermore, Zhang, who had taken Beijing in 1926 and came to head China's internationally recognized government, soon started losing ground as his troops were pushed back by Chiang Kai-shek's decidedly antiwarlord and anti-imperialist Northern Expedition. Zhang thus became less reliable to the Japanese as someone who would, or indeed could, stand for the de facto independent Manchuria they sought. On June 4, 1928, as Zhang, bested by Chiang, made his way back to Mukden from Beiping, Colonel Kōmoto Daisaku 河本大作 (1883–1955) of the Kwantung Army led a few junior officers in blowing up the Old Marshal's train just before he arrived at his destination. Yosano Akiko, who was in Mukden and about to set off on her tour of Fushun, recalled the "faint, strange noise" of the distant explosion and the "fear, the threat of danger, and chaos" that swept the city after.[14] The mortally wounded Manchurian warlord succumbed to his injuries a few days later. The plan had been to replace Zhang with a more pliable puppet whom Japan could bend to its will. Instead, Zhang was succeeded by his son, Zhang Xueliang, who quickly threw his support behind Chiang Kai-shek and China's reunification under Nationalist rule.[15]

Three years later, undeterred elements within the Kwantung Army tried

13. Yamamuro, *Manchuria under Japanese Domination*, 9–29.

14. Yosano, *Travels in Manchuria and Mongolia*, 119.

15. For more on Kōmoto Daisaku and the assassination of Zhang Zuolin, see Orbach, *Curse on This Country*, 161–92.

a slightly different tactic. This time they met with success. On September 18, 1931, Lieutenant Kawamoto Suemori and several of his subordinates set off explosives along a small stretch of Mantetsu's railway line on the outskirts of Mukden. The plot, engineered by Kawamoto's superiors, Colonel Itagaki Seishirō and Lieutenant Colonel Ishiwara Kanji, involved blaming the Chinese troops stationed nearby for this act and then using it as casus belli to invade Manchuria. Although the explosion did little actual damage to the tracks, the Japanese instigators went ahead with the plan, ordering attacks on the Chinese garrison in what they claimed was reprisal for the act of sabotage. This episode, the Mukden Incident, marked the beginning of Japan's invasion of Northeast China. The fighting lasted exactly five months, and by February 18, 1932, Japanese forces had occupied the entire region and set up the client state of Manchukuo. Fushun did not hold out very long, falling in just a day after the first shot was fired.[16]

The Manchukuo state, freshly formed, soon directed its attention to the project of industrial development. At the center of this process (and, indeed, all policy processes in Manchukuo's early years) was the Kwantung Army—likened by historian Yamamuro Shin'ichi to the "lion head" of the complex and conflicted "Chimera" that was Manchukuo, with the emperor system as its "sheep body" and the modern Chinese state its "dragon tail."[17] In March 1933, the Manchukuo government announced the Economic Construction Program of Manchukuo (滿洲國経済建設綱要), its core economic charter drafted by Mantetsu's Economic Research Association and approved by the Kwantung Army.[18] This charter outlined a comprehensive program for transport expansion, currency reform, agrarian increase, and industrial growth. Amid its array of plans, from constructing railways to establishing industrial districts, the charter contained two key clauses concerning energy. First was a call for "rationalizing the production and supply of coal" so as to "supply the public with an ample quantity of the fuel at reduced prices." Second was an injunction that the largely coal-burning electricity producers should "provide the country with a sufficient supply of power at low cost."[19] Harboring developmentalist desires consistent with the demands of carbon technocracy, Manchukuo's leaders

16. *FKZ*, 126.
17. Yamamuro, *Manchuria under Japanese Domination*, 8.
18. Hara, "1930-nendai no Manshū keizai tōsei," 19.
19. Minami Manshū tetsudō kabushiki gaisha, *Economic Construction Program*, 11–12.

embraced a grandiose economic vision that rested on the profligate use of cheap and available coal.

Kubo Tōru, who became Fushun colliery manager not long after the establishment of Manchukuo, turned to Daoist imagery to describe coal's centrality to Manchuria's industrial transformation. Referencing a famous passage from the Daoist classic the *Daodejing* in which the elusive Way is likened to the hub that allows a wheel to function, Kubo described the coal-mining industry as such a "hub" from which multiple "spokes"—such as the steel, machinery, chemical, and ceramics industries—extended, a compelling visualization of carbon technocracy and the industrial order it was supposed to support.[20] Coal production rose in the first few years, from 10.2 million tons in 1933 to 13.6 million tons in 1936, with coal from the Fushun colliery consistently accounting for around three-quarters of total output. Coal consumption grew slightly faster, increasing by over 50 percent in the same period, from 6.1 million tons to 9.6 million tons.[21] Much of this coal went, as in Kubo's metaphor, from the hub to the spokes, powering Manchukuo's industrialization efforts. From the 9.6 million tons of coal consumed in Manchuria in 1936, an estimated 1.3 million tons were used for electricity and gas production, 1.6 million tons for the steel industry, and 2 million tons for the railway.[22] In all these areas, there had been modest gains in the preceding years. Electricity output more than doubled, from 662 million kilowatt hours to 1.6 billion kilowatt hours; pig iron production increased by just under 50 percent, from 434,000 tons to 633,000 tons; and Mantetsu oversaw, by 1939, around 6,200 miles of railway track, more than twice as much as before Manchukuo's founding—some of these taken over from Chinese and Russian operators, others laid new.[23] These coal-fueled advances were to serve as the material foundation for subsequent industrial development.

Ample though Manchuria's coal deposits were, it is still somewhat interesting that the Manchukuo economy relied so heavily on this resource, not only as a direct fuel, as in the driving of steam locomotives, but also as a feedstock for electricity generation. Manchuria as a region contained, after all, vast potential for hydroelectric power. But while the Manchukuo state sponsored a number of important dam-building projects toward the end of the decade, for most of the 1930s, electric power in Manchuria came

20. Kubo, *TōA no sekitan*, 1.

21. Schumpeter, "Japan, Korea, and Manchukuo," 409.

22. Schumpeter, 410.

23. Schumpeter, 384–86; Mitchell, *Japan's Industrial Strength*, 103, 106; "Survey of the Coal Mining Industry," 71.

mainly from thermal plants.[24] One reason for this could have been the large amounts of capital required for dam construction. The Yalu Hydropower Company, established in 1937 to oversee construction of the Sup'ung and other dams along the river running between Manchukuo and Korea, boasted a hefty initial capitalization of one hundred million yen.[25] According to one Mantetsu progress report, however, there was really little need for developing hydroelectric resources given "the almost inexhaustible coal deposits in Manchuria."[26] The imaginary of fossil fuel abundance here, as it often would be in places where carbon technocracy held sway, was such that it minimized the pursuit of alternatives.

The intensification of coal-fired industrialization reoriented the Manchurian economy, with effects that could be felt back in the Japanese home islands. Channeling more Manchurian coal toward use within the region meant that less was available for export to Japan. Japanese coal producers' prior fears of being undermined by Manchurian coal would be allayed by this shift in distribution. In 1935, Fushun exported 2.3 million tons of coal to the Japanese home islands—500,000 tons less than the year before.[27] Agriculture remained the mainstay of the economy, with soybeans as Manchuria's main export. Since 1907, the Manchurian soybean was a major agricultural product sent and sold in escalating volumes to Japan and beyond as food, feed, and, perhaps most importantly, fertilizer—the bacteria in symbiosis with the roots of this legume were particularly adept at fixing to soil the nitrogen so necessary for plant growth. Manchuria was to be the world's biggest producer of soybeans, accounting for 52 percent of global output in 1925.[28] Still, the industrial development that had been limited to Japanese concessions deepened in those areas and extended beyond.[29] This heightened industrialization would, in turn, bolster the Kwantung Army's plans for a militarized Manchuria. "Today, Manchuria's main function is no longer to put its soya beans into every Japanese messpot," one commentator, writing in 1935, remarked, "but to supply the Empire with the sinews of war and to serve as Nippon's front trenches in the war with the Bolsheviki."[30]

24. On Japanese dam building in Manchuria and Korea, see Moore, *Constructing East Asia*, 150–87.
25. Moore, "'Yalu River Era of Developing Asia,'" 121.
26. South Manchuria Railway Company, *Sixth Report*, 60.
27. "Fushun Coal Exports to Japan."
28. Minami Manshū tetsudō kabushiki gaisha, *Report on Progress*, 116–17.
29. On soybeans in Manchuria, see Wells, "The Manchurian Bean"; and Christmas, "Japanese Imperialism and Environmental Disease," 821–25.
30. Aurelius, "Manchukuo."

Coal under the Managed Economy

As one of the institutional structures fashioned to support Japan's turn toward fascist imperialism in the 1930s, Manchukuo's managed economy stood on three legs: statism, militarism, and anticapitalism.[31] The Economic Construction Program of Manchukuo that sketched out the client state's industrial development envisioned an enlarged role for the government in the economy, fixated as it was with fostering military might and constraining capitalistic forces. "In order to avoid the baneful effects which capitalism when unbridled may exert," the charter's fundamental policies began, "it is necessary, in constructing our national economy, to apply a certain amount of national control and to utilize the fruits of capital so that a sound and lively development in all branches of the people's economy may be realized."[32] Manchukuo's military planners pinned much of the blame for the global economic crisis and the broader socioeconomic malaise that had stricken Japan on capitalism and its profit imperative. They therefore sought to create a new economic order in Manchuria in which the development of resources and industries would be guided less by the invisible hand of the free market and more by the visible hand of the technocratic state. In particular, they were interested in bringing "principal enterprises bearing upon national defense, public utility, or public benefit" under public control.[33] The coal-mining industry was to be one of these "principal enterprises."[34]

The managed economy that materialized in the nascent state of Manchukuo had emerged as a paradigm among Japanese thinkers and policy makers in the midst of the depression. Partly in a bid to revive flagging financial and commercial conditions that had begun after World War I and that the 1923 great Kantō earthquake had exacerbated, finance minister Inoue Junnosuke 井上準之助 (1869–1932) pushed forward deflationary policies and returned Japan to the gold standard in 1930. These measures could not have been undertaken at a more inopportune time. Coinciding with the New York stock market collapse and falling prices worldwide, they sent Japan spiraling into an economic crisis that historians Nakamura Takafusa and Odaka Kōnosuke characterize as "an unprecedented national

31. Young, "When Fascism Met Empire."

32. Minami Manshū tetsudō kabushiki gaisha, *Economic Construction Program*, 4.

33. Minami Manshū tetsudō kabushiki gaisha, 5.

34. On the military technocrats and their vision for Manchukuo, see Mimura, *Planning for Empire*, 41–69. On the tensions between soldiers and capitalists and the eventual "uneasy partnership" between the two groups in Manchuria, see Young, *Japan's Total Empire*, 183–240.

disaster."[35] To proponents of the managed economy, this crisis exposed the failings of laissez-faire market liberalism. At the same time, the Kwantung Army's takeover of Manchuria—which some leaders regarded as a way out for the embattled Japanese economy—placed Japan more squarely in the path of potential conflict with other powers, especially following its exit from the League of Nations.

The managed economy promised a rational mobilization of capital, resources, and labor for economic revival and preparation for the impending war. While the increased involvement of states in their economies was a global phenomenon in the aftermath of the depression, the extent to which economic statism in Japan had been oriented toward military buildup was probably matched only by the case of Nazi Germany.[36] In Japan, the first significant manifestation of the managed economy was the Major Industries Control Law of 1931, which legalized and, indeed, encouraged the formation of cartels or trusts among firms in major industries in an effort to eliminate excessive market- and price-damaging competition by coordinating prices and distribution volumes and undertaking other measures to regulate production.[37]

To economist Kojima Seiichi 小島精一 (1895–1966), one of the managed economy's staunchest ideologues, the state's promotion of cartel formation had yielded some successes for the Japanese coal industry. In 1932, the leaders of Japan's large coal-mining firms that made up the Coal Mining Alliance founded a joint-stock company, the Shōwa Coal Corporation (昭和石炭株式會社), to coordinate coal production and sale. Individual coal producers would submit to the company their data on consumer demand, which were hitherto kept as commercial secrets. The company then used the data to derive estimated demand and determine production goals. Although the individual coal producers were still in charge of selling their own products, Shōwa Coal set the terms of sale. This system proved more effective than the previous one that used fines and monetary rewards to incentivize compliance with production targets.[38] By 1937, the company extended control over almost 90 percent of national coal sales. Furthermore, since the Coal Mining Alliance also coordinated with Mantetsu on Fushun coal imports, approximately 80 percent of total coal supply (factoring in

---

35. Nakamura and Odaka, "The Inter-war Period," 39. On Inoue Junnosuke and Japan's return to the gold standard, see Metzler, *Lever of Empire*, 199–239.

36. On the rise of economic interventionism around the globe following the depression, see Patel, *The New Deal*. On the Nazi economy, see Tooze, *The Wages of Destruction*.

37. Gao, *Economic Ideology*, 74.

38. Kojima, *Nenryō dōryoku keizai*, 21–23.

both domestic and imported sources) was under this production agreement. The cartel became particularly effective in maintaining quotas across the industry. In the past, producers could just pay a special levy if they wanted to send more coal than they were allotted to market. Conversely, there was no penalty if they were not able to meet their quota. Now, the cartel completely forbade producers to exceed quotas, and, in the event that colliers could not meet their quotas, it allocated the shortfall to other producers who had the extra capacity.[39]

Kojima argued, however, that this system of autonomous industrial control of coal was not without its shortcomings. To begin with, this arrangement, as he saw it, placed too much power in the hands of the zaibatsu, the large conglomerates that tended to guard entrenched interests, protecting older mines that had high capital demands while blocking the development of newer ones. For example, Shōwa Coal kept allocating much of its production quota to the long-established Jōban colliery, even though this producer consistently failed to meet its targets. In contrast, it assigned the more recently established Taiheiyō colliery only two-thirds of its actual capacity of sixty thousand tons. Perhaps more egregiously, the cartel failed to consider the different types of coal when it determined overall production quotas. This, as Kojima pointed out, meant that there was no assurance, for instance, that there would be enough metallurgical coal to fuel increasing iron output targets. Given these problems, it was necessary, he concluded, for the state to exert more control over the coal industry.[40] The outbreak of war in 1937 served as immediate impetus for accelerating industrial development and the attendant extension of control that Kojima and his fellow proponents of the "managed economy" recommended. Shōwa Coal was dissolved in October 1940 and replaced by the Japan Coal Corporation (日本石炭株式會社), formed as a national policy company to direct the coal industry from the center.[41]

Experiments in statist control over coal and other industrial sectors had taken place a few years earlier in Manchuria. A pillar of the managed economy in the early years of the Manchukuo state was its system

39. Kojima, 81–85.
40. Kojima, 87–90.
41. Samuels, *Business of the Japanese State*, 84–90. As Richard Samuels observes, there were limits to the degree of control that the Japan Coal Corporation was able to exert from the onset. It served primarily as a "tunnel company" linking producers to consumers. Still, it is significant that this model of state control would come into being as much as large firms were able to profit from its workings.

of "special companies."[42] This system had been introduced by the Kwantung Army with the express intention of minimizing capitalistic competition and keeping out the zaibatsu, which military planners accused of long engaging in profiteering at the expense of national prosperity. Operating under an overarching premise of "one industry, one company" (一業一社), army technocrats set up these special companies as government monopolies over "principal enterprises." In managing their respective industries through comprehensive plans, these special companies would, they believed, rationalize the process of industrial development and, in so doing, play an important role in helping the army realize its strategic objectives in the region.[43] This was certainly what the army technocrats had in mind when they organized a special company to master Manchuria's coal-mining industry.[44]

The Manchurian Coal Company (滿洲國石炭株式會社; "Mantan" [滿炭] for short) was founded in April 1934, with the Manchukuo government and Mantetsu each committing eight million yen to its establishment. The hope for this new enterprise, which was ostensibly a Japan-Manchukuo joint venture, was that it would, in placing coal mining under government control, "secure rational exploitation, cheap supply of fuel, development of productive industries, and increase in exports."[45] This was, advocates averred, to be facilitated by the power of planning in standardizing coal-mining knowledge and practice, which would, in theory, allow Mantan to apply efficient and cost-effective methods of production throughout its many mines.[46] As a company pamphlet made plain, such enhancements in imperial extraction were intended mainly for metropolitan benefit, to "plentifully supply the Japanese industrial sector with cheap and good coal."[47] Starting out with several sizable collieries under its control, including Fuxin, Xi'an, and Hegang, the company was initially slated to take control of all the coal mines in Manchuria. Under its second director, Zhang Zuolin's assassin, Kōmoto Daisaku, the company extended its reach across the region. Kōmoto became known as the Manchurian "king of coal."[48]

42. Minami Manshū tetsudō kabushiki gaisha, *Economic Construction Program*, 5.

43. Mimura, *Planning for Empire*, 59–60, 100.

44. For more on the evolution of the managed economy over the Manchukuo years, see Yamamoto, *"Manshūkoku" keizai shi*, 27–73.

45. South Manchuria Railway Company, *Fifth Report*, 90.

46. South Manchuria Railway Company, 90; Kojima, *Nenryō dōryoku keizai*, 69–71.

47. Manshū tankō kabushiki gaisha, *Manshū tankō*.

48. Hara, "'Manshū' ni okeru keizai tōsei," 257–64.

In spite of its best efforts, Mantan would not succeed in snatching for itself the biggest prize of all, Mantetsu's Fushun colliery.[49] Mantetsu was understandably reluctant to relinquish its most profitable enterprise apart from the railway. Yet it argued for retaining the Fushun coal mines not on the basis of their value to the company but on technical grounds. In this effort, Kubo Tōru was the leading voice. Kubo admitted that while certain industries—like the telegraph, the telephone, and the railway—were suited for complete integration, this arrangement was not appropriate for the coal industry. The fundamental reason, he argued, lay in the geophysical variance of coalfields:

> In coal mining, the object on which one works is the coal seam. But coal seams differ in infinite ways: they may be buried deep or located near the surface of the earth, thick or thin; or it could be an issue of angles of incline, faults and folds, the amount of impurities in the coal, the quality of the coal itself, or of gas, coal dust, naturally occurring fires, the hardness or softness, strength or weakness of the rock, presence of water, ground pressure and so on. Hence, in order to mine these coal seams, we have a wide range of equipment, materials, labor, and business management practices. And, because of this, there cannot be just some general regulation of operations across the board.[50]

Proponents of nationalizing the coal industry under a single system lacked basic technical understanding and actual industry experience, Kubo, the consummate technocrat, complained. The particularities and contingencies of coal presented challenges to attempts at standardization. This ancient rock had many qualities, he pointed out, that made it difficult to subsume under the regime's simplifying impulses. Other coal-mining industry leaders echoed Kubo's dissenting opinion. They, too, regarded government officials and military leaders as deficient in the knowledge and experience needed to tackle the challenges facing the industry.[51] Here was a point of tension within carbon technocracy. If technical expertise and scientific rationalization were tenets that those who called for industrial control tried to appeal to, then those same principles could and would be used against their centralizing efforts.

But even in his rejection of complete standardization, Kubo envisioned

---

49. Mimura, *Planning for Empire*, 100–101.
50. Kubo, *Tōa no sekitan*, 11.
51. *CMWE*, 15.

a strong, technocratic role for the state, not as a direct manager involved in the nuts and bolts of running mines but as an overseer "controlling production quantity, quality, and price."[52] As he saw it, this arrangement would lead to the better management of the coal-mining industry, which was ultimately not for the "indulgence of selfish desire" or the "pursuit of profit" but for the proper stewardship of the nation's coal resources as part of the "moral duty of cheaply supplying the products of the mines to society."[53] In this, Kubo shared Manchukuo's technocratic planners' anti-capitalist bent and their conviction that energy should be inexpensive and widely available. Fissures in the energy regime of carbon technocracy in this instance lay mainly over issues of scientific and technological opinion. The notion that the state should be deeply involved in planning and managing the coal sector was not called into question.

Fueling the National Defense State

By 1936, military concerns prompted Manchukuo's leaders to commit to creating a national defense state in which society and economy would be fully oriented toward a looming war. A year prior, army planners had drafted the Strategic Program for Manchurian Development (滿洲開發方策網要), advocating a transition into a second phase of economic development for Manchukuo. This phase, as Ishiwara Kanji and other like-minded men of empire understood it, involved building a national defense state upon the industrial foundation laid in the first phase. Persistent concerns about the Soviet threat came to the fore, prompting an even greater focus on developing military- and war-related industries. This draft plan was followed by several other guidelines drawn up by the Kwantung Army, Mantetsu's Economic Research Association, and the Manchukuo government. In their preoccupation with war readiness, these guidelines contained provisions for reforming auxiliary sectors from transportation to agriculture, setting specific targets for military-industrial output, and privileging self-sufficiency—in what was referred to as "the dogma of local procurement" (現地調弁主義).[54] These guidelines would greatly inform the Manchurian Industrial Development Five-Year Plan (滿洲産業開發五ケ年計畫), provisionally drawn up by the fall of 1936. The coal industry was one of the major areas designated for development.

52. Kubo, *TōA no sekitan*, 13–14.
53. Kubo, 7.
54. Hara, "1930-nendai no Manshū keizai tōsei," 57–61.

Six percent of the drafted comprehensive budget of 2.4 billion yen went toward coal, for a targeted increase in total production by over 50 percent to eighteen million tons.[55]

As the five-year plan started coming together on paper, the Manchukuo government, thinking of implementation, enlisted the services of "reform bureaucrats" (革新官僚), a circle of officials who wanted to integrate the principles of the managed economy into the crafting of policy in Japan and its empire during the wartime era. Unlike army planners, they were not opposed to drawing on the capital and expertise of large corporations to support their technocratic vision. Among these reform bureaucrats, few exerted as much influence as Kishi Nobusuke 岸信介 (1896–1987). Kishi—who was later imprisoned as an accused "Class A" war criminal and still later made prime minister of Japan—became one of the chief engineers of the Manchukuo economy. Invited by the client state's leaders, Kishi arrived in Manchuria in the fall of 1936. Manchurian industrial development had stalled in part because of shortages in capital. Kishi's first task, then, was to attract investments to the region. To that end, he reached out to the so-called new zaibatsu. These conglomerates, which focused on more scientifically and technologically involved heavy and chemical industries as compared to their older counterparts, had come to prominence in the interwar years. By the 1930s, they were flourishing in the shadow of military expansion.[56]

One of these new zaibatsu was Nissan, which was headed by engineer and entrepreneur Ayukawa Yoshisuke 鮎川義介 (1880–1967), a close relative of Kishi's. At Kishi's urging, Ayukawa relocated Nissan to Manchuria to form the core of a new consortium that would reorder the Manchurian economy, the Manchurian Industrial Development Company (滿洲工業開發株式會社; "Mangyō" [滿業] for short). Financed half by the Manchukuo government and half by Nissan at 450 million yen each, Mangyō was designed to consolidate control over Manchurian industry. Ayukawa had been critical of the former special company system in which, as he put it, "a car is a car, a plane is a plane, coal is coal"—that is, industries were being developed in isolation. The new economic structure, as he imagined it, would be modeled after a pyramid, with industries placed under a centralized command and subjected to large-scale, comprehensive

---

55. Zhao, "Manchurian Atlas," 117.

56. Johnson, *MITI and the Japanese Miracle*, 131; Mimura, *Planning for Empire*, 94. On reform bureaucrats, see also Moore, *Constructing East Asia*, 188–92. As Louise Young has noted, though, the old zaibatsu remained the primary source of direct investment in Manchuria over the course of this period. See Young, *Japan's Total Empire*, 214–16.

planning.[57] Manchukuo's technocratic leaders turned the management of Manchuria's special companies, including Mantan, over to Mangyō. They also forced Mantetsu to sell off almost all of its heavy industrial subsidiaries (sixty-eight companies for 550 million yen), which were then given to Mangyō to run. As before, however, Mantetsu was unwilling to part with its Fushun colliery—the only industry aside from the railway that would remain in its hands.[58]

The Manchurian coal industry by the late 1930s was hence divided between Mangyō and Mantetsu. Both were slated for ambitious increases in output, as Manchukuo's planners concerned themselves with fueling the five-year plan and the national defense state. The finalized version of the plan, launched in January 1937, targeted annual coal production to reach 25.5 million tons by the end of the five years, more than double the output of 1936. Of the 25.5 million tons, they expected 15 million to come from the Mantan mines under Mangyō and as much as 10 million from Fushun.[59]

Throughout the 1930s, the Fushun colliery had been further expanding its coal-mining operations. By 1934, not only had Fushun's open pit grown to gargantuan proportions, but its shaft mines also now sank much farther into the earth, so much so that according to one account, "the deepest of the shafts could hold three Washington monuments end to end."[60] Between 1933 and 1935, Fushun's annual output had increased from about 7 million tons to just under 8 million tons. Then within a year, coinciding with the five-year plan's launch, output went up by 1.5 million tons, reaching Fushun's prewar height of 9.59 million tons.

Much of this surge in production could be attributed to the expansion of workings at the Longfeng mine. Opened in 1917, the Longfeng mine underwent major refurbishment in the mid-1930s. As the coal in its upper reaches had been depleted through slope mining, Longfeng turned to shaft mining, which allowed miners access to the coal embedded deeper down.[61] Located on the eastern edge of the coalfield, the Longfeng mine covered an area with deposits totaling a quarter of a billion tons and seams from sixty-five to a hundred feet thick. The coal here, "being fairly adhesive," according to one assessment, was suited for making coke, most of which fired the furnaces of the Shōwa Steel Works in nearby Anshan.[62]

57. Mimura, *Planning for Empire*, 103–4.
58. Mimura, 101, 105.
59. Kojima, *Nenryō dōryoku keizai*, 72–73.
60. Scherer, "Manchoukuo Down to Date."
61. *FKZ*, 180.
62. "The Ryuho Coal Mine," 56.

Serving as the centerpiece of the refurbished mine and the symbol of its mechanized modernity was the massive new winding tower that stood above its western shaft, a veritable monument to the energy regime's ever-expanding vision. Because shaft mining involves tunneling straight down into the earth to access deposits, lifting men and materials up and lowering them down requires a great deal of power—certainly more than in slope mining, in which those transport processes typically take place along an incline. As such, winding towers are installed above shaft mines with assemblages of hoists, motors, and other mechanisms to do the work of lifting and lowering. In hoisting, in coal mining on the whole, and indeed in almost all industrial endeavors, Japanese engineers of the 1930s took Germany as the gold standard. One Mantetsu account noted that the Longfeng winding tower, at over two hundred feet, was of comparable scale to those at the largest German coal mines, such as Koenigsborn, Hannibal, and Minister Stein. In terms of its cage winding system, however, Longfeng's tower was unparalleled. It was able to hoist 12.2 tons in a single lift and 650 tons per hour, outperforming those at Koenigsborn and Minister Stein, which were able to pull 11.7 and 8.4 tons each lift and 293 and 420 tons every hour, respectively. "In size or capacity the tower of Ryuho [Longfeng] commands the undisputed supremacy over all others," the account's author bragged.[63]

This monstrous tower drew much debate from Mantetsu's engineers when it was being designed. Two main points of contention were how the coal would be brought to the surface—by skip or by cage—and the type of hoisting system to use—the drum method or the friction method. (In the drum method, the haulage rope conveying the load was wound around a drum; in the friction method, the haulage rope passed around but was not attached to a wheel, and the load was then offset by tail ropes and counterweights.) Eventually, the company decided to go with the cage and friction system. Skips tended to result in a higher proportion of coal being crushed, reducing the output's value, and the drum method required huge drums of eighty tons that were difficult to manufacture. The result was "a giant elevator shaft" that could lift in one hoist eight coal cars or, if used to transport personnel, as many as 120 miners.[64] The tower itself, "well lighted through the wide and numerous glass windows," shielded the hoisting apparatus from the harshness of Manchuria's long cold season, preventing the "slip-

63. "The Ryuho Coal Mine," 59–60.
64. "The Ryuho Coal Mine," 57.

4.1. The Longfeng mine's winding tower. Standing at about 220 feet, it was tied with a tower at the Koenigsborn coal mine in Germany as the tallest in the world. If the open-pit mine was the primary symbol of Fushun's coal-mining enterprise, this tower was arguably its second.

ping of the lines due to the formation of frosty impedimentary coating over the cable lines in the winter."[65]

As was characteristic of the colliery and extensive coal mining in general, the entire setup at Longfeng consumed much energy even as it was involved in the production of the same. The tower's generators had a capacity of 4,025 kilowatts, "the greatest in the world" and almost four times that of those driving the winding facilities at the massive German Koenigsborn coal mines. Considerable amounts of electrical power were needed for "the smooth and efficient functioning of so great a mechanism."[66] According

65. "The Ryuho Coal Mine," 57.

66. "The Ryuho Coal Mine," 60, 63. One small irony in all of this was that as much as Mantetsu boasted of Fushun's Longfeng mine surpassing its German counterparts in this way or the other, many of the key pieces of machinery it used here were in fact imported from Germany. These included parts of the winding mechanism and winding and coal pit cages made by Demag, electrical equipment by Siemens-Schuckert, and other materials such as coal cars and platforms by other German firms.

to one account, Kubo Tōru was so proud of this powerful and productive new shaft mine that he had "under his safe keeping in his room, the first lump of black coal brought up from the pit."[67] Buttressed by Longfeng's refurbishment and other improvements in operations across the colliery, Fushun remained in this period the single largest supplier of coal to the national defense state.

At the same time, Fushun's importance to Japanese militarists as a source of energy also stemmed from its production of liquid fuels. This was not only from its shale oil industry, whose crude oil output grew steadily, doubling in the half decade from 1930 to 1935 to over 65,000 tons, but also because of inroads in manufacturing synthetic petroleum.[68] In 1939, Mantetsu established in Fushun a plant for producing synthetic petroleum through coal liquefaction, the conversion of coal into oil. Chemically, hydrogen content is one primary difference between the two, with coal containing on average about half as much hydrogen as oil. Theoretically, it was thus possible to turn coal into oil by adding hydrogen.[69] How exactly to best accomplish this was a puzzle many scientists around the world strove to solve.

Two main methods of coal liquefaction had emerged as commercially viable by the early twentieth century. The first was the Fischer-Tropsch process, developed by German chemists Hans Fischer and Hans Tropsch of the Kaiser Wilhelm Institute in 1925. In this method, coal is gasified to produce carbon monoxide and hydrogen, which thereafter are reconstituted into a liquid fuel. The second method was the Bergius process, named after German chemist and future chemistry Nobel laureate Friedrich Bergius. With this method, coal is crushed into a fine powder and mixed into a paste with heavy oil recycled from an earlier reaction. The paste is then put into contact with hydrogen in the presence of an organic catalyst, triggering a conversion into liquid form.[70] In 1936, the Japanese government introduced several laws, subsidies, and tax exemptions to incentivize industrial concerns to establish synthetic petroleum plants. Under such provisions, Mitsui, for instance, purchased from Germany the rights to use the Fischer-Tropsch process and the necessary equipment to carry it out in a plant at its Miike coalfield.[71]

---

67. "The Ryuho Coal Mine," 65.

68. "Mining Industry of Manchukuo."

69. Speight, *Chemistry and Technology of Coal*, 435.

70. Department of Justice, War Division, Economic Warfare Section, "Report on Fushun, Part 3," 1–2. On Friedrich Bergius and his synthetic petroleum process, see Stranges, "Friedrich Bergius."

71. Schumpeter, "Japan, Korea, and Manchukuo," 436.

Experiments with synthetic petroleum by Japanese scientists stretched back almost two decades earlier, spurred, as investigations into shale oil were, by how World War I rendered oil an indispensable resource. The main challenge was not so much making the reaction happen but doing so on an industrial scale. The Tokuyama Fuel Depot led these efforts. Its naval scientists initially sent samples to German laboratories to have trials run on Japanese coal using the Bergius process. When those trials yielded poor returns—only 30 to 35 percent of the coal was liquefied—these scientists commenced their own experiments in-house with Fushun coal. By 1934, fuel experts at Tokuyama were able to liquefy up to 67 percent of the feed-stock coal in experimental reactors they modified to allow for finer coal powder and a better utilization of exhaust heat.[72]

Mantetsu's foray into synthetic petroleum began in 1928 with experiments at its Central Laboratory in Dairen.[73] As with their work on shale oil, Mantetsu scientists collaborated with counterparts at Tokuyama and similarly tapped into transnational technoscientific currents as they proceeded with their coal liquefaction research. In 1932, the company dispatched fuel researcher Abe Ryōnosuke 阿部良之助 (1898–1980) to Germany in 1932 to purchase coal hydrogenation equipment.[74] Abe returned with machinery in tow, which he then set up at the Dairen facility for tests intended to complement those underway at Tokuyama. The priority remained the same as before: scaling up production from laboratory to plant. In 1935, Abe and two colleagues once again traveled to Germany, this time to observe operations at existing synthetic petroleum plants. In August 1936, Mantetsu, now under president Matsuoka Yōsuke 松岡洋右 (1880–1946), secured from the Japanese government a license to build a coal liquefaction plant in Fushun. The following June, the navy and Mantetsu held a meeting at Tokuyama, with Kubo Tōru as one of the company's representatives. There, both parties agreed that Mantetsu's experiments at the Central Laboratory should be replicated at the proposed plant in Fushun. Construction was completed in two years, and the plant commenced production not long after.[75] The method Mantetsu used here was a variant of

72. Department of Justice, War Division, Economic Warfare Section, "Report on Fushun, Part 3," 3.

73. Another notable effort within the Japanese empire to develop synthetic petroleum through coal liquefaction was at the nitrogen fertilizer enterprise Chōsen Chisso Hiryō in colonial Korea. See Molony, *Technology and Investment*, 226–32.

74. On Abe Ryōnosuke, see Katō, *Manekarezaru kokuhin*.

75. Department of Justice, War Division, Economic Warfare Section, "Report on Fushun, Part 3," 4–5.

the Bergius process, a direct liquefaction process purportedly perfected in its Dairen laboratory. At the time of the plant's launch in 1939, Mantetsu proclaimed the promise of synthetic petroleum, not only because of "the relatively high degree of research and experimentation completed in this line" but also because of "the abundance and extensive distribution of coal supply." As we shall soon see, neither research breakthrough nor bounty of available coal should have been taken as a given.[76]

Primarily by coal but also by shale oil and synthetic petroleum, Fushun proved so pivotal to powering industrial Manchuria and the Japanese national defense state that it caught the attention of Japan's foes, who spotted in its centrality a vulnerability that might be exploited. "Destruction of the hoists and the winding towers for both the open pit and shaft mines at Fushun would paralyze coal production," an American government economic warfare report from 1943 projected. "And the effect of stoppage of coal production at Fushun," it quickly added, "would strike at every industry within a 400 mile square from Fushun to Mukden to Dairen to Antung [Andong]." This assessment was accurate. Fushun, at that point, still supplied 70 percent of the coal extracted in Manchuria. The Shōwa Steel Works, the greatest in the region, counted on Fushun for almost three-quarters of its coal and coke. Mantetsu relied entirely on Fushun for its railway fuel. And Fushun's main power plant, which was the primary source of electricity for surrounding industrial centers, also ran on coal mined on site. This network of energy dependence, which had been cultivated under carbon technocracy and reinforced by the national defense state, seemed strikingly fragile. If just the power plant were incapacitated, "this would mean," the report noted, "stopping operations in the coal mines, since all transportation and mining equipment in the mining district is completely electrified; it would mean a stoppage of iron and steel production at the Shōwa and Honkeiko [Benxihu] Iron and Steel Works; and it would mean a stoppage of munition and machinery production in Mukden."[77] As the war waged on, Japan would feel the true depth of this dependence most as coal shortages arose.

Empire of Energy

Delivering on its promise of resource richness, Manchuria dominated coal output in the Japanese empire in the years leading up to and into the war.

76. "Coal Liquefaction," 18.

77. Department of Justice, War Division, Economic Warfare Section, "Report on Fushun, Part 1," n.p.

In 1938, half of the thirty-two million tons of coal produced outside of the home islands came from this region.[78] The Manchurian coal industry both represented and set the pace for developments in coal mining across Japan's colonies. Although varying in their quantity and quality of coal resources, these other territories all endured the extractive endeavors of the Japanese state, the coal mined from beneath their occupied soil used to fuel local industrialization or sent to light furnaces back in the home islands or elsewhere in the empire. The circulation of coal between these sites of extraction would, over the years, be subjected to attempts at increased coordination, particularly by technocratic planners who started to see these as parts of one larger imperial energy system. To get a sense of this interconnected empire of energy, let us now briefly survey the coal-mining activities in the Japanese colonial possessions of Taiwan, Korea, Karafuto, and North China.

Coal mining in Taiwan developed over the decades of Japanese rule, which began when the Qing empire ceded the island to Japan in the 1895 Treaty of Shimonoseki. The Dutch had known of the presence of coal on the island when they seized it in the seventeenth century, mining this resource to use in smelting iron and to sell abroad. The Qing largely halted these operations, only restarting them after the Opium Wars, when demand for bunker fuel increased with the influx of ships calling on the newly opened Chinese treaty ports. Taiwan's coal-mining industry remained small in the first two decades of the Japanese takeover. Aside from shortages of capital and expertise, the fact that the colonial government had set aside the island's most valuable coalfield in Keelung for naval procurement meant that coal-mining entrepreneurs were left working poorer deposits. There was also a problem of economic geography: Taiwan's coal mines were in the north of the island; the sugar factories that were coal's main consumers in the south. Until the north–south railway line was completed in 1908, the sugar industry ended up importing most of its coal.[79]

Taiwan's coal-mining industry did, however, expand from World War I onward. As the region saw increases in industrial activity, Japanese authorities took a greater interest in extracting Taiwanese coal, which found ready markets in the home islands, South China, Hong Kong, and Southeast Asia.[80] The total output of Taiwanese coal rose steadily in the period that followed. Production was 1 million tons in 1921, 1.7 million tons in 1925,

78. *CMWE*, 14.
79. Chen, "Development of the Coal Mining Industry," 182.
80. Chen, 183–87.

and 1.9 million tons in 1927.[81] In the late 1920s and the early 1930s, Taiwanese coal mining suffered setbacks, in part because of the global depression but also because of Chinese boycotts of Japanese goods in response to the Ji'nan and Mukden Incidents.[82] Nevertheless, by the mid-1930s, the industry had recovered, and the mines that weathered those political and economic storms further mechanized their operations to increase the quantity and quality of their output.[83]

Korea, which Japan had taken over as a protectorate in 1905 and formally annexed in 1910, possessed sizable coal deposits. About half of this coal was anthracite of high quality, burned to heat homes and drive naval vessels.[84] Korean coal was concentrated in the northern part of the peninsula, and P'yŏngyang emerged in this period as a center of anthracite extraction. According to one estimate in a 1921 report by the Bank of Chosŏn, "the veins are 32 miles in length and 7½ miles in width, and the amount of coal they contain is estimated at 200,000,000 tons." The colonial government began working mines in the area in 1907. A decade later, annual output was over 150,000 tons, 90 percent of which went to the Tokuyama Fuel Depot to be "used by the Navy as a substitute for Cardiff coal."[85] Colonial resources gained particular purchase amid mounting concerns over self-sufficiency. Other sites would also be excavated in those years, similarly upon government initiative. In 1930, the colony's total coal production amounted to almost 900,000 tons. A striking feature of Korean coal consumption was that because much of domestic output consisted of anthracite, large quantities of bituminous coal had to be imported in order to power industry and the railway for which the higher-ranked coal was less suited. Of the 1.6 million tons of coal Korea consumed in 1930, 640,000 tons came from local production, 510,000 from China, and 420,000 from Japan.[86]

Karafuto was a Japanese territory in the southern portion of Sakhalin, the almost six-hundred-mile-long island north of Hokkaido and east of the Far Eastern Krai. Japan had obtained this half of the island from Russia through the 1905 Portsmouth treaty and held it until 1949. Across Sakha-

81. Bradley and Smith, *Fuel and Power in Japan*, 24.

82. The Ji'nan Incident of May 1928 was a violent conflict between Chiang Kai-shek's National Revolutionary Army and Japanese troops stationed in Ji'nan, the capital of Shandong Province.

83. Chen, "Development of the Coal Mining Industry," 188.

84. "Chōsen shi dai kōgyō"; Bradley and Smith, *Fuel and Power in Japan*, 24.

85. Bank of Chosen, *Economic History of Chosen*, 188.

86. Bradley and Smith, *Fuel and Power in Japan*, 25.

lin, coal was abundant, its quality, according to one assessment, "better than the average Japanese coal and slightly inferior to the best, but not equal to coal from Fushun in Manchuria."[87] In the fourteen thousand square miles that made up Karafuto, the Japanese colonial government actively mined this energy resource. The coal industry grew immensely from the late 1910s through the 1920s. In 1916, yearly production was 7,000 tons; in 1930, it was nearly 650,000 tons.[88] North of the border, Japanese interests also ran several coal mines along the thickly forested coast under concession from the Soviet government, to whom they paid royalties of 5 to 8 percent of the extracted value.[89] Although Japanese imperialists intended from early on that Sakhalin be one source of coal for Japan and Korea, the sea around the island froze between November and April, meaning that supply was at best seasonal.[90] Still, by 1935, Karafuto was sending at least a third of its over 1.5 million tons of annual output to fuel enterprises in other parts of the empire.[91]

North China, with its rich endowment of coal, had attracted Japanese industrialists and empire builders for decades. Since the late nineteenth century, Japanese capital had been flowing into the region for the development of its coal mines. The interest here, as in Manchuria, was in securing a continental coal supply that could help meet demand for both metallurgical coke and ordinary fuel back in the home islands.[92] A notable enterprise in this regard was the Sino-Japanese Luda company in Shandong, formed in 1923 to run mines that the Japanese had seized from the Germans during World War I.[93]

The allure of increased control over the region's resources deepened with time. Soon after the founding of Manchukuo, Mantetsu and the Japanese garrison stationed in Tianjin sent researchers to conduct extensive surveys of coal and other natural resources in North China. One main impetus behind their investigations, a Mantetsu report openly claimed, was "the promotion and expansion of imperial economic power."[94] These surveys subsequently served as the basis for various Japanese plans to develop North China's strategic resources, of which coal, especially for coking and the production of

87. Great Britain Foreign Office, Historical Section, *Sakhalin*, 33.
88. Bradley and Smith, *Fuel and Power in Japan*, 27.
89. "The Significance of Saghalien."
90. Bradley and Smith, *Fuel and Power in Japan*, 28.
91. *CMWE*, 43.
92. Wright, *Coal Mining in China's Economy and Society*, 126.
93. On the history of Luda, see Wright, "Sino-Japanese Business in China."
94. Cited in Nakamura, "Japan's Economic Thrust into North China," 224.

synthetic petroleum, was deemed particularly promising. Possessing more coal than even Manchuria, this region, by one estimate, accounted for 55 percent of China's total reserves.[95] As Japan's presence in North China became more pronounced, punctuated by its installation of the East Hebei puppet regime in 1935, these coal resources were increasingly exploited at the whim and for the wants of the Japanese imperial state.[96]

With the growth of the Japanese empire during the interwar period, carbon technocracy came to increasingly define the development and interdependence of its constituent parts. Japan's coal imports reached new heights in 1933, stimulated by the expansion of iron and steel production and, as such, by the need for coking coal, most of which came from the continent. Of the 3.5 million tons of coal arriving in Japan's ports that year, 2.4 million tons were from the Fushun colliery.[97] Imports continued to rise in the years after, with Japan's colonies remaining the main suppliers. But as the preceding sketches demonstrate, the flows of coal were not necessarily just from colony to metropole. They were also between colonies and, at times, from metropole to colony as well—the Japanese export of bituminous coal from Kyushu to Korea being a good example of this.[98] These crisscrossing circulations of coal undergirded Japan's empire of energy, and when the Japanese military assemblage embarked on its warpath in the late 1930s, it relied on such networks to fuel the industrial engines that kept it going.

Warscape of Intensification

The start of the Second Sino-Japanese War in the fall of 1937 saw statist pressures to ramp up coal extraction across Japan and its empire. Although it was not unusual for most industries to strive for increases in production from year to year, the extent of the targeted increases was remarkable. The Cabinet Planning Board (企畫院), the powerful central government agency formed in October 1937 to manage the wartime economy, set lofty goals for coal output. For example, Manchuria had witnessed an increase in coal production from 13.6 million to 14.3 million tons between 1936 and 1937 (an increase of 5.1 percent), but the target then set for the following year of 1938 was 17.5 million tons (a would-be increase of 22.4 percent).[99] The

95. Nakamura, 232–33; Paine, *The Wars for Asia*, 39.
96. Paine, *The Wars for Asia*, 39–40.
97. Bradley and Smith, *Fuel and Power in Japan*, 6.
98. *CMWE*, 14.
99. "Survey of the Coal Mining Industry," 69; *CMWE*, 46.

extent to which coal was important to the war effort was evident not only in overall consumption increases but also in the greater proportion being consumed by war-related industries. For instance, the iron and steel industry, which was key to munitions manufacturing, used 4.1 million tons of coal in 1933, which was 12.9 percent of national consumption, but in 1940, it used 11.4 million tons, which represented 18 percent of the total.[100] By one estimate, coal accounted for over two-thirds of the energy consumed by Japan's war economy in 1943. This was the case even though hydroelectric power, the next most important source of energy, underwent a more than sixfold increase in generation capacity over the preceding two decades.[101] The Japanese war machine hungered for coal, its cravings intensifying as the conflict prolonged and its needs becoming increasingly impossible to meet.

I refer to this situation as a warscape of intensification, a play on energy historian Christopher Jones's idea of "landscapes of intensification." Jones, writing about expanding systems of energy transport in the late nineteenth- and early twentieth-century American mid-Atlantic, coined this term to capture how sociotechnical infrastructures like canals, pipelines, and wires were marked by "synergistic feedback loops" through which they "stimulated and sustained the ever-increasing consumption of coal, oil, and electricity."[102] A warscape of intensification was similarly characterized by cycles of cause and effect that drove an escalating demand for energy. Warfare is an incredibly energy-intensive activity.[103] This was particularly evident in wars waged at least partly for the seizure of energy resources, as Japan's invasion of East and Southeast Asia ended up being. For even if expansionist belligerents managed to wrest control of the resources they sought, they almost always found themselves having to pump more energy into maintaining and developing extractive sites and transportation systems to get those resources out of the ground and to wherever they would be consumed. This dynamic brought into being warscapes of intensification in which aggressor states were then motivated to expand farther for access to even more resources and to mine presently held deposits with greater ferocity. The resultant arrangement, while possibly impressive in geographic extent—again, as Japan's wartime empire was—was vulnerable insofar as the flows of energy that held it together, moving in increasing

100. *CMWE*, 14.

101. *CMWE*, 9.

102. Jones, *Routes of Power*, 2, 6–10, 189, quotations on 2, 189.

103. For one reflection on the relationship between war and energy resources, see Smil, "War and Energy." For an account that looks at the militarized environment of war through the lens of energy at its most capacious extent, see Muscolino, *Ecology of War*.

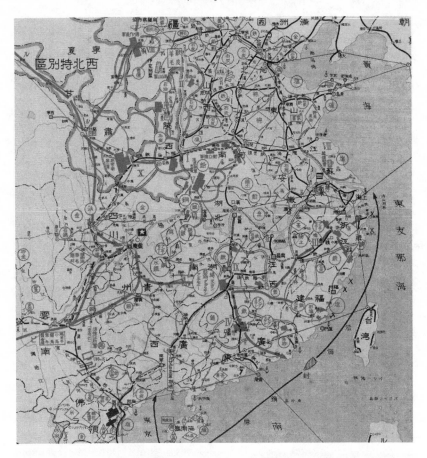

4.2. Excerpt from a Japanese map dated March 1941 and titled *Shina kōNichi senku oyobi shigen kōtsū mō yōzu* [Map of China's anti-Japanese war zones and network of resources and transportation lines]. While its full extent covers almost the entirety of China and the states that it borders, only North, East, and Central China and, interestingly, French Indochina are populated with data regarding resources and transportation lines. This excerpt shows most of the China portion of the map. The Nationalist government's wartime capital of Chongqing is marked by the flag with a star just left of center. One may spot the coal resources identified by looking out for the Japanese characters for coal (石炭). These are largely concentrated in the north. Other resources labeled here include antimony, gold, iron, lead, silver, and tin as well as cotton, kaolin, salt, tea, timber, and wood. (Image courtesy of the Harvard-Yenching Library, Harvard University.)

volumes and across greater distances via sea lanes and land routes, could be disrupted. Moreover, the intensification of extraction, even if initially successful, was seldom sustainable. Warscapes of intensification often unraveled in a downward spiral.

In Manchuria, the call for steep increases in coal production was worked into the five-year plan. Japanese leaders had further increased output targets for coal, like those of other resources, after the outbreak of war. According to the original plan, coal capacity was supposed to hit 27.1 million tons by 1941; in the revised plan of 1938, it was set at 34.9 million tons.[104] If securing domestic self-sufficiency had been a priority for Manchukuo's planners in the preceding period, the war redirected the Manchurian economy toward supplying the home islands, more openly subordinating it to the demands and desires of the metropole.[105] In the case of coal, this was not only about extracting the resource and sending it directly to Japan proper but also about fueling other local industries that produced materials the home islands needed, such as iron and steel. As a Mantetsu report described it, "The plan for increased output of coal has been made the basis, or the nucleus of, the Five Year Industrial Development Plan."[106] Fushun was at the center of this initiative for intensified industrialization. "The Fushun colliery," another Mantetsu report from 1937 vaunted, "has adapted to the needs of the current political situation, and, taking 'coal mining in service of the nation' as its mission, has determined plans for the greatest coal production and has exerted much effort in increasing extraction."[107] Fushun's output had, in fact, hit its prewar peak the year before and was in the middle of a slow start to what would soon be a quick decline.

Elsewhere in the empire, too, colliers furiously pursed the intensification of coal mining. In Taiwan, the colonial government established a statist corporation tasked with instigating the expansion of mining while controlling prices and distribution.[108] These efforts met with some early success. In 1939, Taiwan's actual output of 2.6 million tons exceeded the planned target of 2.4 million tons.[109] The coal-mining sector in Korea also witnessed an upswing around that time, its development a key part of the peninsula's five-year plan modeled on Manchuria's and launched in 1939. In South P'yŏngan Province, for instance, a new railway line was laid to facilitate access to the anthracite-rich coalfields to the north in Kangdong, Samtŭng, and Taesŏng.[110] Korean coal production increased by almost 80 percent within two years, from 3.4 million tons in 1938 to 6.1 million

104. Nakagane, "Manchukuo and Economic Development," 145.
105. Beasley, *Japanese Imperialism*, 216.
106. "Survey of the Coal Mining Industry," 74.
107. Mantetsu, "Tōkei nenpō," 398.
108. Chen, "Development of the Coal Mining Industry," 189.
109. *CMWE*, 47.
110. Park, *Colonial Industrialization*, 153; "Korea Starts Industry."

tons in 1940.[111] Similarly, in Karafuto, the Japanese set up a new governmental board to look into raising production in 1939, lifted a ban on new mines in 1940, and established a large statist mining company in 1941. Together, these measures spurred a jump in coal output. In 1935, Karafuto produced 1.5 million tons of coal; in 1941, its production peaked at 6.5 million tons.[112] North China, which Japan managed to formally seize early in the war, was also made to suffer the strains of intensified extraction. Coal production in this region more than doubled, surging from just under 10 million tons in 1938 to a staggering 24.9 million tons in 1942.[113] Finally, when Japan invaded Southeast Asia after its attack on Pearl Harbor in December 1941, it also began exploiting a number of important coalfields in the region, from the Hongay coal mine in Indochina to the Bukit Assam coal mine in the East Indies.[114] It is partially a testament to coal's ubiquity beneath the earth's surface that across Japan's Greater East Asia Co-Prosperity Sphere (大東亞共榮圈), both mining and empire expanded simultaneously, mutually reinforcing each other and giving form to a warscape of intensification.[115]

The Japanese metropole's reliance on the colonies for coal deepened by 1940 as domestic production struggled to supply Japan's growing industrial demand. "Coal, the lifeblood of modern industry, has begun to flow sluggishly in the veins of the Japanese economic body," one contemporary commentator wrote.[116] The previous summer, a terrible drought hit the nation—"said to be," by one account, "without parallel in the last 170 years of Japanese history."[117] As a result, thermal power plants had to burn more coal to make up the decrease in hydroelectric power generation, and when the elevated demand could not be met, energy shortages ensued.[118] Coal scarcity had been a point of concern for at least a year by that point. For example, in order to conserve coal, Tokyo had put a ban on morning baths—public bathhouses only opened their doors at noon.[119] In April 1938, Kubo Tōru gave a talk to the Osaka Industrial Association, where he made a case for further exploiting continental coal resources, par-

111. *CMWE*, 14.

112. *CMWE*, 43.

113. *CMWE*, 48.

114. "More Driving Power."

115. On the Greater East Asia Co-Prosperity Sphere, see Yellen, *Greater East Asia Co-Prosperity Sphere*.

116. Bloch, "Coal and Power Shortage," 39

117. Bloch, 43.

118. Bloch, 12; Barnhart, *Japan Prepares for Total War*, 143–44.

119. "Public Bath Rate-Cutting."

ticularly in North China, so as to meet Japan's shortfall.[120] It was in the dry spell of 1939, however, that the problem became severe. This situation was exacerbated by the fact that the cartelistic Shōwa Coal Corporation provided thermal power plants with coal of such low quality that they had to burn more coal than they normally would to generate the same amount of electricity, and their operations were thus compromised.[121] In a bid to deal with this and other issues related to energy shortages, the state's grip on the coal industry tightened. This was most apparent in the reorganization of Shōwa Coal into the quasi-governmental Japan Coal Corporation, which now held a complete monopoly over coal sales and distribution, further shoring up the dominance of carbon technocracy within the Japanese imperial state.[122]

Still, it had become evident to many observers of the industry that Japanese coal mining had run into a rut. A postwar assessment saw this as partly an outcome of "the drive for increased production at any cost." The picture painted was of a warscape of intensification that appeared progressively irrational: "Each year, more and more low grade, ill-equipped and poorly engineered mines were brought into production. And each year the galleries of the high-grade mines were pushed deeper and deeper." Aside from quantity for quality, another trade-off here was between speed and sustainability, as "the pressure for immediate production demanded sacrifice of sound long-range development of such resources as were available." Most egregious in this regard was the forgoing of maintenance necessary for the mines to continue to be workable. "Absence of this developmental groundwork," the assessment concluded, "was to prove one of the obstacles to achieving greater production during the war years."[123] Eventual overextension seemed inevitable.

In the midst of this pattern of intensifying production, Japanese coal mines also experienced shortages of skilled labor and mining materials. In terms of labor, the coal-mining industry regularly experienced high turnover because of the work's danger and difficulty, coupled with relatively low pay. This turnover posed a particular challenge since it generally took about six months to two years for a miner to become proficient in the use of mining machines. The problem only worsened after the onset of the war, as experienced miners were conscripted to serve in the military. As for

120. "Meeting Discusses Shortage of Coal."
121. Hein, *Fueling Growth*, 39.
122. *CMWE*, 13. Japan Coal was capitalized half by the government and half by private coal-mining companies.
123. *CMWE*, 12.

mining materials, steel (for cables, pit props, and machines), cement (for construction), and rubber (for conveyor belts) were often in limited supply, and shortages became more pressing as the war dragged on.[124]

Together, these factors conspired against the expansion of coal mining that Japanese war planners had sought, bringing the warscape of intensification to a breaking point. Japan's coal production peaked in 1940 at 57.3 million tons. Even then, the industry had been struggling, as evidenced in the decline in mining efficiency, defined as average output per miner per year. That year, the figure was only 173 tons (as compared to 227 tons in 1933), and it continued its downward slide in the years that followed. That the Japanese coal industry managed to deliver growth in absolute output while mining efficiency consistently fell points to the considerable amounts of labor, capital, and resources that were pumped in at diminishing marginal returns to make whatever gains possible.[125] Given this situation in the home islands, Japanese technocrats looked more readily to the colonies—especially Manchukuo, North China, and Karafuto—for the coal they needed to fuel Japan's wartime industries.

Yet as much as Japanese planners regarded it as a primary source of energy for the Japanese economy, the Manchurian coal sector was not without problems of its own. Coal production in Manchuria had been increasing in the early years of the war, but it was doing so at a much slower pace than planners had projected. In 1938, output was fifteen million tons, almost a million tons more than the year before. However, this did not even meet the original five-year plan's target of eighteen million tons for that year, let alone that of the more ambitious revised version. According to the plan, Mantan's mines were supposed to account for most of the increase in coal production, while the Fushun colliery, as the other more established major producer in the region, was expected to see moderate rises in output.[126] In 1938, Fushun achieved 95 percent of its targeted output, but Mantan only reached 68 percent of its goal.[127] Then and after, industry observers consistently spoke of Manchurian coal in terms of shortages, as the gap between planned and actual output continued to widen.[128]

The prospect of coal scarcity in Manchuria went beyond the towering heights of industry and into the *kangs* and stoves of ordinary women and men. Coal was, after all, not just an industrial fuel but a household one as

---

124. *CMWE*, 12.
125. *CMWE*, 11–12, 26.
126. Mitchell, *Industrialization of the Western Pacific*, 85–86.
127. Iguchi, *Unfinished Business*, 183.
128. "Transfer of Mining Equipment."

well. As with war mobilization across the rest of Japan's empire, the home was understood as a site of sacrifice for the needs of the frontline. It was this logic (and the increased industrial demand for coal and the corresponding sense that coal was in short supply) that motivated some technocrats to agitate for an intensification of household fuel frugality. Mantetsu itself put out a manual in the late 1930s with instructions on how to burn coal economically in the home. "On account of the steady rise in coal consumption in Manchuria and this year's coal shortage," it began, "there have been calls for the shortening of periods during which heating is used and the enforcement of 'low-temperature living' [低温生活] . . . being mindful of the use and conservation of coal, which is the driving force of national defense and production, is the duty of all citizens on the home front."[129] As with their counterparts in colonial Korea, who similarly promoted "low-temperature living" as part of an agenda for wartime forestry conservation, Manchukuo officials launched a campaign asking that ordinary subjects embrace the cold for the sake of fueling the nation.[130] In Manchuria, the campaign for "low-temperature living" was strongly backed by the Concordia Association (滿洲國協和會), Manchukuo's state-sanctioned organization for the mobilization of the masses.[131] Drawing on transnational medical discourses that linked overheated building interiors to increased instances of illness, advocates claimed that it also benefited one's constitution to scale back on coal use, recommending that individuals instead do more exercise, wear thicker layers, or train themselves by rubbing cold water over their bodies in order to better combat the frigidity for which Manchurian winters were infamous.[132]

At the same time, that household fuel consumption even figured in efforts to address coal scarcity underscores an unintended consequence of statist controls. To begin with, the widespread use of coal in Manchurian homes was somewhat unplanned. Locals had, in the past, typically burned sorghum stalks as fuel for their heating and cooking needs. However, much of the land originally used for growing sorghum was gradually turned over to the cultivation of soybeans, "the ears of kaoliang [sorghum] giving place to bran blossoms on the vast fields all along, and in Changchun." But the shift from sorghum to soybeans meant that farmers and the wider rural population had less of their conventional fuel, and many had to buy and

---

129. Minami Manshū tetsudō kabushiki gaisha, *Katei ni okeru sekitan*, 1.
130. On "low-temperature living" in colonial Korea, see Fedman, "Wartime Forestry."
131. On the Concordia Association, see Duara, *Sovereignty and Authenticity*, 73–76.
132. Smith, "'Hibernate No More!,'" 138–41.

burn coal instead.[133] Later on, the price of coal had been kept relatively low because technocratic planners were committed to ensuring a steady supply of this main energy resource so essential to industry. As a result, coal soon became the cheapest fuel option, which prompted households across Manchuria to switch from firewood, charcoal, and sorghum stalks to coal for their heating and cooking needs. This, in turn, then increased the demand for coal, reduced the availability of this resource for industrial use, and exacerbated the sense of a coal problem.[134] It is little wonder, given these dynamics, that carbon technocracy tended toward eventual exhaustion.

On the supply side, there were several reasons why Manchurian coal mines—and Mantan's in particular—may not have been able to meet increases in demand and elevated production goals, but lack of capital was not one of them. Mantan received a princely capitalization of eighty million yen from the government and Mantetsu. To match the revised planned output, the parties forked out five times more than what they had initially budgeted. But this enterprise ended up directing most of its considerable financial resources to securing new mining zones and adding to existing facilities rather than expanding productive capacity.[135] As in the home islands, various insufficiencies also hindered the industry's growth here.

At a roundtable on Manchurian economic development held in the spring of 1938 at the Yamato Hotel in the Manchukuo capital of Shinkyō (Changchun), Mantan head Kōmoto Daisaku spoke of shortages of labor, technical expertise, and machinery holding back the business of coal mining. Labor had taken a hit after the war started, as Chinese migrants who made up the bulk of the workforce began facing difficulties crossing over to Manchuria. Yet that was an issue that had, he claimed, already been largely resolved. As for technical expertise ("the engineers who worked for the great enterprise of Manchurian and Mongolian development"), Kōmoto was also hopeful. Whatever Mantan lacked, he reasoned, could be made up by seconding experienced engineers from Fushun and other established mines and by having the company itself rapidly train its less experienced technical personnel. In his assessment, it was the lack of machinery (or, more precisely, the rubber and copper that went into making machines) that constituted the most serious problem for which there was no ready solution.[136]

---

133. Minami Manshū tetsudō kabushiki gaisha, *Report on Progress*, 116–17; "Trade with Manchuria."

134. Kubo, *Tōa no sekitan*, 103.

135. Mitchell, *Industrialization of the Western Pacific*, 86.

136. "Kensetsu tojō no Manshū keizai."

As it turned out, all three problems worsened over the course of the war. Labor shortages became acute, a challenge that Japanese imperialists sought to resolve by using "special workers," as the next section will discuss. The shortfall of technical personnel also intensified. Kubo, writing in 1941, noted a drop in the number of students enrolled in mining engineering courses across Japanese institutes of higher learning, including at institutes specializing in this field of study, such as Kyushu Imperial University—a result, he suggested, of the "danger, dirtiness, and toil" of the mining engineer's job in relation to his paltry remuneration.[137] As for machinery, whatever deficiencies that had existed before only deepened. On the one hand, machines were being worn down as the war waged on. On the other hand, machine manufacturers back in the home islands, on which colonial industries depended a good deal for supplies, started slowing down or even halting production. For instance, after 1942, Hitachi, Mitsubishi Electric, and Yasukawa Electric, the three main producers of mining machinery, reportedly fulfilled almost no orders.[138] Moreover, transport issues, especially once the Allied blockade began in 1941, meant that even if materials were available, they might not have been able to make it to Manchuria.

Fushun felt such challenges keenly. In 1940, the colliery released a report assessing its third year of the five-year plan in which it tried to account for its failure to meet output goals. One main cause it identified was the trouble it had in importing equipment. The report pointed out that because the mines had "become greatly mechanized," their functioning was contingent on a regular flow of machinery inputs, especially from Japan proper, since demand for these inputs quickly overtook the ability of the colliery's machine and tool factories to produce them. The embrace of the machine, as was consistent with carbon technocracy's "rationalizing" impulses, may have been intended to reduce reliance on labor (something that, as we saw in chapter 2, did not come to pass), but it ended up cultivating new kinds of dependencies and, correspondingly, vulnerabilities. "If there is a delay in replenishing stocks of machines or a scarcity of service parts, it directly affects the mining of coal and other preparatory work," the report noted. Furthermore, in Fushun, there were also shortages of various materials that "not only directly caused the decline in production, but also exerted a greater [negative] influence on future output." In 1940, the col-

---

137. Kubo, *Tōa no sekitan*, 58, 61. On the shortage of engineers in the wartime period and attempts to address this problem through the expansion of education, see Nishiyama, *Engineering War and Peace*, 17–23.

138. *CMWE*, 17.

liery obtained only 63 percent of the sixty-five thousand tons of ordinary rolled steel that it needed for operations and similarly insufficient amounts of other key resources such as copper for electrical wires and cement for construction. In addition, as in the home islands and in Manchuria generally, Fushun often lacked sufficient labor, and the report estimated that the colliery as a whole needed about six thousand to ten thousand more miners a month.[139]

Systemic shortages of materials and men were compounded by calamitous episodes that further crippled production. As before, disaster lurked in the subterranean spaces of extraction. Here in Fushun, as in metropolitan coal mines, the push to produce had led operators to compromise oftentimes already lax safety standards, as "more and more workers were taken off maintenance work, and men and materials were diverted to actual digging."[140] Among the other factors the 1940 report offered to explain Fushun's underperformance was a gas explosion at the Longfeng mine.[141] Two months after this report, another gas explosion occurred at the Wandawu mine, killing thirteen miners.[142] An assessment by Mantetsu's Research Section later that year brought up these two incidents as evidence of increases in the dangers of the work environment at Fushun. Because ongoing efforts at expansion mostly involved going deeper into the ground, operations were subject to greater gas-related risks with which existing facilities were unable to cope. "Based on current ventilation equipment, areas that may be regarded as dangerous are aplenty," the assessment reported. "Now is the time to try to employ big blowers to improve the circulation of air," it then recommended.[143]

The pattern here had become almost too predictable. Pernicious productivist pursuits generated new threats for which the solutions imagined were primarily technological. In the end, though, even this assessment struck a pessimistic note. The challenge of the deteriorating work environment, along with material shortages and the inexperience and insufficiency of workers, made raising output at Fushun "difficult." Were these challenges to be left unaddressed, the assessment concluded, "one can expect there to be a continued fall in production."[144] Those who oversaw the Japanese wartime economy did not share such a cautious outlook, however,

139. "Sangyō kaihatsu gonen," 401–3.
140. *CMWE*, 18.
141. "Sangyō kaihatsu gonen," 403.
142. "Bujun tankō bakuhatsu."
143. "Mantetsu chōsa bu," 404–6.
144. "Mantetsu chōsa bu," 404–6.

and expected Fushun and other overworked coal mines across the imperial warscape of intensification to further escalate their extractive efforts in the interest of bolstering Japan's energy regime.

Squeeze

In 1942, the Manchukuo state launched its second five-year plan. Like the preceding plan, this one charted out ambitious increases in output for various industrial products. Where it parted ways with the first plan was its absolute commitment to autarky. While self-sufficiency had been an aspiration before, it was now a necessity. The Allied powers had restricted and then cut off exports of oil and other strategic resources, which precipitated Japan's plunge into the Pacific War with the bombing of Pearl Harbor in December 1941. This plan was thus explicit in stating that there would be no technological, material, or capital inflows from beyond the economic bloc made up of Japan, Manchukuo, and China.[145] Coal, which had been given top priority under this arrangement, was targeted for an 80 percent jump in production by 1946. Even more so than in the first plan, Mantan was expected to bear the bulk of this leap. Three of its largest mines—Fuxin, Hegang, and Mishan—were slated for output increases of 1.7, 3, and 2.5 times, respectively.[146] The plan once again marked Fushun for comparatively modest growth since it was deemed more developed. Its target for 1942 was 7.2 million tons, just about half a million more than the year before. The idea was that its output would gradually climb to around 8 million tons by 1945.[147] With production on the decline since 1937, there was much work to be done.

Over the next few years, the Japanese empire squeezed its sites of extraction, inflicting violence on miners and their environments, as carbon technocracy's limits were exposed through a swirl of overreach and exhaustion. Japanese authorities regarded more workers as necessary for bringing production up to meet the demands of war, and in Fushun, as in other parts of the empire, they attempted to secure this supply through forced labor.[148] Referred to as "special workers" or "guided workers" (輔導工人), these laborers were mostly prisoners of war from North China or men seized specifically to work in the mines.[149] The first of these "special workers" were

145. Zhao, "Manchurian Atlas," 129.
146. Zhao, "Manchurian Atlas," 130; "Manshū sekitan zōsan."
147. MSZ, vol. 4, no. 2, 410, 412–13.
148. Kratoska, "Labor Mobilization."
149. Wang, "Manshūkoku" rōkō, 286.

forcibly brought to Fushun in 1940. In a June 1941 memo to the colliery manager, Fushun's general affairs section noted that in the half year since the previous November, 564 "special workers" had arrived at the mines "in order to ensure labor resources." According to the office, there was an urgent need to begin counterespionage work soon since many of these workers were former Communist soldiers or operatives who could still wage a "secret war" through "intelligence reports, strategies, and propaganda" from within. Adding to this concern, the report emphasized, was that their number would soon be in the "several thousands."[150] According to one calculation, around 40,000 "special workers" would labor and, in many cases, die in Fushun by the end of the war.[151]

In Manchuria, Fushun was not alone in its use of forced labor. Industrial enterprises such as the collieries of Fuxin, Beipiao, Xi'an, and Benxihu, the Shōwa Steel Works in Anshan, munitions factories in Fengtian, and the Fengman hydroelectric dam all employed "special workers." To support Manchuria's wartime industry, Japanese imperialists sent an estimated one hundred thousand "special workers" to the region each year.[152] Beyond Fushun and Manchuria, forced labor provided much of the energy for the workings of the wartime empire, extending from industrial production to infrastructural construction. The most infamous of these projects was the "Death Railway" that Japan built between Thailand and Burma, expending more than one hundred thousand civilian and prisoner-of-war lives in the process.[153] The coal industry in Japan proper also relied on the "'enforced collective immigration' of Korean contract laborers," as a postwar report phrased it, in order to make up for "the shortage of husky Japanese for work in the mines." In 1941, 17.7 percent of coal miners in Japan were Korean and, to a lesser degree, other non-Japanese; by 1945, this proportion increased to 36.4 percent.[154]

The regulatory mechanisms set up earlier to manage ordinary workers adapted well to this climate of even more merciless control. Upon arrival in Fushun, "special workers" were processed like regular ones. They had their fingerprints taken and their identity cards made. Whereas the interest before had been to monitor worker mobility, the intention here was in completely restricting it, preventing these "special workers" from running

150. "Bujun tankō sōmukyoku," 593.
151. Fu, "Zaixian de lishi," 7.
152. Wang, "Manshūkoku" rōkō, 290–92.
153. Rivett, Behind Bamboo.
154. CMWE, 12 and 30. On Korean forced labor in the Japanese empire, see Naitou, "Korean Forced Labor."

away. Other precautions included housing them together in big buildings surrounded by electric fences and watchtowers, distinguishing them from regular workers by dress—leaving them in their old military attire, putting the character "special" (特) on the back of their shirts, or drawing a red circle on their hats—and ensuring that they were under military escort each time they entered or emerged from the mines.[155]

Still, in spite of these measures, a good number of "special workers" managed to escape. According to a 1943 report, over fifteen thousand men had fled from Fushun in just the first half of the year.[156] If they were lucky, they had assistance. Wang Jizhou, a "special worker," was fortunate to have befriended Yan Shuting, an electrician. Because of the nature of Yan's job, in which he went around performing maintenance work on the colliery's electrical systems, he enjoyed a freedom of movement unlike most others. Wang, who impressed Yan as someone "with culture and foresight," convinced the latter to help him break out. Yan and a coworker disrupted a telephone line connected to the guarded factory where Wang worked. Claiming that they needed to service this line, they gained access to the factory. The two electricians then met up with Wang, put him into a uniform similar to theirs, and snuck him out of the site and, a while later, out of Fushun.[157]

There was little wonder why "special workers" risked trying to flee. Forced to work for more than ten hours a day and fed poorly (usually acorn flour), many of them wasted away.[158] Illness and disease were rampant, becoming challenges that Japanese officials tried to keep in check through means often harsh and cruel. At stake, after all, was a potential blow to the productivity they so cherished. Wang Keming, a former "special worker" in Fushun, recalled that when he came to Fushun late in the summer of 1941, he was first taken to get "disinfected" (消毒). "Getting disinfected was not the same as being given a shower," he clarified, "but it was having one's body sprayed by a kind of acidic substance with fumes that when inhaled caused one to choke and shed tears." According to Wang, Japanese authorities went to the most extreme lengths to contain any possible spread of infectious diseases: "If they found anyone who had contracted an infectious disease, they would douse them in oil and burn them alive."[159] In a postwar written confession, Kashiwaba Yuichi, who headed Fushun's

155. Fu, "Zaixian de lishi," 9.
156. Wang, "Manshūkoku" rōkō, 291.
157. "Yan Shuting huiyi."
158. Fu, "Zaixian de lishi," 8–9.
159. "Wang Keming huiyi."

police department at the time, admitted to overseeing the setup of Fushun's quarantine operations in 1942. He recounted an instance in which Japanese authorities at a quarantine station at Yongantai, having taken stock of the facility's "inadequate supplies" and the "extremely weak physical condition" of the men housed there, decided to have the infirm brought to the machinery repair plant and flung into its blast furnaces.[160] Because diarrhea was regarded as a sign of infectious disease, when Wang or his compatriots had loose stools, they would quickly cover up their waste with soil to avoid being discovered and killed. Historian of medicine Warwick Anderson has termed the fixation on fecal management in colonial public health regimes—that ultimately served to mark distinctions between colonizer and colonized—"excremental colonialism."[161] Here, Wang and other "special workers" found themselves having to try and outmaneuver excremental colonialism taken to its most odious extreme. Later, a doctor who came from Shinkyō to survey conditions at the colliery diagnosed these bouts of diarrhea as having resulted from severe malnutrition and not infectious disease.[162]

It is perhaps impossible to truly quantify how much squeezing labor out of workers both "special" and regular at Fushun and other mines in Manchuria and beyond may have contributed to productivity. If it did, such measures were sorely inadequate. Coal output continued to decline in Manchuria as it did in the rest of the Japanese empire. As the war dragged on, preexisting problems only got worse, and worker shortages persisted despite the use of forced labor. One reason was that agricultural laborers, who had occasionally been directed toward mine work in the past, were needed in the fields, as food scarcity had also arisen. Japanese authorities attempted other measures to deal with the shortfall. On the one hand, they tried limiting outflow by introducing legislation in 1944 that fixed mining labor in place, prohibiting miners from leaving to take up other forms of work that was perhaps more profitable and almost certainly less perilous. Enforcement was an issue, however, especially when there were powerful actors like the military who sought to requisition mining labor to build airfields and other infrastructure. On the other hand, Japanese authorities tried encouraging inflow by rallying and recruiting labor for the mines through nationalistic campaigns that bore slogans such as "coal mining is also on the front line." Through one of these campaigns, several thousand

160. "Kashiwaba Yuichi."
161. Anderson, *Colonial Pathologies*, 104–29.
162. "Wang Keming huiyi," 22.

students signed on as part-time workers. Behind the problem of worker shortages, though, was the issue of efficiency, tied as this was to experience and skill. The work force steadily grew over these years. In Japan proper, miners numbered 338,000 in 1941 and 416,000 in 1944. But efficiency continued to drop as more and more inexperienced and unskilled workers were brought on board. At only 164 tons per worker per year in 1941, efficiency would further fall to 119 tons by 1944.[163]

As for equipment and materials, the other area in which shortages seemed relentless, the situation, too, was increasingly dire. Aside from machinery deteriorating through wear and tear and the stoppage of production by the major mining equipment makers, one problem on this front was that machine repair shops within many mines themselves started being repurposed for armament manufacturing, removing the pivotal maintenance function when it was most needed. Materials vital to operations continued to be insufficient as well. "In general," a postwar assessment observed, "allocations of materials were far below the industry's requirements; moreover, actual receipts, especially of steel products and cement, rarely measured up to the local allocation figures." Furthermore, military officials often "made in-roads on even these meager supplies," prompting those who ran the mines to turn to the black market for many essential inputs.[164]

The conundrum of wartime coal supply was that it involved not only extraction (complicated as that already was) but also transportation—a point that became painfully clear with the onset of the Allied blockade. After Pearl Harbor, Japan's shipping lanes, through which coal and other resources moved from colonies and recently occupied areas back to the home islands, came under attack, particularly from American submarines. At the beginning of the Pacific War, Japan had about 6.5 million tons of registered merchant shipping; at the end of the war, capacity had plunged to 1.5 million tons, countless vessels having sunk to the bottom of the ocean.[165] By 1943, Japanese plans for supplying the industrial demands of the metropole began focusing less on extracting resources from the far ends of the expanding empire (such as Southeast Asia) and more on mobilizing resources from sites nearby. In order to further limit exposure to Allied assault on the open waters, Japan started to rely more on land transport. Manchurian coal bound for Japan had previously gone through the port of

163. CMWE, 17.
164. CMWE, 17.
165. Milward, War, Economy, and Society, 317.

Dairen, where it was loaded onto ships that crossed the Yellow Sea. Now, it went first by rail to Korean ports and then by ship to the home islands via the more enclosed Sea of Japan. By 1944, as much as three-quarters of the Manchurian coal sent to Japan reached it by means of this "Korean relay." Still, Japanese imports of coal from its colonies steadily declined. Between 1939 and 1944, coal from Manchuria fell from 848,000 to 561,000 tons, while coal from North China, which came to make up almost half of imports, fell from 3.04 million to 1.52 million tons.[166] Moreover, the blockade also affected movement of coal among the Japanese islands. Shipments from coal mines in Hokkaido and Kyushu to the industrial heartland in the main island of Honshu dropped from 20.7 million tons in 1941 to 11.1 million tons in 1944.[167] By July 1945, the available coal was almost half of its wartime peak.[168]

As the coal supply shrank in the last few years of the war, the patterns of use shifted to reveal the embattled empire's priorities. Most major consumers of coal, including producers of iron and steel, chemicals, and electric power, had to scale back their usage between 1941 and 1944. In that period, only the railways, the machine manufacturing industry, and the liquid fuel industry witnessed consumption increases.[169] Coal continued to go toward the production of synthetic petroleum, as precious liquid fuels became even more dear with the war's progress. However, the effectiveness of this enterprise was suspect. Although Japanese engineers boasted that they had perfected the hydrogenation process and bragged about the quality of liquefied coal they were able to produce at Fushun and elsewhere, Allied petroleum chemists questioned these claims, surmising that Japanese improvements involved applying lower temperatures to the process, which produced a "higher quality of oil and a greater yield of oil per unit of bituminous coal" but at a slower rate per unit. This was, these scientists concluded, "just another example of Japanese industrial boasts without foundation."[170] Japan's petroleum shortage persisted despite efforts to supplement the paltry amounts extracted with synthetic fuel and shale oil.[171] What is more, even as the quantity of coal dwindled, so did its qual-

166. *CMWE*, 35, 46.

167. *CMWE*, 21.

168. *CMWE*, 3.

169. *CMWE*, 24.

170. Department of Justice, War Division, Economic Warfare Section, "Report on Fushun, Part 3," 11.

171. The limited success of Japan's wartime coal liquefaction enterprise seems to have been not so much a result of the lack of organizational coordination that Walter Grunden identifies as a major factor that hindered developments in "Big Science" projects from nuclear bombs

ity. Into the Pacific War, the grade of coal determined its price, as set by the Coal Control Association (石炭統制會), which had been formed by the Japanese government in November 1941 to oversee the industry's expansion. However, in October 1943, the association decided to standardize the price regardless of grade, which quickly led to a decline in quality as producers increasingly went for the low-hanging fruit of easily mined, poorer-quality coal.[172]

Fushun struggled in the crushing grip of the Japanese wartime state. It consistently failed to meet the targets outlined under the second five-year plan. In 1943, the third year of the plan, for instance, Fushun's goal was to produce 7.45 million tons of coal. Actual output was 5.13 million tons, which not only fell short of the mark but was even less than the 6.36 million tons from the year before. A report by the colliery explaining this poor showing expressed "distress" over the "unfavorable conditions," which included the usual lack of materials, equipment, and labor. What really dealt a blow to production, though, was an outbreak of cholera. The disease brought operations at Longfeng to a halt, and when suspected cases also started appearing at Ōyama and Laohutai, workers at those mines were placed under quarantine.[173] Mantetsu may have tried to prevent diseases at this extractive site by monstrous methods like spraying workers with harsh chemical disinfectants or murdering those among them who were suspected of being sick, but its long-standing practice of packing workers tightly in close quarters coupled with the likely deterioration of sanitary systems amid the warscape of intensification gave rise to an environment in which tiny comma bacilli, whose only interest lay in reproduction, could thrive and, in so doing, thwart carbon technocracy's oversized plans.[174]

At the open pit, colliers faced problems wrought by the continuing excesses of the imperial wartime regime. There was, the same report noted, a tendency from early on for authorities to treat the open pit like "a natural storehouse of coal"—"any time something cropped up, it could be forced

---

to aeronautical weapons. Because of the long-standing cooperation between two of the main players in fuel research, Mantetsu and the navy, coal liquefaction appeared to have had a firmer foundation for the necessary mobilization of expertise and resources. Rather, the primary problem may have been, as Anthony Stranges contends, the Japanese impatience with scaling up from laboratory to plant, no doubt spurred by the exigencies of war. See Grunden, *Secret Weapons*; and Stranges, "Synthetic Fuel Production."

172. *CMWE* 15–16.

173. "1943-nen gensan," 419.

174. For a history of cholera, see Hamlin, *Cholera*. For another exploration of how patterns of life and labor within certain environments transformed by empire can produce the diseased bodies of colonized subjects, see Derr, *The Lived Nile*, 99–126.

to do its part." Because of this tendency, when the demand for coal suddenly increased after the Mukden Incident, Fushun was "forced to complete excessive coal production tasks." With the open pit, this often meant exceeding the stripping ratio between overburden removed and coal mined that was necessary for sustainable production. "Because of factors such as difficulties in importing machinery and repair parts, the lack of human resources, and the lengthening of time taken to transport [debris] by rail to the dump site, the overburden stripping work, when compared to what was planned, was frequently put off," the report explained. The situation was, it concluded, not only "irrational" but would also result in "the decline of coal production capacity."[175] Fushun's output continued to plummet in the following years, dropping to 2.5 million tons in 1945, almost a quarter of its wartime peak. When the war concluded, the Coal Capital was but a shadow of its former self, exhausted by the demands of wartime mobilization and the limits of carbon technocracy.

Conclusion

Japan surrendered on August 15, 1945, bringing an end to its war and its empire. Just under a week earlier, the United States had dropped an atomic bomb on the port city of Nagasaki. As the bomb detonated, its plutonium core was compressed, setting off a fission chain reaction that turned matter about the size of a baby's fist into the energy equivalent of about four thousand tons of coal, released in an instant.[176] Almost everything within a half mile from the explosion was obliterated. Death and destruction stretched out for many miles beyond. The hills surrounding Nagasaki were "scorched by the flash radiation," giving them, according to one report, "the appearance of premature autumn."[177] A few days before, the Americans dropped a first atomic bomb on the city of Hiroshima some 260 miles away. Its core was different (uranium, not plutonium) but its effects were similarly devastating. Nicknamed Fat Man and Little Boy, respectively, the bombs the Americans released over Nagasaki and Hiroshima were anthropomorphized expressions of human inhumanity. An estimated quarter of a million people died in these two bombings and their long aftermath.[178]

175. "1943-nen gensan," 419–22.
176. My estimate here is calculated from the energy released by Fat Man as being eighty-eight terajoules and the energy contained within a ton of coal being on average (with the caveat that values can differ greatly based on the type of coal) twenty-two gigajoules.
177. Manhattan Engineer District, "Atomic Bombings of Hiroshima and Nagasaki."
178. Barnaby, "Effects of the Atomic Bombings," 2.

After the war, the United States Strategic Bombing Survey conducted a study on coals and metals in Japan's war economy. Formed in 1944 to investigate the impact of the American aerial bombardment of Germany so as to provide findings for assessing air power and its future development, the survey extended its scope of research to Japan and the Pacific as soon as the war wrapped up. The researchers concluded that coal had a more important role in the Japanese war economy than earlier recognized. However, because sites of coal production were dispersed, attacking those sites directly would not have been practical. That said, the transportation lines through which Japan's coal traveled had been extremely vulnerable. Therefore, the researchers reasoned, while the blockade was effective, a concentrated bombing campaign on sea and rail routes would have further hit coal supply and paralyzed Japanese industrial activity.[179]

In their survey of efforts at mining coal across the Japanese empire, the American researchers detailed the constellation of difficulties that compounded one another and undermined the entire extractive enterprise. At the same time, though, they seemed all too ready to relate to the underlying logic that drove that venture.

> Coal resources in Karafuto, Korea, Manchukuo, and North China were enormous, and the prewar production of each region was capable of substantial expansion. Had they been given freedom from hostile interference and sabotage, labor, the necessary equipment and materials, and sufficient transport from mine to port and thence to consumers in Japan proper, the Japanese would have had an almost unlimited supply of coal for the taking. It is small wonder that they succumbed to that lure as an attractive alternative to the harsh and unpopular measures which would have been required to squeeze the maximum of production from their domestic resources, particularly in view of their inescapable dependence on imported coking coal.[180]

That these American researchers, too, would appreciate the "lure" of the "almost unlimited supply of coal for the taking" points to the wide reach of carbon technocracy for which idealizing endless energy was central. This is perhaps not surprising given how entrenched the prodigious consumption of coal was in developmentalist notions of progress across industrial societies. What is slightly startling is the researchers' suggestion that this supposedly unlimited supply would have been for the taking if this broad

179. *CMWE*, 9–10.
180. *CMWE*, 13.

range of conditions, including freedom from wartime disruptions, were met. Such a hypothetical disregards the fact that "hostile interference and sabotage" were responses to Japan's extractive endeavors—endeavors that required increasing amounts of energy to support an empire built in part on the very premise of securing such power. In many ways, this warscape of intensification that took shape under Japanese imperial leadership exposed the vulnerabilities of carbon technocracy, pushing the energy regime to consume itself—in terms of both carbon and human resources— more quickly than it could produce. Furthermore, the colonial coal from Manchuria and elsewhere could never truly be free "for the taking," as the American researchers suggested. It was, after all, mined through the ruthless exploitation of Chinese, Korean, and other non-Japanese labor, upon whom measures deemed too "harsh and unpopular" for Japanese miners were readily inflicted.

One morning in the spring of 1948, Kubo Tōru smoked his last cigarette. He had been arrested by the Chinese Nationalists soon after they moved into Manchuria. Along with ten others, he was made to stand trial for the Pingdingshan massacre. At the time of the atrocity, Kubo had been serving as Fushun's deputy manager but was, as the accusations went, the one in charge of its operations because the manager, Godō Takuo, had been occupied with concurrent appointments, including heading the iron and steel works in Anshan. Following months of court proceedings, seven of the accused were deemed culpable in the affair and sentenced to death. Kubo was among the condemned. Before he was taken to the execution grounds, Kubo requested a cigarette, seemingly resigned to his fate. As he finally stood upon those grounds, he reportedly pointed to the back of his head and said, "Over here. Please do a good job." Two shots and he was dead.[181]

Kubo's execution, like the other trials and sentences of Japanese war crime suspects, was meant to provide closure to this chapter of Japanese imperial rule—a violent end to an era of carbon technocracy distinguished by its own violence.[182] Chinese Nationalists sought justice for the most heinous of atrocities, such as the Pingdingshan massacre. In this case, though, whether someone like Kubo should have taken the fall was and remains an open question (no members of the Kwantung Army were persecuted for this atrocity). Nevertheless, if their goal was to mark the end of Japan's

181. Kume, "Heichōzan jiken," 81.
182. On postwar war crimes trials of Japanese in China, see Kushner, *Men to Devils*.

imperial extraction in Fushun, the Nationalists could not have found someone more representative than Kubo. He had spent most of his adult life in this mining town, from the time he first set foot in it as a student and was so moved by "the magnitude and inexhaustibility of its coal seams below" and "the extensiveness and newness of the facilities above" that he would return and stay.[183] The intensive extraction undertaken by Japanese imperialists did have its own afterlife, though. The decade that followed witnessed the Chinese Nationalists attempt to revive it and the Chinese Communists succeed in its resurrection.

183. Kubo, *Bujun tankō*, 1.

FIVE

Nationalist Reconstruction

Xie Shuying 謝樹英 (1900–1988) was despondent. It was October 1, 1947, exactly a year since the Chinese Nationalist government entrusted its National Resources Commission with running the Fushun coal mines, and the situation on the ground looked far from promising. Reviving production as one war ended and another began was an uphill task. Xie, a fuel specialist with a mining engineering degree from the Technische Hochschule of Berlin, had fallen into his role as colliery manager rather abruptly. The fact that his predecessor had quit a month into the position may have signaled the challenges that lay ahead. In his message to Fushun's employees on that first anniversary, the main theme was "striving in the midst of adversity" (從艱苦中奮鬥). Prior practices of intensive coal extraction—first by the Japanese wartime state and then by Soviet troops who occupied Fushun in the interregnum between the Manchukuo regime's collapse and the Nationalist takeover—had downplayed or completely neglected necessary maintenance work, undermining present and future exploitation of the ever-essential energy resource. Material shortages, runaway inflation, and mounting tensions with the Chinese Communists in the region all served to exacerbate matters. Still, the Fushun colliery, as "a center of coal and electricity," was too important to not make an effort, and Xie tried to be upbeat. He concluded his address—which was otherwise laden with lamentations—on a rousing note, encouraging Fushun's workers "to serve the country with blood, sweat, and zeal."[1]

The National Resources Commission for whom Xie worked was the highest institutionalized form of carbon technocracy under the Nationalist state. Founded in 1932 as the National Defense Planning Commis-

---

1. Xie, "Cong jianku zhong fendou."

sion (國防設計委員會), this government agency was staffed mainly with scientists and engineers who conducted surveys and drafted plans for developing human and material resources in preparation for an anticipated war with Japan. In 1935, when the commission was renamed, it was further empowered to directly undertake the actualization of those plans. In this, it joined other government agencies, such as the National Reconstruction Commission (建設委員會) and the National Economic Council (全國經濟委員會), which, too, had been pursuing industrial plans of their own (sometimes in opposition to one another).[2] After the Second Sino-Japanese War broke out in 1937, the National Resources Commission subsumed the functions of these other agencies to become the primary architect of statist industrialization. When that war ended in 1945, it was tasked with taking over Japanese assets in China, and it was in this capacity that the commission came to operate the Fushun colliery. Given how central fossil fuels were to driving the military-industrial assemblage the National Resources Commission aspired to create, expanding the coal-mining industry became one of its enduring interests. Geologist and longtime commission chairman Weng Wenhao 翁文灝 (1889–1971), in reflecting on coal mining and China's industrial future in 1932, argued for more factories to be developed in coal-rich areas. "The things that industry needs most," he reasoned, "are raw materials and motive power [原料與動力]."[3] Sites of extraction like Fushun promised abundant supplies of such things.

The Nationalist regime ended up holding on to Fushun for but two and a half years. As time passed, the challenges Xie highlighted only became more pronounced. Still, the National Resources Commission's experience in Fushun was more than just a tale of insurmountable adversity. It was part of a longer narrative in which the Nationalist state tried to extend control over the Chinese coal-mining industry in its pursuit of "reconstruction." Although the Chinese word *jianshe* (建設) translates more directly to "construction," it was always rendered into English as "reconstruction" within these contexts (as seen, for example, in the official English name of the National Reconstruction Commission). This may have been a product of historical happenstance. Yet it was nevertheless consistent with the regime's ambitions to not only build up the economy but, in doing so, remake China into a strong and modern nation worthy of the past it so

2. On the origins of the National Resources Commission, see Kirby, *Germany and Republican China*, 91–95. On the competing interests among the Nationalist state's economic factions, see Coble, *Shanghai Capitalists*, 208–60.

3. Weng, "Zhongguo meikuangye de eyun," 6.

glorified.[4] Planning was deemed key to this project. Minister of Finance T. V. Soong 宋子文 (1894–1971), writing in 1931, envisioned "the creation of a really effective planning organization which would guide the productive forces of the country . . . and rigidly map out the essential ends which for a given course of years each of the different components is obliged to pursue."[5]

Sun Yat-sen 孫中山 (1866–1925), the "father of the Chinese nation," deserves much credit for the Nationalists' penchant for planning. In 1920, he put forward his *Industrial Plan* (實業計畫), a wide-ranging proposal for China's industrial transformation that rested heavily upon the state's mobilization of science and technology.[6] It would be the clearest articulation of his developmentalist vision and the blueprint on which his successors drafted their economic policies. Coal occupied an important place in this plan, both as an industry to be developed and a resource essential to developing other industries. It should be, Sun contended, mined extensively and priced inexpensively so as to ensure a ready supply for industrial applications. At the same time, Sun, whose capacious political thought included socialistic undercurrents in its attention to "people's livelihood" (民生), suggested that this bountiful and cheap coal might also contribute to meeting the public's quotidian needs and that the miners involved in its extraction should receive "high wages."[7] His successors were generally less concerned about these auxiliary ideals. Their involvement in the coal sector would be informed by the logics of carbon technocracy in which the interests of ordinary individuals were often eclipsed by the demands of industry and the state.

This chapter traces how carbon technocracy emerged out of and was shaped by crisis in Nationalist China from the 1920s to the 1940s. It begins with a case of "coal famine" (煤荒) in the early 1930s, which brought into relief Chinese industries' dependence on Fushun and other foreign coal and the relative underdevelopment of domestic coal resources. In response to these challenges, the Nationalist regime took up measures ranging from regulating competition to running its own mines, the beginnings

4. For a reflection on this issue of translating *jianshe* as "reconstruction" in the context of rural reconstruction, see Merkel-Hess, *The Rural Modern*, 15.

5. Cited in Young, *China's Nation Building Effort*, 293.

6. Sun, *Shiye jihua*. The *Industrial Plan* was first published in English as Sun, *International Development of China*. I use that earlier version here.

7. Sun, *International Development of China*, 156. On ideas about "people's livelihood," see Zanasi, *Saving the Nation*, 33–38. On how the Nationalist state under Chiang Kai-shek co-opted and reinterpreted Sun's understanding of "people's livelihood" in its discourses promoting frugality, see Zanasi, *Economic Thought in Modern China*, 183–89.

of what would be a growing presence of the technocratic state within this key industry. The chapter then follows the activities of National Resources Commission engineers who coordinated the development of coal mines in the interior during the war with Japan and the takeover and management of the Fushun colliery and the industrial Northeast thereafter. This account demonstrates how the conditions of war both helped consolidate the ideal of carbon technocracy under the institution of the National Resources Commission and, conversely, compromised the actual workings of that energy regime.

The Chinese Coal Industry Undermined

The coal shortage that hit Shanghai in 1931 coincided with the first hint of the coming winter. As the crisp late October winds blew in from the north, there was talk on city streets of dwindling coal supplies and the onset of a "coal famine."[8] The center of China's industrial activity in those prewar years, Shanghai was a voracious consumer of coal, most of which came by ship to this treaty port. "As a traveler approaches Shanghai, steaming up the Whangpoo [Huangpu] River," one contemporary observer wrote, "he will note nearly a dozen coal wharves back of which are a score or more of coal storage yards piled high with coal."[9] The city did, in fact, hold stockpiles of coal that fall. In one estimate, these totaled at least a month's worth. The problem was that half of that cache was from Japanese-owned mines, Fushun included.[10]

Difficulties in weaning itself off Japanese coal revealed the extent to which China's industry relied on foreign fuel. Chinese popular sentiment toward Japan was at a record low following the Japanese invasion of Manchuria that began a month earlier with the Mukden Incident. Across China, capitalists and consumers alike called for a boycott of Japanese goods, as they had on multiple past occasions in response to Japan's infringements on China's national sovereignty.[11] The boycott that ensued seemed

8. Zhong, "Da ke zhuyi zhi meihuang"; "Shanghai Faces Coal Shortage."

9. Bacon, "Coal Supplies of Shanghai," 196.

10. "Shanghai Faces Coal Shortages." The figure given for total available coal was 240,000 tons. With coal consumption in Shanghai averaging 8,000 tons a day, this volume of coal was expected to last the city exactly one month.

11. The post–Mukden Incident boycott was the ninth against Japan since the beginning of the twentieth century. For more on its broader impact on the Japanese business community in China, see Wilson, *Manchurian Crisis*, 173–75. Japan was, however, not the only country to be targeted by Chinese consumer nationalism. For more on antiforeign boycotts in Republican China, see Gerth, *China Made*.

5.1. Coal yard by the British Cigarette Company buildings along the Huangpu River, Shanghai. This photograph dates to 1931, the same year as the post–Mukden Incident "coal famine," and is attributed to Jack Ephgrave. (Image courtesy of Adrienne Livesey, Elaine Ryder, Irene Brien, and Historical Photographs of China, University of Bristol, https://www.hpcbristol.net.)

the most successful by far, its efficacy enhanced by the global depression: Japanese imports fell by two-thirds in 1931.[12] Among the Japanese goods blocked from Chinese markets was coal, including (and especially) coal from Fushun. In Shanghai, the local coal trade association firmly resisted fresh imports. It also required all coal dealers to register existing stocks of Japanese coal, which were then seized and stored away—neither for sale or use.[13] This ban on Japanese coal—which typically accounted for no less than one-third of Shanghai's coal cache—depended on a corresponding increase in the supply of Chinese coal. When domestic sources proved unable to meet the shortfall, a coal famine seemed all but imminent.[14]

The Nationalist regime had, in fact, anticipated this problem. Soon after hostilities erupted in Manchuria, H. H. Kung 孔祥熙 (1881–1967)—the banker, businessman, and brother-in-law to Chiang Kai-shek who then

12. Zhang, Zhao, and Luo, *Jingji yu zhengzhi*, 227.

13. "Shanghai shi meiye." Violations of the ban by Chinese merchants were punishable by a fine of a few hundred yuan or a suspension of business operations for a number of days. See, for example, "Guomei jiuji jian you banfa."

14. Bacon, "Coal Supplies of Shanghai," 209; "Shi shanghui jiuji meihuang."

served as minister of industry—submitted a proposal to the Executive Yuan, the government's primary executive organ. Centered on the prevention of such a coal famine, Kung's proposal began with the premise that troubles with local coal transport had rendered major Chinese cities reliant on Japanese coal. "If Japanese coal were to be cut off," he continued, "then factories across various regions would be forced to shut down, disrupting public order—the consequences are almost unthinkable." His proposal consisted of four measures to guard against that outcome by providing support for the domestic coal industry: (1) allocating rolling stock (railway vehicles) for coal transport; (2) extending loans for mine capitalization through national banks; (3) deploying local government militia to protect mining sites; and (4) ordering increases in the production of coal of all types.[15] The proposal's provisions pointed to some of the main problems holding the industry back—capital, security, and transport. It also marked the Nationalist state's increasingly interventionist stance toward coal mining.

It was, after all, not for lack of domestic coal reserves that China relied so heavily on Japanese coal as to have Shanghai face the shortage of 1931 when those imports were boycotted. By contemporary counts, China ranked third globally in coal deposits.[16] China's wealth in coal had been repeatedly delineated through geological surveys since the nineteenth century. Most notable was the work of German geologist Ferdinand von Richthofen, who had famously contended in 1870 that "the world, at the present rate of consumption of coal, could be supplied for thousands of years from Shansi [Shanxi] alone"—a claim that Chinese elementary and middle school textbooks, for instance, would cite for decades.[17] While the geologists who came after him deemed Richthofen's estimates "exaggerated," there was little disagreement that China's coal resources were nevertheless considerable to the point that one might imagine them inexhaustible.[18] The China Geological Survey (中國地質調查所), which Weng Wenhao helped found in 1921, provided what was arguably the most reliable calculation of Chinese coal deposits in the prewar years at over two hundred billion tons.[19]

15. "Shiye bu guanyu yufang meihuang."

16. Hou, *Zhongguo kuangye jiyao, di si ci*, 96.

17. On Ferdinand von Richthofen's geological surveys in China, see Shen, *Unearthing the Nation*, 29–35; and Wu, *Empires of Coal*, 33–65. Quotation from Richthofen, *Baron Richthofen's Letters*, 29, as cited in Shen, 31. On Richthofen's claim being reproduced in Chinese textbooks, see Weng, "Zhongguo dixia fuyuan," 7.

18. "Early Estimates."

19. Weng, "Zhongguo dixia fuyuan," 9.

The problem, then, lay in accessing this abundance. More specifically, it concerned the development of domestic mines that were Chinese owned. Coal mining in China, which had been steadily expanding since the end of the nineteenth century, remained, as in those early years, an industry dominated by foreign and Sino-foreign firms. Including Fushun and Kailuan, which together made up around half of the domestic total from large mines, these firms accounted on average for three-quarters of the coal that China's large mines churned out between 1896 and 1936.[20] A main challenge Chinese mines faced was capitalization. Modern coal mining required large amounts of capital, much of which went toward the mechanization that enabled extraction of greater depth and wider scale.[21] Mines that were wholly or partially foreign owned tended to be better capitalized than all but a handful of Chinese mines.[22]

Chinese mines had actually been faring quite well in the decade or so before the establishment of Chiang Kai-shek's Nationalist government in Nanjing in 1928. This was a product of two developments that were favorable to Chinese industry in general. The first, as in the case of Japan, was the reduction of European imports and economic activities during World War I, which created opportunities for Chinese producers to enter hitherto foreign-dominated local markets. The second was the continued extension of China's railway system as part of what seemed to be a mutually reinforcing loop: more coal was needed to smelt more iron and steel to make more trains and tracks, which resulted in a better internal movement of coal and other commodities, not to mention an increased demand for locomotive fuel.[23] As a result of these changes, many Chinese coal-mining companies made sizable profits, and a good number of them expanded their operations, raising significant capital by reinvesting their gains or securing loans

20. Wright, *Coal Mining in China's Economy and Society*, 118.

21. It bears mentioning that it is not a general principle that mechanization would necessarily lead to the rapid growth of an industry. See Mokyr, "Editor's Introduction," 14–15.

22. The amount of capital required for the excavation of a modern coal mine varied based on a host of factors, including the size and shape of the deposits, the depth to which operations were extended, and the targeted daily output. For example, the Ministry of Industry had planned in 1931 to develop an anthracite coal mine in Anhui, and in order to have a working that yielded two thousand tons a day, the initial expense was calculated at 3,273,000 yuan. See "Text of Kung Memorandum." Small coal mines in rural China were also capital-intensive operations that reflected changes in the larger coal industry in the 1920s and 1930s. For a compelling account through the example of a local Confucian elite who got involved in the business of such small coal mines, see Harrison, *Man Awakened from Dreams*, 113–35.

23. This may be taken to be another instance of the "synergistic feedback loops" that Christopher Jones has identified in the dynamics between energy and transport that set in motion patterns of continued growth. See Jones, *Routes of Power*, 75.

from banks.[24] This represented the height of pre–World War II Chinese coal production.

The chaos surrounding the Nationalists' seizure of power brought those advances to a halt. On June 6, 1926, Chiang Kai-shek launched the Northern Expedition, a campaign in which he set out to defeat the Beiyang government and the various warlord cliques in a bid to reunify the country under Nationalist rule. The ensuing fighting disrupted railway lines, curtailing or completely cutting off the flow of coal to urban markets.[25] For instance, the Liuhegou mine in Henan Province, which supplied the central city of Hankou, was hard hit when the Ping–Han railway transporting most of its coal stopped running for much of 1926 and 1927 because the crucial part of the line that crossed the Yellow River had been damaged in battle. This enterprise, which had been doing well enough to also boast an iron foundry and a cement plant, could not sell its coal and incurred almost three million yuan in debts before the end of the decade.[26] In addition, belligerents on all sides further impaired coal circulation by frequently commandeering rolling stock for military transport. Chinese coal mines buckled under the weight of war, their conditions worsened by additional taxes and other types of "squeeze" exacted by the revolving cast of competing authorities. Many were forced to close or suspend work.[27]

The war highlighted again how essential transportation was to the coal industry. As before, commentators decried the difficulties involved in getting coal from mine to market. The geographical divide between the large coal mines in the North and the centers of industrial activity in the South became more pronounced as the demand for coal in those expanding industrial centers grew.[28] Dominated by the railway, the movement of coal in early twentieth-century China took several other forms, often involving transfers from one form to another along the way. For shorter distances, it was not unusual to rely on human or animal power. Well into the 1930s,

24. Wright, *Coal Mining in China's Economy and Society*, 22–23, 97–102. For one account of the industrial boom in China during and after World War I, see Bergère, *Golden Age of the Chinese Bourgeoisie*.

25. Zhang, Zhao, and Luo, *Jingji yu zhengzhi*, 221–22.

26. Wright, *Coal Mining in China's Economy and Society*, 151; "How the Chinese Coal-Mining Industry"; Gao, "Ping–Han yanxian," 7–9. To put a figure on it, while the Ping–Han railway transported 146,000 tons of Liuhegou coal in 1926, it transported only 40,000 tons in 1927.

27. "How the Chinese Coal-Mining Industry"; "Chinese Civil War Deals Blow." As one example of "squeeze," the Liuhegou mine was forced to supply ten thousand tons of coal for military use every month in 1928. See Shanghai shangye chuxu yinhang diaocha bu, *Mei yu meiye*, 187.

28. Weng, "Zhongguo meikuangye de eyun," 6.

caravans of camels hauled coal from the Mentougou mine to Beiping during the bitter chill of the winter months.[29] Where there was access to rivers, canals, or the coast, water transport was another option. In this regard, shipments by sea tended to be more reliable than via inland waterways. For one thing, fluctuations in river or canal depth due to rainfall patterns and the redirection of water for irrigation purposes made riverine transport unpredictable. Another factor was the susceptibility of inland water routes to theft and banditry and the added cost of security to guard against such incursions.[30] As much as two-thirds to three-quarters of the coal produced in China each year from the mid-1910s to the mid-1920s was, however, carried by the railway.[31] This heavy reliance on rail transport meant that line interruptions from ripped up tracks to requisitioned rolling stock were particularly debilitating for the coal industry and those who depended on it.

The closure and suspension of mines and the corresponding decline in domestic coal production brought about by the disorder of the Northern Expedition opened up fresh opportunities for foreign coal on the Chinese market. Aside from Japanese and Indo-Chinese imports, the Sino-British Kailuan and Japanese Fushun collieries also stood to gain. Kailuan and Fushun had rail access to nearby ports—Qinhuangdao and Dalian, respectively—from which their coal could be shipped south along the coast to Shanghai and the rest of the lower Yangtze Delta region, thus avoiding the transport difficulties that accompanied the internecine war.[32]

The Fushun colliery had, in fact, been looking into the possibility of expanding its sales in South China. In a 1925 report, its general affairs bureau recognized that while "as a coal market, the industrial district along the Yangtze River had a great future," the "shocking strength" of Kailuan coal in Shanghai and beyond made Fushun coal's chances of seizing a greater share of that market seem dim.[33] The Northern Expedition bettered its odds by reducing Chinese competition. For a period, Fushun also benefited from Kailuan's transport troubles. Kailuan had gotten into a dispute with the Bei–Ning railway over its refusal to pay demurrage fees for loaned rolling stock. In response, the latter would "withdraw all the locomotives operated by the [Kailuan] administration, on the pretext that they needed

29. "Beiping luotuo." Each camel could carry about six hundred catties of coal (one catty is about a pound and a third).

30. Ting, *Coal Industry in China*, 72–74.

31. Wright, *Coal Mining in China's Economy and Society*, 84–85.

32. Wright, 53–54.

33. Abe, *Bujun tan no hanro*, 17–20.

repairing," dealing a blow to the Sino-British colliery's sales.[34] For example, the Shanghai power company—the city's largest coal consumer—was originally under contract to buy from Kailuan, but the colliery's inability to deliver the goods prompted the company to turn to Fushun instead.[35] Fushun's exports to the Yangtze Delta region grew almost tenfold, from 103,000 tons in 1921 to 993,000 tons in 1930.[36] By the time of the Mukden Incident, Fushun coal became so vital to the regional market that supply reductions could give rise, as they did, to fears of a coal famine.

The Nationalist government quickly organized a response to the crisis. On November 17, 1931, not long after a coal shortage was pronounced, H. H. Kung convened a meeting with delegates from the Ministries of Railways, Communications, Military Affairs, and Finance, the banking sector, and the coal industry to discuss measures for mitigating the crisis.[37] This meeting resulted in the formation of the National Coal Relief Commission (國煤救濟委員會), made up of representatives from the parties in attendance and tasked with following through on resolutions decided in their discussions.[38] Although the coal industry, like other industries, boasted a number of trade groups and associations, this commission was different in that it was established upon government initiative and included officials in addition to industrialists.[39]

Of the propositions put forward at the coal famine relief meeting, the most pressing concerned securing more rolling stock. Even after the chaos surrounding the Northern Expedition died down, Chinese coal mines still struggled with transportation issues. At an earlier convention on commercial rail transport hosted by the Ministry of Railways, coal industry leaders lamented that the railway system was so "unpredictable in its operations" that there were "mines without access to a single train for stretches of time," resulting in "coal being stockpiled as high as a mountain on the supply side" while "those on the demand end encounter shortages." This led, they contended, to losses both for coal producers—with "miners worrying about their livelihoods" and "mine owners watching their capital

34. "Coal Shortage Threat"; "Local Industrialists Are Alarmed."
35. Leiyu, "Minguo 19 niandu Dongbei meikuangye," 12.
36. *Bujun tankō tōkei nenpō.*
37. "Shi bu zhaoji jiuji meihuang"; "Jiuji meihuang."
38. "Guomei jiuji weiyuanhui cheng."
39. All earlier associations were composed almost entirely of industry men. An example of this was the National Coal Development Commission (國煤發展委員會), established in 1928. See "Meitan ye shangxie fenhui."

waste away and their businesses fail"—and for coal consumers, who were forced to rely on foreign coal, which, as the accusation went, was prone to "opportunistic price hikes."[40]

After the coal crisis meeting, the Ministry of Railways made arrangements to set aside more railroad cars for coal transport on the Ping–Han and Jin–Pu railways—the two main lines running from the North to the South—with the express aim of ensuring a steady flow of coal to Shanghai and other surrounding areas threatened by shortage.[41] In the name of coal famine relief, the ministry also established a separate commission to oversee coal transport. The rationale, as one official explained, was that because the freight cost of coal was "minuscule" relative to its weight, railways would, if going by the "logic of business," find it more profitable to move other goods instead. The purpose of the new commission, then, was to directly control the allocation of vehicles on each and every route to make sure that they were carrying enough coal to alleviate the problem of paucity.[42] For a while, these statist efforts at resolving the coal famine by bolstering transport appeared to be working. At the very least, large Chinese mines were able to send considerably more coal to the Yangtze Delta region. For instance, the Zhongxing mine in Shandong Province, which used the Jin–Pu railway to convey its coal, saw its exports to Shanghai jump more than threefold from 29,700 to 98,500 tons between 1931 and 1932.[43]

What really ended the crisis, though, was military conflict. The January 28 Incident, as it came to be called, was, to a large degree, the culmination of Sino-Japanese tensions that had been steadily escalating since the Mukden Incident. On the premise that Chinese authorities had failed to provide timely and satisfactory restitution for the recent bout of popular anti-Japanese activities—from boycotts and confiscations of Japanese goods to attacks on Japanese persons and property—Japanese naval forces stationed in Shanghai launched an "intervention" in early 1932 that turned into over a month of fierce fighting.[44] This short war took a considerable toll on Shanghai's industrial capacity: 896 new factories valued at 68 million yuan were leveled and other production facilities suffered damages totaling as much as a billion yuan. The demand for coal fell as a result.[45]

40. Tiedao bu quanguo tielu shangyun huiyi banshichu, *Quanguo tielu shangyun huiyi*, 153.
41. "1,000 Cars Will Transport Coal."
42. "Tiedao bu jiuji meihuang."
43. Hou, *Zhongguo kuangye jiyao, di wu ci*, 120.
44. Jordan, *China's Trial by Fire*, 10–43.
45. Jordan, 197–98.

Although not at all in the manner that Chinese industrialists had hoped, the specter of scarcity was thus dispelled.

The post–Mukden Incident coal famine scare brought to light the extent to which China's industries relied on foreign fuel. It also underscored how much domestic coal production lagged behind industrial demand. Fushun coal's success in China was both a consequence and a cause of the Chinese coal sector's troubles. Accordingly, state planners and industry leaders saw the need to address structural challenges that had held back advancements in coal mining, particularly in the area of transportation. This was not the first time the Nationalist government involved itself in coal affairs, but it marked the beginning of Nanjing's heightened intervention in the industry.

Developmentalist Visions

The interest that the Nationalist state took in matters of coal was grounded in its larger commitment to industrial transformation.[46] Nanjing was, as historian William Kirby has contended, "consumed with the industrial metamorphosis of *national* life, planned," as this was, "by a central—and centralizing—government."[47] Its ideological backbone was to be found in Sun Yat-sen's 1920 *Industrial Plan*. Sun was, of course, not the first to view industrialization as the key to China's national salvation. Many Chinese thinkers beginning at least with the "self-strengtheners" of the late Qing regarded industry as the source of national wealth and power.[48] What set Sun's vision apart was its immensity of scale and the enlarged scope of government needed to see it through. He sought to bring about nothing less than a "second industrial revolution" in China that would "increase the productive power of man many times more than the first one."[49] Concretely, the plan involved an ambitious agenda for road and rail construction, dam building, electrification, urban redesign, and the mechanization of economic activity, all financed largely by foreign capital (which Sun was eager to attract) and carried out by scores of engineers and other experts

46. It should be noted that there were forms of industrialization pursued by private actors that were distinct from those of the developmental state. One example is the "vernacular industrialism" epitomized by the activities of maverick entrepreneur Chen Diexian. See Lean, *Vernacular Industrialism in China*.

47. Kirby, "Engineering China," 137 (italics in original).

48. Schwartz, *In Search of Wealth and Power*, 16–17.

49. Sun, *International Development of China*, ii.

brought into the apparatus of government.[50] Coal, in powering the machines involved in these processes, proved crucial to his plan.[51]

Sun described coal as the "sinews of modern industries," a biological metaphor possibly inspired by his medical training (Sun had been educated as a physician). It conveyed well how that fossil fuel held together the industrial system while allowing it to move forward. His plan called for copious amounts of coal and iron—the "two great essentials to modern industrialism"—and for cultivating self-sufficiency in those twinned resources. It would not do, Sun argued, to just rely on imports. In his analysis, current producer countries lacked the capacity to supply the necessary materials for Chinese development. To begin with, their resources were already all going toward the post–World War I reconstruction of Europe. Furthermore, the rapid depletion of those resources had also prompted these countries to start thinking about conservation for their own futures, making them unlikely to commit to meeting the huge projected material demands of China's industrial endeavors even if they could. The coal that China needed was to come from China itself. Like other boosters of Chinese coal, Sun celebrated its abundance ("China is known to be the country most rich in coal deposits"), but—in the same breath—bemoaned its underdevelopment ("yet her coal fields are scarcely scratched"). As part of the first program (among six) within his plan, he pushed for the large-scale development of the "unlimited iron and coal fields of Shansi [Shanxi] and Chili [Zhili]." And as with most other Chinese coal boosters, the sheer wealth of this resource fueled in Sun ideas of inexhaustibility and intentions for extensive extraction that aligned with the developmentalist premises of carbon technocracy.[52]

Because coal was so indispensable to industrialization, it was, as Sun saw it, a "necessity of civilized community." Like their Japanese contemporaries, Sun and many after him took the amount of coal produced and consumed as one index for their country's level of development. Calculated per capita and placed alongside corresponding numbers for other polities, which were almost always much higher, China's coal production and consumption figures seemed like yet another quantification of the country's backwardness. Sun contended that if China were as "equally developed" as the United States, then, "according to the proportion of her population," Chinese coal output should be four times that of the Ameri-

50. Kirby, "Engineering China," 137–39.
51. Sun, *International Development of China,* ii–iii.
52. Sun, 13–14, 155–56, and app. 2, xi.

cans.[53] A 1930 article in the *North China Industry Quarterly* pointed out that each Chinese person was using a mere 0.06 tons of coal per year, which was one-seventy-sixth the annual per capita coal usage in the United States, one-eighth that in Japan, and one-sixty-fifth that in Britain—an indication, the author concluded, of just how much China's industrial sector was lagging behind those of other countries.[54] That such coal figures, like most economic indicators, failed to truly capture the progress they ostensibly gauged did not stop state planners from taking them as tangible measures of development.[55] At this point in the fossil fuel age, coal consumption, seemingly concrete in its quantifiability, was understood as desirable in and of itself, and Sun sought to stimulate its increase.

Sun also expressed interest in how coal might help improve the lot of ordinary Chinese. Aside from backing the development of mines, Sun supported bringing coal prices down "as low as possible" in order not only to "give impetus to the development of various industries" but also to "meet the demands of the public."[56] By his rough estimates, those who lived in the countryside devoted about 10 percent of their working hours to collecting firewood, and those who lived in towns and cities spent around 20 percent of their living expenses on the same fuel. He bid the former switch to coal and the latter to electricity and gas—both produced through burning coal—for if coal were indeed cheap, this could mean savings in terms of time and money.[57] In fact, for much of the interwar period, household use consistently accounted for the largest proportion of coal consumption. The China Geological Survey's report of 1926 noted that 43.3 percent of coal went toward household use, while 32.6 percent powered industry, 8.4 percent fueled transportation, 8 percent drove operations at mining sites, and 7.7 percent was exported.[58] To promoters of industry, however, this distribution, like the low per capita coal consumption, demonstrated the "extent of [China's] industrial underdevelopment."[59] Carbon technocracy placed a premium on increases in coal consumption, but often only for narrowly industrial ends.

Weng Wenhao shared Sun's insistence on the importance of develop-

53. Sun, 155–56.

54. Wang, "Mei zhi yongtu," 12.

55. On the adoption of modern economic indicators leading to reductionistic visions of progress, see Cook, *The Pricing of Progress*.

56. Sun, *International Development of China*, 156.

57. Sun, 150–51.

58. Xie, *Zhongguo kuangye jiyao, di er ci*, 99–100.

59. Wang, "Mei zhi yongtu," 12–14. According to Wang, in Germany at the time, only 6.1 percent of coal went toward household use.

ing China's coal resources. A pioneer in Chinese geological research with a doctorate from the University of Louvain in Belgium, Weng was appointed by Chiang Kai-shek to be chairman of the National Defense Planning Commission at the agency's founding in 1932. To his eye, the "most fundamental fact that needed to be considered in regard to the prospects of the Chinese coal mining industry" was the much-repeated north-south divide: "the North's coal deposits are exceedingly abundant; the South's coal supplies exceedingly in want." Apart from calling for the improvement of transport channels to facilitate the "southward shipping of northern coal," Weng proposed developing industrial zones in North China, closer to existing or potential sites of coal extraction.[60] The Nationalist state attempted such strategic siting, though primarily in Central and East China, where it had a firmer footing. One infamous episode involved the Ministry of Industry initiating a project with German partners to construct a massive coal and steel complex, financed in part by German loans, employing German machines and knowhow, and using iron ore from Anhui and Hebei and coal from Anhui. Right before the deal was concluded, however, it came to the attention of the would-be signatories that Anhui coal was not suitable for coking.[61] Weng was among those who criticized the proliferation of such plans in the late 1920s and early 1930s. "Every day I hear about many plans, but their content is far removed from reality; . . . they do not address themselves to the real situation of our own nation," he grumbled.[62] Weng was, nevertheless, not opposed to planning per se. In coal production, as in other industrial activities, plans may deliver on their promise, he maintained, were they grounded in technical expertise and coordinated from the center.[63]

Chinese public discourse, accepting as it was of the coal-fired development espoused by Sun and others, was conflicted in its assessment of the Fushun colliery that exemplified this ideal. On the one hand, the Japanese enterprise was deeply abhorred. Many Chinese commentators castigated Fushun for all manner of ills, most of all the abuses it inflicted on the multitude of Chinese workers laboring in its coal mines. For example, the *Mining Weekly* (鑛業週報), the main periodical for China's extractive industries, frequently ran articles on accidents and other more intentional injustices that bore such titles as "No Lows to Which the Fushun Colliery Would Not

60. Weng, "Zhongguo meikuangye de eyun," 6.
61. Kirby, *Germany and Republican China*, 83–85.
62. Kirby, 84–85, cited quotation by Weng Wenhao on 85.
63. Kirby, 85, 97, 206–7.

Stoop" or "Fushun Colliery Trifles with Chinese Lives and Properties."[64] On the other hand, the Fushun colliery was greatly admired, largely on account of the magnitude of its modern operations and the magnificence of its open-pit mine. Writing in 1932, Hou Defeng 侯德封 (1900–1980), a leading geologist with the China Geological Survey, noted that "Fushun, with its massive output, has long ranked first among the coal mines in China."[65] In spite of its moral failings (or, arguably, because of them), Fushun represented an actualization of carbon technocracy and the industrial modern ideal toward which the Nationalist regime, in all its striving at reconstruction, so earnestly aspired.

The Nationalist State in the Business of Coal

In the years between the Mukden Incident and the Second Sino-Japanese War, the Nationalist state continued to extend its reach into the business of coal, involving itself in crises from which it tended to emerge with a stronger presence within the industry. The National Coal Relief Commission may have been established in response to the coal shortage of 1931, but when it convened again in late 1932, it was to address the opposite dilemma: the dumping of Fushun and Japanese coal.[66]

Coal from Fushun and Japan began flooding Chinese markets as Japanese products started once again flowing steadily into China with the lifting of boycotts against Japanese goods after the short Shanghai war. Fushun coal had been stockpiling because of recent restrictions on its sales in Japan (chapter 3) and its general overproduction over the preceding years. To rid itself of this accumulated reserve, Mantetsu chose to grossly undercut its Chinese competitors. In 1932, Fushun coal was sold for 7.7 Japanese yen per ton in Mukden, about 40 miles west of Fushun, while it went for 7.5 yen in Shanghai, more than 750 miles away as the crow flies. The company absorbed the transportation costs, which included additional railway fees as well as payments for the use of port facilities and for shipping.[67] As for Japanese coal overall, export prices to China had been slashed so as to offset the escalating value of the gold-backed yen from 1930. Although this was initially insufficient to keep Japanese coal prices competitive, the yen's devaluation after the Japanese government took it off the gold stan-

64. "Wu er bu zuo"; "Fushun meikuang wanhu."
65. Hou, *Zhongguo kuangye jiyao, di si ci*, 228.
66. Li, "Wai mei qingqiao," 45–47.
67. Li, 45–47; Zhang, *Shiye jihua yu Guomin zhengfu*, 150.

dard at the end of 1931 resulted in those prices plummeting, and Japanese coal upped its market share in Shanghai considerably.[68] Chinese coal producers were hard hit. The National Coal Relief Commission wrote to the Ministry of Industry in desperation: "The dumping tactics now used by Japanese coal have already won over sellers [i.e., Chinese coal merchants] and are also welcomed by consumers. It is hard to imagine a future for national coal!"[69]

Once again, the Nationalist state stepped in, welcomed by coal industrialists who sought relief in regulation. The National Coal Relief Committee and the Ministry of Industry initially proposed to address this issue by increasing the production and hence competitiveness of domestic mines. In addition to calling for transportation reform, they suggested bringing in technical personnel to install and upgrade mechanical facilities and reorganize operations to reduce production costs and increase output.[70] However, upon further reflection, they decided that it would take too much to effect these changes within a short period of time and that a more immediate solution would be to lower domestic freight rates and raise foreign import tariffs.[71] Regaining tariff autonomy was one of the major early accomplishments of the Nationalist government, which then raised tariffs both to secure more revenue and to protect Chinese industries. Under the new tariff schedule introduced in May 1933, tariffs on coal increased from around 1.6 to 3.5 Chinese yuan per ton.[72] This appeared to have the desired effect. Japanese coal imports fell. While recognizing the statehood of Manchukuo, established in February 1932, would be a contentious issue in China throughout the existence of this short-lived client state, Chinese government officials quite readily treated it as a foreign entity when it came to trade, and the new tariffs were extended to Fushun coal as well. Its imports, too, dipped accordingly.[73]

But with what appeared to be the resolution of the dumping crisis came other problems for which the Nationalist regime would proffer statist solutions. Competition between Chinese coal producers, who now contended less with foreign firms than with each other, intensified. This competition drove prices and profits so low that the government started expressing con-

68. Wright, "Nationalist State and the Regulation," 134–35.

69. "Guomei jiuji weiyuanhui cheng."

70. "Guomei jiuji weiyuanhui cheng," 486.

71. Zhang, Zhao, and Luo, *Jingji yu zhengzhi*, 263–65; "Caizheng, shiye, waijiao."

72. Wright, "Nationalist State and the Regulation," 135.

73. Zhang, Zhao, and Luo, *Jingji yu zhengzhi*, 265. On tariffs in Nationalist China, see Boecking, *No Great Wall*.

cern about the state of the coal mining industry, which was also suffering from sluggish demand amid the industrial slowdown of the mid-1930s.[74] At the same time, the interventionist position that Nanjing had taken in relation to the economy became more pronounced, particularly with Chen Gongbo 陳公博 (1892–1946) as the minister of industry between 1932 and 1935.[75] Chen advocated for a greater degree of "control" (統治) in coal mining, as he would in almost all economic areas. In his assessment, the Chinese coal sector had to deal with several issues, including "irrational" management leading to high expenses regardless of output volume, "unscientific" technical expertise because engineers were far removed from actual mining conditions, and persistent problems of capitalization and transportation.[76] State control promised to address these issues by coordinating resources within the industry. While previous meetings between government and business interests might be regarded as instances of such coordination, the most extreme attempt to coordinate the industry from the center was a proposed coal cartel that did not come to pass.

In 1936, the Ministry of Industry tried to initiate a national coal cartel to curtail the competition blamed for impeding the industry's development. That summer, Chen's successor, Wu Dingchang 吳鼎昌 (1884–1950), called a meeting to pitch this cartel to representatives from major Chinese and Sino-foreign coal-mining companies, the Ministries of Finance, Industry, and Railways, the National Reconstruction Commission, and the National Resources Commission.[77] Although the participants endorsed this plan and elected a committee to draft the regulations, several key actors—namely the Japanese-controlled coal mines in Shandong, the Kailuan Mining Administration, and the Huainan colliery run by the National Reconstruction Commission—refused to back any collusive agreements. As these producers had been faring well within their regional markets, they regarded the fixing of existing market shares that was part of the proposed regulations an impediment to their current and future growth. The cartel consequently fell through. Of interest here is how one of the

74. Zhang, Zhao, and Luo, *Jingji yu zhengzhi*, 265. On the mid-1930s silver crisis and economic depression in China, see Coble, *The Shanghai Capitalists*, 140–60; and Shiroyama, *China during the Great Depression*.

75. On Chen Gongbo, the Nationalist Left, and their vision of a *minzu* economy, see Zanasi, *Saving the Nation*. On commercial schools as vectors for the transmission of new ideas about the economy that undergirded the rise of the regulatory state, see Yeh, *Shanghai Splendor*, 30–50.

76. Chen, *Si nian congzheng*, 52–53.

77. This abortive coal cartel of 1936 is the central event in Wright, "Nationalist State and the Regulation." My narrative here draws from his account, especially 139–46.

main parties that foiled the Ministry of Industry's ambitious plan was none other than a colliery run by another economic arm of the state.[78]

The Nationalist government's operation of its own coal mines had in fact begun with the National Reconstruction Commission, which was created in 1928 and chaired by entrepreneur Zhang Renjie 張人傑 (1877–1950). Zhang, an early supporter of Sun Yat-sen and one of the Nationalists' "four elder statesmen," described the "sole aim" of the commission as "the realization of the various Reconstruction projects advocated by the late party leader." To that end, "it will adopt scientific methods in the administration of various affairs," he declared.[79] Among the first of such affairs was developing southern coal mines. This would, Zhang insisted, not only offer a corrective to the perennial problem of the north-south imbalance but also provide fuel for the commission's two power plants.[80] The National Reconstruction Commission swiftly acquired the Changxing colliery in Zhejiang and started excavating the Huainan coal mines in Anhui. Under its management, Huainan grew rapidly into a formidable coal-mining enterprise, the largest in East China. In 1931, its output was 31,000 tons. By 1936, production was 584,000 tons, almost twenty times more. Several factors led to this increase, including the employment of more technically trained work supervisors, the waiving of certain local taxes, and the construction of the Huai–Nan railway, which allowed coal to be transported to the river port of Yuxikou on the Yangtze River, where it could then be sent downstream to Shanghai and its environs.[81] Given Huainan's success, there was little incentive to join a cartel that limited its production just so that other mines could catch up. Historian Tim Wright regards this episode of the abortive cartel as an example of a weak Nationalist state unable to exert its will on China's economic affairs.[82] One can also read this episode as an indication of competing economic units within the government that shared Sun's developmentalist vision but were divided over how to realize it while protecting their own institutional interests along the way.

The National Resources Commission, as another one of these economic agencies under the prewar Nationalist state, also considered the flourishing of the coal-mining industry an important part of the larger industrialization plan to which its leaders were committed. As was consistent with its

78. Wright, "Nationalist State and the Regulation," 142.
79. Chang, "Reconstruction Program for 1930."
80. Zhongguo Guomindang zhongyang weiyuanhui dangshi weiyuanhui, *Zhang Jingjiang xiansheng wenji*, 120; Tan, "Nanjing Guomin zhengfu shiqi," 107.
81. Tan, "Nanjing Guomin zhengfu shiqi," 108, 110–12.
82. Wright, "Nationalist State and the Regulation," 146–48.

overall transformation from research institute to managerial bureau, the commission began by surveying the industry to see what might be done to assist it and later went on to take part in running mines. A central figure in these endeavors was an engineer by the name of Sun Yueqi 孫越崎 (1893–1995).

Sun was a Zhejiang native with a degree in mining and metallurgy from Peking University. His choice of study had been influenced by his father, who had opened and operated a gold mine in the northernmost province of Heilongjiang and advised him against becoming an official ("a high-level hoodlum," in his phrasing) and (perhaps with the intention of having his son come help with his business) to consider mining engineering instead. Not long after graduating, Sun did indeed go to the Northeast, where his father arranged for him to observe operations at several industrial sites, including the Fushun colliery. In the end, he was, however, determined to find his own way. In 1923, Sun got a job with a Sino-Russian company set up to excavate coal mines in Muling far to the east of Harbin.[83] He remembered "riding a snow sleigh through remote mountains and thick forests to reach the godforsaken" place where the site of extraction was to be established.[84] A few years later, Sun caught the eye of Weng Wenhao, who was visiting the Muling colliery on China Geological Survey business. Purportedly "admiring [Sun's] work ethic and accomplishments," Weng recommended him for a job as chief engineer at a coal mine in Hebei. At the time, Sun was just about to leave for further studies in the United States and hence "politely declined."[85] Between 1929 and 1932, Sun completed graduate work in mining engineering at Stanford and Columbia Universities and practicums at coal mines in England, France, and Germany. Through these transnational exchanges he honed the expertise and credentials that later secured him a leadership position in the government's newest and unquestionably most technocratic economic agency.[86]

Upon returning to China in November 1932, Sun received an invitation from Weng to join the National Defense Planning Commission.[87] In 1933, under the commission's orders, Sun conducted surveys of the coal mines along the important Jin–Pu railway. Soon after, Weng sent him to the near bankrupt Zhong-Fu colliery in Henan to help reorganize opera-

83. Sun Yueqi keji jiaoyu jijin, *Sun Yueqi zhuan*, 4–7, 24–25, 34, 38–39.
84. Cited in Xue, "Sun Yueqi yu 20 shiji," 103.
85. Sun, "Wo he Ziyuan weiyuanhui," 14.
86. Sun Yueqi keji jiaoyu jijin, *Sun Yueqi zhuan*, 58–74.
87. Sun, "Wo he Ziyuan weiyuanhui," 14.

tions.[88] Putting on several hats in succession, from government representative to chief engineer to general manager, Sun proved instrumental in turning things around for the failing enterprise, beginning with correcting engineering missteps such as mining beyond the recommended recovery ratio, which had compromised safety and crippled production. Zhong-Fu recorded a loss of more than 950,000 yuan in 1934. By 1936, its profits exceeded 1.5 million yuan, placing it just behind Kailuan and Zhong-xing as the third-largest coal producer in Nationalist China.[89] Sun's work at Zhong-Fu and other mines served him well as war broke out and he became central to statist efforts at developing the coal-mining sector and strengthening the apparatus of carbon technocracy in the Chinese interior.

Fueling the War of Resistance

On July 7, 1937, Chinese and Japanese troops exchanged fire at the Marco Polo Bridge to the southwest of Beiping. Initially, this episode seemed like yet another in a string of minor incidents between China and Japan that had sparked and quickly died down in the years following Manchukuo's establishment in 1932. However, with breaches in ceasefires, the buildup of units on both sides, and directives from Tokyo to stay on the offensive, the conflict exploded, escalating from a skirmish to a full-blown battle. It marked the beginning of the Second Sino-Japanese War. As the fighting persisted and the Imperial Japanese Army took Shanghai and marched on Nanjing, the Nationalist government beat a quick retreat, first to Wuhan and then, when it became apparent that this, too, would fall, Chongqing, from which it held on tensely to China's Northwest and Southwest.[90]

In its move inland, the Nationalist government brought along not only its administrative apparatus but also the machinery and labor from many factories and mines. To help coordinate this industrial migration, it founded the Industrial and Mining Adjustment Administration (工鑛調整處), which, together with the National Resources Commission, oversaw the relocation of hundreds of thousands of tons of industrial equipment and tens of thousands of workers.[91] Coal mines were included

88. Chiang Kai-shek had taken a personal interest in the revival of this enterprise, which was a Sino-British concern. See "Oong Ordered to Reorganize."

89. Xue, "Sun Yueqi yu 20 shiji," 104–5.

90. For an account of this opening phase of the war, see Mitter, *China's War with Japan*, 73–167.

91. Eastman, "Nationalist China during the Sino-Japanese War," 562–63; He, "Kangzhan shiqi guoying meikuangye," 31–35.

in this shift. Although their deposits were unmovable, the machinery with which those seams were worked could be taken apart, crated, and lugged onto trucks, trains, and boats for shipment inland. A mine was, after all, more than its resources. It was also constituted by its labor and equipment. As war spilled across North and Central China, the coal mines in these regions were in danger of falling into enemy hands. Many, in fact, did. Faced with this prospect, Sun halted operations at Zhong-Fu and arranged for its mining equipment to be ferried away. Zhong-Fu's technical assets—which included inclined shaft hoisters, centrifugal fans, electric pumps, steam locomotives, alternating current generators, and boilers—were later reinstalled in coal mines in the interior, most notably at the Tianfu colliery in the wartime capital of Chongqing.[92]

With this industrial migration, the interior's carbon resources would be opened up to exploitation on a hitherto unprecedented scale. Many coal mines were either established or expanded in this period.[93] Tianfu, formerly a modest operation mined almost entirely by manual labor, became the largest and most productive coal-mining enterprise in Nationalist-controlled China through the influx of machinery and Sun's direction as its general manager. In the past, subterranean coal transport here was done by workers bearing bamboo baskets on their backs. Human effort was also solely responsible for pumping water. Ventilation was inadequate, and without sufficient circulation of air, the excavated subsurface spaces were so hot and humid that miners mostly worked naked. Under these poor conditions, workers faced discomfort and even death—their fatality rate was very high. To facilitate mechanization, Sun had a power station built to supply electricity to the transplanted equipment, which, in turn, helped transform operations at this site. Tianfu's output in 1938 was 55,000 tons. In 1945, it was over 450,000 tons.[94] While the day-to-day work environment also improved, the mechanized site held its own dangers too.

Following their retreat inland, the Nationalists consolidated the state's economic planning efforts under the technocratic National Resources Commission. In January 1938, the Ministry of Industry, the National Reconstruction Commission, and other governmental economic units were combined into the Ministry of Economic Affairs (經濟部). The National Resources Commission, which originally reported to the Military Affairs

92. Xue, "Sun Yueqi yu 20 shiji," 106; Zhongguo jindai meikuang shi bianxiezu, *Zhongguo jindai meikuang shi*, 431.

93. Sun, "Chuanmei chanxiao," 42.

94. Xue, "Sun Yueqi yu 20 shiji," 106.

5.2. The Tianfu colliery, near the Nationalist government's wartime capital of Chongqing. Under the technocratic National Resources Commission, it became the largest coal-mining enterprise in Nationalist-controlled China. This photograph was taken by Joseph Needham when he was in China working for the Sino-British Science Cooperation Office. (Image courtesy of the Needham Research Institute.)

Commission, was placed at the center of this new ministry, with Weng Wenhao at the top as minister.[95] Given a broad mandate to "develop and manage important resources" (開發及管理重要資源) in "foundational industries," "essential mining enterprises," "power facilities," and "any other undertakings designated by the government," the commission pursued plans to build up the interior's heavy industrial capacity.[96] In 1939, it put forward a three-year plan for establishing state enterprises that targeted increases in the output of key resources like coal in an attempt to "strive to grow the economy of the interior so as to meet the needs of both military matters and industrial development."[97] Although Chiang Kai-shek approved less than half of the proposed budget, the commission nevertheless worked with what it was given to enlarge the industrial sector and

95. Kirby, "Technocratic Organization and Technological Development," 29–30.

96. "Jingji bu Zhiyuan weiyuanhui," 96.

97. Zheng, Cheng, and Zhang, *Jiu Zhongguo de Ziyuan weiyuanhui*, 50. The targeted output for coal listed here is 64,000 tons by 1942, which seems too low given that output in 1939 was already 5.5 million. It is more likely that the target set was 6.4 million.

its place within it.[98] The commission's operations expanded in the years that followed. In 1937, it had a staff of under 2,000; by 1945, its employees exceeded 63,000.[99] As historian Cheng Linsun has noted, the National Resources Commission "became the single largest government employer outside the armed forces."[100] The number of industrial enterprises under the commission went up almost fivefold over the course of the war, from 24 at the start to 115 at the end. Of these, 20 were coal mines.[101] Sun Yueqi alone oversaw four, concurrently serving as general manager of three other collieries in addition to Tianfu.[102]

Through the involvement of the Nationalist state, inland China's coal-mining sector grew steadily.[103] The mines wholly or jointly operated by the National Resources Commission only accounted for, on average, 10 percent of total output, but they boasted the largest and most mechanized operations and ended up supporting the multitude of much smaller private mines with technical assistance and recommendations for production reform.[104] Moreover, the Ministry of Economic Affairs also undertook several broader measures to bolster the industry, creating, in 1938, the Fuel Management Administration (燃料管理處) to help coordinate its "planning and preparation of production, transportation, allocation, and storage of fuels in this time of emergency."[105] Such measures included laying down wartime regulations to speed the launch of new mines; extending large sums of credit to mines to support their establishment and subsequent operations; and securing for mines necessary materials and machinery from both local and foreign suppliers, purchasing by preemption and rationing out thereafter. Furthermore, in order to ensure sufficient labor and expertise for coal production, the ministry arranged for miners and technicians in the industry to defer military service into which men were increasingly

98. Zheng, Cheng, and Zhang, 50–51. This plan was preceded by an important one that the National Resources Commission had embarked on in 1936 but that was cut short by the war. On this 1936 plan, see Kirby, "Kuomintang China's 'Great Leap Outward.'"

99. Zheng, Cheng, and Zhang, *Jiu Zhongguo de Ziyuan weiyuanhui*, 112.

100. Cheng, "Industrial Activities," 55.

101. Zheng, Cheng, and Zhang, *Jiu Zhongguo de Ziyuan weiyuanhui*, 107, 109–10. Of the 115 industrial enterprises, 57 were under the sole proprietorship of the National Resources Commission, 41 were joint enterprises between the commission and other private or public entities, and 17 were enterprises that the commission had invested in but was not involved in running.

102. Xue, "Sun Yueqi yu 20 shiji," 106.

103. For an account of how local technocrats pushed for state-led development in Southwest China's mining industry decades before the Nationalist government's retreat into the interior, see Giersch, *Corporate Conquests*, 123–70.

104. He, "Kangzhan shiqi guoying meikuangye," 96.

105. "Jingji bu Ranliao guanlichu," 65.

being drafted.[106] Their battlefield would not be at the frontlines but in the deep recesses of the mine. For some, it was just as deadly. In 1939, an electric cable in one of Tianfu's pits sparked and caught fire, triggering an explosion that killed 102 miners.[107] Coal production in Nationalist-controlled areas was 4.7 million tons in 1938. It peaked at 6.6 million in 1943 before declining as the continuing conflict and related inflationary pressures slowed down further industrial expansion across the board.[108]

In the Chinese interior, the coal mined in increasing intensity over this period supplied the energy needs of the wartime state and society.[109] The coal from state-run mines disproportionately served industrial purposes. According to a 1945 National Resources Commission report, military industry consumed 39 percent of such coal, regular industry and power generation 28 percent, transportation 19 percent, and household use 14 percent.[110] "As the source of motive power, coal is the mother of all other industries," Sun Yueqi proclaimed in 1940.[111] Although the commission began developing hydroelectric power facilities in the highlands of the Southwest, electricity production remained predominantly coal-fired, one reason for this being the considerable cost of dam building.[112] However, when we look at not only state-run mines but private ones as well, the biggest consumer of coal, as before the war, remained the home. Coal, along with firewood, heated food and water and warmed domestic spaces for a population that had swelled as refugees fleeing the Japanese armies rushed into the region.[113] In 1941, the China Geological Survey estimated that over half of the coal consumed in the Southwest that year was for household use.[114] For those who dwelled close to coal mines, these sites occasionally

106. Li, "Shi nian lai," I.11–I.13. On conscription under the Nationalist regime during the Second Sino-Japanese War, see van de Ven, *War and Nationalism in China*, 255–58; and Xu, *Soldier Image and State-Building*, 59–63, 68–70.

107. Zhongguo jindai meikuang shi bianxiezu, *Zhongguo jindai meikuang shi*, 432.

108. On wartime inflation in Nationalist China, see Young, *China's Wartime Finance and Inflation*.

109. On energy in China during the Second Sino-Japanese War more generally, see Muscolino, *The Ecology of War*.

110. Wu, "Kangzhan qijian guoying meikuang," 149.

111. Sun, "Chuanmei chanxiao," 43.

112. Zhu, "Shi nian lai zhi dianli shiye," J.23–J.24. For more on electricity generation in Nationalist-controlled China during the war, see Tan, "Revolutionary Current," 64–112.

113. There are no firm figures for number of people who fled to Southwest China during the war, though it is likely that it was in the millions. See Eastman, "Nationalist China during the Sino-Japanese War," 565.

114. Jin, *Zhongguo kuangye jiyao, di liu ci*, 21.

provided additional relief. Zeng Yongqing, a housewife residing on the outskirts of Chongqing in those years, recalled taking shelter in a nearby abandoned coal mine during the frequent Japanese bombing of the wartime capital and the "horrible experience" of standing and waiting in "at least a foot of dirty black water" while air raids continued overhead.[115]

Complementing its efforts to expand production, the National Resources Commission also sought more efficient consumption by enhancing the quality of coal extracted and the uses to which it could be put through technoscientific means. In spite of the comparative lack of preexisting infrastructures for scientific and technological work in the Chinese interior, the Nationalist regime managed to capitalize on the Chinese experts who had retreated inland with the government, the equipment they managed to bring along with them, foreign scientific and technological assistance, and the urgency of war to engender advances in science and technology.[116] In this regard, the National Resources Commission was, according to Joseph Needham—the British biochemist and later historian of science who spent four years in wartime China fostering Sino-British scientific cooperation—one of the two "greatest organizations in China concerned with applied sciences" (the other being the Army Ordnance Administration).[117]

In 1938, Weng Wenhao founded the Mining and Metallurgy Research Bureau (礦冶研究所), which was housed in a facility near the Tianfu colliery and staffed almost entirely by National Resources Commission experts.[118] The bureau undertook dozens of surveys of coal and other resource industries across the interior provinces, but its primary work was in coal research, with a focus on ameliorating existing local methods of coal washing and coking to suit the properties of Sichuanese coal.[119] The coal in Sichuan tended to come in thicker seams, allowing it to be mined more easily and in large quantities, but it also had high contents of both ash and

115. "An Abandoned Housewife: Zeng Yongqing, born in 1916 in Chengdu, Sichuan Province," in Li, *Echoes of Chongqing*, 111.

116. Needham, "Chinese Scientists Go to War." For some of the ways in which wartime conditions shaped science and the work of scientific practitioners in the Chinese interior, see Reardon-Anderson, *The Study of Change*, 293–318; Schneider, *Biology and Revolution*, 94–104; Greene, "Looking toward the Future"; Barnes, *Intimate Communities*, 21–51, 120–58; and Brazelton, *Mass Vaccination*, 55–100.

117. Needham, "Chinese Scientists Go to War," 57. On Joseph Needham in wartime China, see Mougey, "Needham at the Crossroads."

118. On the Mining and Metallurgy Research Bureau, see Lei, Fang, and Qian, "Kangzhan shiqi houfang yejin ranliao."

119. He, "Kangzhan shiqi guoying meikuangye," 106–9.

sulfur, rendering it unsuitable for making metallurgical coke.[120] Following a series of trials, the bureau's fuel specialists redesigned troughs and other coal-washing apparatuses so that more ash and sulfur could be removed and more cleaned coal recovered (a result of better calibrating how the machines regulated the volume and speed of water flow). They also improved upon coke oven design to better their odds at successfully coking the feedstock and to shorten the time for the whole process.[121] The bureau then proceeded to introduce these techniques and technologies to collieries and coke-consuming industries in the interior, with Tianfu as a pilot. In 1942, Nationalist-controlled China produced three hundred thousand tons of coke, three times more than the annual average in the 1930s. This mass of manufactured fuel would then fire the metallurgical processes so crucial to the Nationalists' efforts at wartime industrialization.[122]

Amid severe petroleum shortages, coal was also used by Chinese scientists to develop synthetic fuels, as their Japanese counterparts in Fushun and Tokuyama did. Foreign and Chinese geologists had been exploring for petroleum across China since the turn of the century, discovering deposits of varying sizes, a few of commercial potential.[123] The provinces of Shaanxi and Sichuan had some production in the 1930s, but this was on an extremely small scale, and China's petroleum products—gasoline, kerosene, fuel oil, and lubricating oil—were largely imported, mostly from the United States and the Dutch East Indies.[124] After the war began, Japan's occupation of the coast and subsequent closure of the Burma Road, by which the Chinese interior was supplied overland through British Burma, meant that Nationalist China lost access to many essential imports, including petroleum.[125] While the Allied forces were able to airlift some supplies from India to China over the stretch of the Himalayas they called "the hump," the National Resources Commission led the way in seeking more self-sufficient solutions to the problem of petroleum.[126]

To that end, the commission opened oilfields and experimented with

---

120. Yu, "Bensuo liang nian lai ximei lianjiao," 8.

121. For more details on these innovations, see Lei, Fang, and Qian, "Kangzhan shiqi houfang yejin ranliao," 276–79.

122. Lei, Fang, and Qian, 279–80. Weng Wenhao called smelting the "basic heavy industry," the first among those targeted for development in China's wartime industrialization. See Wong, "China's Economic Reconstruction."

123. On petroleum exploration in China in the first half of the twentieth century, see Shen, Unearthing the Nation, 166–73.

124. Heroy, "Petroleum," 123–33.

125. Smyth, "China's Petroleum Industry," 189.

126. Mitter, China's War with Japan, 181.

liquid fuel substitutes. In Yumen, Gansu Province, geologists dispatched by Weng Wenhao identified promising petroleum deposits. Sun Yueqi then oversaw the development of this site into one of sizable oil extraction while maintaining his management of the four collieries.[127] "In 1944, when the United States air force flew from Chengdu to bomb Tokyo and Japanese-occupied sites such as the Tangshan coal mine and Linxi power plant of the Kailuan colliery, the fuel that ground services used was from the Yumen oilfield," Sun proudly recalled decades later as he reflected on Yumen's wartime contributions.[128] As for liquid fuel substitutes, these included power alcohol derived from fermenting molasses or grains, gasoline cracked from tung oil and other vegetable oils, and liquid fuel manufactured from distilling coal.[129] To produce liquid fuel from coal, the commission set up a plant in Jianwei in western Sichuan, where the region's abundant high-quality coal could be processed through low-temperature distillation into a variety of petroleum products, including high-octane aviation fuel.[130]

In boosting its technical capabilities in the coal and other industrial sectors, the National Resources Commission often looked to the United States. This was a pivot away from the prewar period, when the commission collaborated closely with German counterparts to draw up and carry out plans for China's industrialization.[131] As a provision through the lend-lease arrangements, in which the United States government committed aid to China and other Allied nations, the commission sent teams of technicians and engineers across the Pacific for training.[132] Yu Zailin 俞再麟 (1908–1947), a fuel specialist who had been key to the Mining and Metallurgy Research Bureau's work on coal washing and coking, was on the first team that traveled to the United States in 1942.[133] Yu spent a total of two years there, beginning with a three-month stint at the Coal Research Laboratory at the Carnegie Institute of Technology in Pittsburgh before moving on to

127. Ministry of Information of the Republic of China, *China after Five Years of War*, 96. On the opening of the Yumen oilfields, see Bian, *Making of the State Enterprise System*, 61–63. Another important attempt at wartime petroleum exploration and development by the Nationalist state was in Xinjiang. See Kinzley, *Natural Resources and the New Frontier*, 124–27.

128. Sun, "Wo he Ziyuan weiyuanhui," 25.

129. "Science and Life in War-Time China," 51–52. On liquid fuel substitutes in wartime China, see Reardon-Anderson, *The Study of Change*, 297–300.

130. Chinese Ministry of Information, *China Handbook*, 484.

131. Kirby, *Germany and Republican China*.

132. Kirby, "Planning Postwar China," 226–27.

133. There were two other major study trips during the war, one in 1943 and the other in 1944. See He, "Kangzhan shiqi guoying meikuangye," 110–11.

practicums at a Pennsylvanian coal and coke company and an Illinoisan steel firm. He then traveled through Pennsylvania, Maryland, Alabama, Illinois, and Ohio, where he observed operations at coal, coke, iron, steel, and machine-making facilities.[134]

If they were not full-fledged technocrats before, the Nationalist experts who participated in this training would emerge from it with a heightened confidence in the primacy of technoscientific solutions to problems of industry and economy. In his report thereafter, Yu noted how coal-washing operations in American industries had been expanding over the past decade to deal not only with increasing coal output through mechanization but also with coal's resultant decreasing quality. "Lately, all collieries in the United States, in their pursuit of greater production, have been using machines in their operations, no matter if this is mining, loading and unloading, or transportation, and, as a result, much gravel is being mixed into the coal," he wrote. As he saw it, coal washing was an effective fix to problems of quality whether due to coal's character or its means of production and processing. For Yu, this technology's appeal lay in its purported ability to overcome all the imperfections of nature and the sullying effects of mechanized extraction: "No matter what the quality of the coal might be, as long as it is washed, it can be made into coke and used for smelting." China would do well, he advised, to follow the American example and more extensively embrace coal washing, especially in the interest of supporting the "necessary development" of the iron and steel sector amid the projected expansion of industry after the war.[135]

By thinking ahead to peacetime possibilities, Yu was by no means alone; Nationalist technocrats had been planning for China's reconstruction in anticipation of war's end since at least 1940. The postwar economic structure they envisioned was increasingly defined by state-centered management and control, in effect an extension of the system they had been putting together in the interior over the course of the war. Representatives from the United States, wartime China's closest ally, balked at this development. American diplomatic officials were invested in shielding American firms in China from regulatory pressures, which these firms had previously been able to avoid because of extraterritorial provisions but which they would no longer be able to ignore after the United States relinquished these ju-

---

134. National Resources Commission correspondence (April 23 and 24, 1942), in Cheng and Cheng, *Ziyuan weiyuanhui jishu renyuan*, 1:87–90.

135. Yu, "Ximei gongye."

dicial concessions in 1943.[136] Still, the Nationalist government stayed its statist course. Essential to its plans was the territorial integration of Manchuria into China. The region's industrial might, cultivated over decades of Japanese rule, was understood as necessary for buttressing the recovering Chinese economy. In the Nationalists' developmentalist discourse, the Fushun colliery and its steel-manufacturing counterpart in Anshan were the bedrock of Manchurian industry upon which China's industrial future would be built. To Weng Wenhao, for instance, failing to "recapture" these important enterprises from Japan would be "tantamount to denying China its opportunity to fully industrialize."[137] Seizing these sites after Japan's surrender and the collapse of its empire turned out, however, to be far from a straightforward affair.

Taking Over the Industrial Northeast

Japan was on the cusp of capitulation when the Soviet Union invaded Manchuria. Early in the morning of August 9, 1945, just three days after the United States dropped the first atomic bomb on Hiroshima and a matter of hours before it released the second over Nagasaki, the Soviet Union declared war on Japan and launched an offensive on the Japanese client state of Manchukuo. Japan surrendered on August 15, bringing World War II in Asia to a close. Just a day earlier, the Nationalist and Soviet governments signed the Sino-Soviet Treaty of Friendship and Alliance, in which the Soviet Union agreed to "render to China moral support and aid in military supplies and other material resources, such support and aid to be entirely given to the National Government as the central government of China" in "the common war against Japan."[138] In return, the Nationalist government assented to a postwar plebiscite for the independence of Outer Mongolia, which was nominally under Chinese suzerainty and which Stalin had deemed of strategic importance to the Soviet Union.[139] During the negotiations over the terms of this treaty, one issue that arose was the Soviet withdrawal from Manchuria, in anticipation of the Red Army's invasion of the region upon its declaration of war on Japan. When pushed on this matter by Nationalist representative T. V. Soong, Stalin contended that "3 months

---

136. Kirby, "Planning Postwar China," 222–26, 229–31.
137. Weng, "Zhanhou Zhongguo gongyehua," 6.
138. "China, Soviet Union," 53.
139. On long-standing Soviet interests in Outer Mongolia, see Elleman, "Soviet Policy on Outer Mongolia."

as a maximum would be sufficient for completion of withdrawal."[140] It would take more than twice as long. Within two weeks of Japan's defeat, Soviet forces occupied all the major cities and communication lines across Manchuria. In the months that followed, Nationalist and Soviet officials engaged in talks about how Manchuria might be handed over to China, even as the Chinese Communists moved into the region and started establishing themselves in the countryside.[141] Manchuria emerged once again as a much-contested site and became, in the words of historian Steven Levine, the "anvil of victory" on which Communist success would be forged.[142]

The Soviet occupation dealt a debilitating blow to the Manchurian economy. Japan, thinking it was there to stay, had invested heavily in the region; the Soviet Union, knowing that it would soon have to leave, extracted mercilessly from it. Soon after their arrival, Soviet troops began systematically dismantling the massive industrial apparatus that the Japanese had set up. They took apart factories and plants, removing their machinery—especially for power generation and transformation—loading it onto trucks and trains, and transporting it back to the Soviet Union as "war booty." A 1947 American intelligence report on Soviet objectives in China contended that this act of industrial looting was to intentionally deindustrialize this region so that it "presents a minimum military threat to the USSR."[143] Others have argued that it was to acquire resources to rebuild industries in the western Soviet Union that had recently been destroyed in the fighting with invading German forces.[144] Either way, as the intelligence report graphically described it, the removal of machinery "cut the heart out of Manchuria's highly developed industrial structure."[145] In June and July 1946, American oilman turned ambassador Edwin Pauley led a mission to survey Japan's remaining industrial assets in Manchuria in order to ascertain the reparations the Allied powers could demand from their defeated foe. In a report he submitted to American president Harry Truman thereafter, Pauley estimated that the material loss from Soviet sacking amounted to as much as 895 million dollars.[146] In an oral history interview decades later, Pauley would remark that Stalin "took everything that was worth a darn out of Manchuria."[147]

---

140. "Record of a Meeting between T. V. Soong and Stalin."
141. On the negotiations, see Gillin and Myers, "Introduction," 7–8, 30–39.
142. Levine, *Anvil of Victory.*
143. "Implementation of Soviet Objectives in China."
144. Levine, *Anvil of Victory,* 68–69.
145. "Implementation of Soviet Objectives in China."
146. Pauley, *Report on Japanese Assets,* 37.
147. Fuchs, "Oral History Interview," 61.

Fushun was not spared. Although the Soviets left most of the equipment in its underground mines "practically intact," they seized "vitally needed power shovels, locomotives, mining cars, and large quantities of maintenance materials and supplies" crucial to the open-pit mine's functioning.[148] The most damage was inflicted, however, not upon the mines themselves but on the power plants that supplied them with electricity. The Soviets took the six best generators from the main plant, leaving behind four, two of which were out of commission. Coal production suffered accordingly. In the Pauley mission's assessment, while there was still enough mining equipment to produce 8,000 tons of coal a day, power shortages kept output at 2,500 tons.[149] A bigger problem was pumping. Because of the colliery's electrification and the mechanization of its processes, the removal of the generators meant that there was "scarcely enough power to operate the pumps much less conduct normal mining operations."[150] This opened up the mines to the threat of flooding, and many of them were inundated in the coming months and years, falling into severe disrepair. Summing up the situation in Fushun, the Pauley mission concluded that the "damage to the Fushun mines exceeded by far the value of equipment removed."[151] Fushun was a technological system held together by energy. Without enough of this energy to maintain its basic functions, the entire system did not merely slow down—it started to come apart.

The National Resources Commission had been preparing for the take-over of Japanese industries since 1942, and its members dispatched to the Northeast after the war, like those sent to other parts of the country in those days of disorder, were at the forefront of such efforts.[152] In Sino-Soviet discussions regarding Soviet withdrawal and the transfer of Manchuria's industrial assets to Chinese possession, the Nationalist government was represented by a team that included influential banker and bureaucrat Chang Kia-ngau 張嘉璈 (1889–1979) and Chiang Kai-shek's son Chiang Ching-kuo 蔣經國 (1910–1988). The younger Chiang, in particular, seemed well suited to the task. An erstwhile Trotskyite fluent in Russian, he was very familiar with the Soviet Union, having spent more than a decade there, including a few years working at Uralmash, the gigantic heavy machinery facility in the Urals nicknamed "the factory of factories."[153] On

148. Pauley, *Report on Japanese Assets*, 78.
149. Pauley, app. 3, Plant Inspection Report 1-L-2, 4.
150. Pauley, 78.
151. Pauley, 78.
152. Zheng, Cheng, and Zhang, *Jiu Zhongguo de Ziyuan weiyuanhui*, 133.
153. Taylor, *The Generalissimo's Son*, 61–62. A third key figure on this team was its leader,

5.3. Nationalist officials hosting a dinner for Edwin Pauley and the rest of the American reparations mission during their visit to Fushun in July 1947. Pauley is seated right beneath the painted portrait of Sun Yat-sen, "the father of the Chinese nation." To his left is Chang Kia-ngau, the banker and bureaucrat who had represented the Nationalist government in Sino-Soviet negotiations concerning the Chinese takeover of Manchuria. (Image courtesy of the Harry S. Truman Library and Museum.)

the ground, this transfer was to be carried out by National Resources Commission personnel.

Sun Yueqi, appointed by the Ministry of Economic Affairs as special commissioner to the Northeast, assumed responsibility for coordinating the takeover. When he first received this appointment in September 1945, Sun, who was based at the Yumen oilfield at the time, placed announcements in newspapers and put up notices all around to recruit technical and managerial staff for the work in Manchuria. Before long, around 1,200 individuals from the commission's various factories and mines were ready to go. Their advance was, however, blocked by Soviet authorities, who had been similarly delaying and restricting Nationalist troop movement into the region. Only in March 1946, when Soviet forces started pulling out,

chairman of the Northeast headquarters of the Military Affairs Commission Xiong Shihui 熊式輝 (1907–1974). See Gillin and Myers, "Introduction," 8.

5.4. Fushun's main power plant after Soviet troops looted its generators. The loss of its two hundred thousand kilowatt capacity threatened operations at the colliery, particularly in terms of the mines flooding because of the lack of electricity to run the pumps. This also resulted in nearby urban areas, most notably Mukden, losing electric service since they too relied on this power plant. This photograph was taken during the American reparations mission led by Edwin Pauley and is attributed to Marlin E. Fenical, United States Army Signal Corps. (Image courtesy of the Harry S. Truman Library and Museum.)

could the commission begin to send personnel in on a large scale to seize industrial Manchuria and, in so doing, to finally secure "the basis," as Sun described it, "of China's present reconstruction."[154]

An earlier Nationalist attempt at industrial recovery—specifically of the Fushun colliery—ended in tragedy. In talks with Soviet officials of the Changchun Railroad (the postwar administrative merging of the South Manchuria Railway and the former Chinese Eastern Railway, which Manchukuo had purchased from the Soviet Union in 1935) in late December 1945, Chang Kia-ngau had discussed an anticipated rise in the demand for coal to fuel increased military transport by rail with the presumed arrival of Nationalist troops and the withdrawal of Soviet ones.[155] Within that context, he thought it essential that the Nationalists claim the Fushun coal

154. Zheng, Cheng, and Zhang, *Jiu Zhongguo de Ziyuan weiyuanhui*, 152; Sun Yueqi keji jiaoyu jijin, *Sun Yueqi zhuan*, 177–78; Sun, "Zhongguo gongye de qiantu," 43.
155. Shen, *Mao, Stalin, and the Korean War*, 213n54.

mines. The man he designated for that job was National Resources Commission engineer Zhang Shenfu 張莘夫 (1898–1946).[156]

Zhang's life began—as it would end—in Manchuria. Born in a village in Jilin Province, he traveled far beyond seeking a purpose larger than himself. As a young man, he won a place at Peking University and, not long after, a scholarship to study in the United States. There, he specialized in mining engineering at the Michigan School of Mines. His choice of subject seemed to stem in part from genuine interest ("I cleaned a coal-cutting machine today and took it apart. It was most interesting," he jotted down in his diary) but mostly from the belief that this would allow him to best serve his country ("If China wants to become prosperous and strong, it must develop its steel industry; otherwise it cannot resist the encroachments of foreign nations.").[157] Like many educated Chinese of his time, Zhang placed much faith in national salvation through science and technology.[158]

Zhang's first job upon returning to China was as an engineer at the Muling colliery. Putting his technical training to use, he thrived in this role. A close friend and colleague from that time remembered him as "full of brilliant ideas" whenever they "discussed engineering improvements" and "never the least bit inattentive to his work" even "when it was bitterly cold at thirty below zero."[159] At Muling, Zhang met and befriended Sun Yueqi. When Sun left for further studies abroad, Zhang replaced him as the technology department's director. Several years later, the two found themselves working together again, this time at Zhong-Fu, where Zhang assisted Sun in the ponderous task of shipping the colliery's mining machinery inland in the early phase of the war. He then joined Sun at Tianfu, where most of that machinery ended up, having been brought in by his old friend to help run this massive enterprise as its colliery manager. When Sun was appointed special commissioner to the Northeast after the war, he approached Zhang to come on board once more as his deputy. When it became apparent that

156. Gillin and Meyers, *Last Chance in Manchuria*, 197. Zhang's granddaughter, journalist and author Leslie Chang, wrote about his life and death in her book on rural–urban migration in contemporary China. See Chang, *Factory Girls*, 120–48. The first character of Zhang's personal name (莘) can also be read as *xin*, and it is rendered as such in many mentions of Zhang in English, but Leslie Chang's account makes clear that it should instead be voiced as *shen*, as his name, which he adopted during his time in the United States, was a truncation of the phrase *shenshen zhengfu* (莘莘征夫) from the predynastic *Discourses of the States*, which Chang translates as "many diligent men drafted into service." See Chang, 138.

157. Cited in Chang, *Factory Girls*, 135.

158. On scientific nationalism among this generation of mostly foreign-trained scientific elite, see Wang, "Saving China through Science."

159. Cited in Chen, "Zhang Shenfu xunguo," 40.

the Soviets would not allow the industrial takeover corps they assembled to enter Manchuria, though, Sun turned to another task, accepting an invitation from T. V. Soong to lead the requisition of enemy assets in North China, while Zhang flew on to the Northeast with his friend Dong Wenqi, who was going to assume his post as mayor of Mukden. Not long after he arrived in the region, Zhang received the request from Chang Kia-ngau to go take over the Fushun colliery.[160]

Zhang Shenfu was most likely excited about this assignment. More than a decade earlier, following Japan's occupation of Manchuria, he had written an article about how China's national defense depended on developing its heavy industries, particularly the "interlocking" metal, fuel, and machinery industries. "Making iron requires coke," he wrote, "making coke requires coal, and machines are made of metals and powered by fuels." In it, he mentioned the Fushun colliery as part of an "unrivaled" nexus of iron and coal production that had "unfortunately" fallen to foes: "Have we not occasion to sigh in indignation?"[161] Now, the opportunity to reclaim this important site had presented itself. There was talk of danger, but Zhang, who, according his wife, "bore the nickname 'energetic Zhang Fei' [a general from the Three Kingdoms period famous for his rash bravery]," was undeterred.[162]

Zhang set off for Fushun by train the morning of January 14, 1946, leading a team of engineers and accompanied by several armed police. Upon arrival in Fushun, the group was received by Soviet troops, who brought them to the old Mantetsu Colliery Club in Yongantai. The guns their escorts carried were summarily seized, and the group was confined to that

160. Xue, *Gongkuang taidou Sun Yueqi*, 120–22; Sun Yueqi keji jiaoyu jijin, *Sun Yueqi zhuan*, 177–78, 196; Dong, "Wangyou Zhang Shenfu"; Chang, *Factory Girls*, 142–43, 146; Gillin and Meyers, *Last Chance in Manchuria*, 182, 195; Sun, "Wo he Ziyuan weiyuanhui," 33–34. There is slight dispute over who should take the blame for sending Zhang Shenfu to Fushun and thus inadvertently to his death. According to Dong Wenqi, Sun Yueqi was the one who was supposed to go to Fushun, but he was afraid of doing so and sent Zhang in his stead. Leslie Chang's narrative follows this explanation. That Sun was given charge of requisitioning enemy assets in North China by T. V. Soong after the Soviets denied his industrial takeover corps access to Manchuria is at least true. As to whether he used this or other business as an excuse to avoid going to the Northeast is less clear. In Chang Kia-ngau's diary, he mentions deciding to send Zhang to Fushun and asking Zhang to go. He does not state that this was in lieu of Sun. Sun's own account of Zhang's murder seems defensive. He claimed that once he realized that the Soviets would not grant his industrial takeover corps passage into the region, he dissolved the corps, and Zhang then left for the Northeast with Dong on his own accord. Still, one wonders, as Chang does, whether Sun felt at least partially responsible for the death of this friend and colleague of almost two decades.

161. Zhang, "Woguo de guofang," 158, 161.

162. Pan, "Zhang Shenfu furen."

compound for the next two days. On the evening of January 16, their Soviet hosts told Zhang outright that they would not be allowed to take over Fushun and that they should leave right away. The group was then ushered to the railway station, where they got on a train bound for Mukden. When the train stopped at Lishizhai, a sleepy station to the west of Fushun, a gang of men boarded the carriage that Zhang and his fellow engineers were on and forced them off. They were then brutally murdered, bayoneted to death in the black and cold of that merciless winter night.[163]

Zhang's death shook the nation. "That Mr. Zhang Shenfu, as one of China's few mining and metallurgy experts, would . . . be murdered in cold blood is of the utmost loss to our country and our people," one newspaper report lamented.[164] In the following months, government officials and engineering associations across China's principal cities organized memorial services that drew tens of thousands. At one held in Mukden, Zhang's bloodstained clothes and a photograph of his bayonet-pierced body were displayed.[165] An army performance troupe, on the premise that Zhang's death was a tragedy "worth singing about and crying over," wrote a traditional opera based on it titled "Regrets upon a Triumphant Return" (凱旋恨), which it then put on for soldiers to spur them "to tread upon the bloodstains of martyr Zhang as they proceed to complete their unfinished mission."[166] These efforts to mourn the murder of a patriot were concurrently attempts to stir up calls for vengeance against those responsible.

As to who had killed Zhang and his compatriots, this was a question of contention. Soviet authorities, when confronted, pointed the finger to "local bandits."[167] The Nationalist government and the Chinese press in general accused the Chinese Communists.[168] In August, half a year later, the Nationalists arrested and then executed Communist company commander Mo Guangcheng, from whom they extracted a confession to having participated in this deed.[169] Back in February, when the news of Zhang's slaying belatedly broke, Chinese public opinion had been turning against the Soviet Union for its continued occupation of Manchuria. Suggestions that the Soviets were culpable in the murder—beginning with the fact that it happened in an area supposedly under their jurisdiction but also given their

163. Gillin and Meyers, *Last Chance in Manchuria*, 223.
164. "Canzao shahai de Zhang Shenfu."
165. "Shenyang gejie zhuidao Zhang."
166. "Zhang Shenfu gushi pingju."
167. Gillin and Myers, *Last Chance in Manchuria*, 261–62.
168. Cao, "Guanyu Zhang Shenfu zhongzhong."
169. "Zhang Shenfu an xiongfan."

support for the Communists—further fueled anti-Soviet sentiments that erupted in demonstrations throughout major urban centers.[170] These demonstrations alone did not prompt the Soviet departure from Manchuria, but they pushed the Nationalist government to harden its position against extending any economic concessions to Moscow, exactly the opposite of what the Soviet troops' enduring presence in the region was intended to achieve. On March 11, Soviet authorities suddenly announced that they would withdraw their forces.[171]

The Soviets left Fushun on March 15. Chinese Communist operatives, led by Wang Xinsan 王新三 (1914–1990), a Fushun native with a degree from the Counter-Japanese Military and Political University in Yan'an, had come two months earlier, days before Zhang Shenfu's ill-fated visit, in a bid to beat the Nationalists in the race to take over the mines. Wang and his comrades pulled out soon after the Soviets departed, on March 21. Wang would later return to Fushun to oversee the colliery after the Communists seized the area in October 1948.[172] For now, though, the Nationalists finally held Fushun.

Although Zhang Shenfu did not manage to take over the colliery, in an indirect way, his death, in helping to hasten the Soviet withdrawal, facilitated the Nationalist control of Fushun. Others perished in similar industrial recovery attempts. Yu Zailin, the coal-washing expert who participated in the National Resources Commission's first training program in the United States, was shot and killed when he went to take over the Beipiao colliery, also in Manchuria. One account would call him "the second Zhang Shenfu."[173] Selected for their technical expertise, these engineers entered exceptionally perilous situations for which they could hardly prepare. Their deaths foreshadowed the difficulties the Nationalists would continue to face in the Northeast even as they took on the herculean task of reviving industry on the ground while the storm clouds of battle gathered above.

The Nationalist Resources Commission in Fushun

The Nationalists found Fushun in disarray. After the Soviets departed, troops from their Fifty-Second Army marched in and secured the site before a team of Ministry of Economic Affairs personnel headed by National

170. "Chungking Students Strike"; "Aiguo hu quan"; Wasserstrom, *Student Protests*, 247–48.

171. "Chiang Assures Nation." On possible reasons why the Soviet Union withdrew from Manchuria when it did, see Levine, *Anvil of Victory*, 78–79.

172. *FMBN*, 331–32.

173. "Zhang Shenfu di er."

Resources Commission engineer Cheng Zongyang 程宗陽 (1892–1977) arrived to restore production. The colliery's capacity, crippled by Soviet looting of machinery, had been further hamstrung by Soviet mining methods. At the open pit, the Soviets, as one account noted, "went all out to just extract coal," and the essential work of "stripping overburden fell to a standstill." In the subsurface mines, they were similarly "only interested in mining coal and not in engineering considerations, to the extent that they even felled the pillars of coal supporting the inclined shafts and haulage ways, which resulted in destroyed tunnels that collapsed in endless succession and obstructed ventilation that led to many areas that could have otherwise been mined being sealed off because of coal-seam fires."[174]

Cheng, a mining expert who had studied abroad at the Massachusetts Institute of Technology and Columbia University and who previously succeeded Zhang Shenfu as Tianfu's colliery manager, had his work cut out for him.[175] One of the first things the new Nationalist management did as recovery efforts began was to rename the Ōyama mine after Zhang Shenfu, a commemorative act to honor this martyred engineer whom Fushun's leaders would hold up as a paragon of perseverance ("not yielding even unto death") as difficult days continued.[176] Half a year later, the National Resources Commission formally took over the colliery with Cheng as its manager. He resigned after just a month. Over the next year or so, his replacement, Xie Shuying, broke his back in his attempt to return Fushun to former productivity.

Securing technical expertise was a priority, and the commission not only brought in Chinese engineers and technicians from other parts of China but, notably, also turned to Japanese experts left in Fushun for their knowledge and skill.[177] The use of Japanese experts here was in keeping with national policy. The Nationalist government, recognizing that many of the factories and mines it was seizing would benefit from continued Japanese expertise, introduced regulations for retaining and using Japanese

174. Xie, "Cong jianku zhong fendou," 230.

175. Wang, "Kuangye zhuanjia Cheng Zongyang."

176. "Jinian Zhang Shenfu xiansheng," 361.

177. There were a few other foreigners recruited to work in Fushun at this time. For example, German electrical engineer Wilhelm Beyer, who had worked at the Benxihu colliery since 1939 and who was stranded in Manchuria after the war, was recruited by the commission to help with the restoration of the heavily damaged Fushun power plant, an experience he described as "fruitful technical collaboration with my numerous Chinese colleagues, with whom I could work in the sense of true comradeship." See Wilhelm Beyer and Chen Zhongxi (n.d.), file no. 003-010102-0058, Academia Historica, Taipei.

personnel in late 1945.[178] According to one tabulation from December 1946, there were around fourteen thousand Japanese technical specialists employed in postwar China, over ten thousand of them in Manchuria.[179] In Fushun, these experts were initially "relegated to mere manual or miscellaneous work," as Yamazaki Motoki 山崎元幹 (1889–1971) observed when surveying the site. Yamazaki, who was Mantetsu's last president, had been taken on first by the Soviets and then by the Nationalists as an industrial adviser. In regard to Fushun's Japanese engineers and technicians, he gingerly suggested that it might be better were they to serve their "original purpose."[180] Employing Japanese experts probably became more attractive as the conflict in Manchuria escalated and the number of Chinese willing to venture into the region for work fell.[181] Either way, it appears that carbon technocracy's immediate demands for technical guidance and support came to override more nationalistic considerations of punishing the vanquished. The commission arranged for many Japanese experts to return to the work they did before, dividing them into three groups based on how soon they were expected to be replaced by Chinese counterparts—six months, a year, or two years. When opportunities for repatriation to Japan emerged, however, most of them took off.[182] (If Japanese experts made a significant contribution to Fushun's reconstruction, it was after the Communist takeover, when a few of them who stayed on assisted in the socialist state's industrial drive. This will be addressed in the next chapter.)

Important as it was, technical expertise was not enough to push Fushun forward. Unrelenting shortages on several fronts—equipment, power, food, capital—held back advances. "Even an ingenious housewife would find it difficult to make a meal without rice," one commentator remarked, recognizing the unenviable position Xie Shuying, as overseer, occupied.[183] In terms of equipment, in addition to the "war booty" carted off by Soviet troops, there were machines that were purportedly "destroyed or seized by

178. Yang, "Resurrecting the Empire," 190–91.

179. Cited in Yang, 205.

180. Yamazaki, Bujun tankō shucchō, 25.

181. For instance, in 1948, Fushun's chief engineer Wang Rulin was promoted to deputy colliery manager but was still expected to do his original job as well, which might have been a cost-saving measure, as the raise Wang was given was less than paying for two positions, but could also have been a result of insufficient qualified personnel. See Wei Huakun to Sun Yueqi (n.d.), file no. 003-010102-2208, Academia Historica, Taipei.

182. Pauley, Report on Japanese Assets, app. 4, Plant Inspection Report 2-B-1, 4; Fushun kuangwu ju meitan zhi bianzuan weiyuanhui, Fushun kuangqu dashiji, 83. On postwar Japanese repatriation, see Watt, When Empire Comes Home, 1–137.

183. Gong, "Fushun meikuang zhi yan'ge," 31.

Communist soldiers as they withdrew."[184] A large number of tools and materials had also been taken by workers and other locals in the chaos of regime transition.[185] At the open pit, increased rockfalls resulted in valuable
machinery—ironically, otherwise used to prevent those rockfalls by stripping overburden—being buried under rubble.[186] Moreover, the difficulties
Mantetsu had faced with importing new equipment after the start of the Allied blockade meant that many of those that remained now were outdated
or worn from use. Among the things in shortest supply were, one commission report detailed, detonator caps, electric picks, all devices for power
generation, rolling stock, and spare parts for various machines.[187] This
problem of material deficiency ran rampant across mines. Other collieries
under the commission reached out to Fushun to ask whether it might send
them equipment for their operations—requests Fushun, unable to keep
up with its own needs, invariably turned down.[188] Fushun itself actively
sought to procure equipment from elsewhere, perhaps most interestingly
purchasing several key devices, like flexible shafts, from former colonizer
Japan through arrangements with the Allied powers who now occupied the
archipelago nation.[189] Where it could, the colliery tried to meet its own
needs. It restarted its machine manufacturing and maintenance operations
and, within a year, was able to produce equipment from files and milling
cutters to electric winding machines.[190] As supplies of timber for pit props
dwindled, the colliery also began looking into afforestation with acacia.
Although acacia yielded wood that was less suitable for this purpose than
the larch that was typically used, it had seeds that were more readily available.[191] In spite of these efforts, the challenge of securing sufficient equipment and materials remained.

As mentioned earlier, the biggest blow that the Soviets dealt Fushun
was in gutting its power-generating facilities, which greatly reduced its coal
output and dragged the colliery into general decline. By one count, Fushun's power-production capacity fell by three quarters as a result of Soviet

184. Gong, 26.

185. Meidu Fushun bianxie xiaozu, *Meidu Fushun*, 55.

186. Cui, "Lutiankuang chang xianzai yu jianglai," 205.

187. Ziyuan weiyuanhui Fushun kuangwu ju, *Gongzuo shuyao* [Overview of work] (May 16,
1947), 8–9, file no. 003-0101301-0235, Academia Historica, Taipei.

188. Bo Zuoyi and Wei Huakun (March 17 and April 13, 1948), file no. 24-12-05-01-01,
Institute of Modern History, Academia Sinica, Taipei.

189. Fushun colliery manager and National Resources Commission (n.d.), file no. 24-12-
05-01-01, Institute of Modern History, Academia Sinica, Taipei.

190. Xian, "Benju jiwu chu gaikuang," 290.

191. Tsuboguchi, "Tan yanghuai zaolin."

sacking.[192] The Nationalists prioritized recovering that lost capacity from the start. Still, power shortages were a common occurrence in this period. One notable incident happened in late 1946, when Sun Yueqi, now deputy chairman of the National Resources Commission, was visiting Fushun and other industrial areas in Manchuria. As testament to the importance of this site of energy extraction, which many still saw as "not only the cradle of Northeast China's revival but also a great center for the rejuvenation of industries throughout the country," Fushun drew a number of high-level Nationalist visitors like Sun. Others included commission chairman Qian Changzhao and Executive Yuan president Zhang Quan.[193] Around the time of Sun's visit, "severe troubles" with the boilers and machinery at the main power plant led to insufficient electricity being generated. As a result, much of the work across the colliery "ground to a halt." In the interest of "safeguarding military supply and civilian use," Sun ordered that, "as there is not enough electrical power, [the colliery] must try its utmost to bring together human power to mine coal [盡量集中人力採煤]."[194]

Even in less urgent times, the labor of workers was integral to the colliery's functions under the commission. This was consistent with the contradiction in carbon technocracy of an enduring reliance on human energy in spite of the impulse to replace it with carbon energy. "In the modern era, the invention of machines has only managed to reduce the waste of human power and has not been able to fundamentally 'automate' [自動] everything," the local Fushun periodical concurred.[195] When the Nationalists took control in April 1946, workers at the colliery totaled sixty-five thousand, a significant drop from the ninety-five thousand right before Japan's surrender but a sizable number nonetheless.[196] Although not subjected to the same degree of direct abuse that characterized the squeeze during the late stage of the preceding war, these workers were exposed to dangers from an unraveling environment. According to one report, major accidents in Fushun's subsurface workings occurred twenty-nine times in 1944, fifty times in 1945, and thirty times in the five months since the Nationalist takeover, with increased incidences of flooding and fire due to the diminished ability to pump and vent the mines.[197]

192. Yamazaki, *Bujun tankō shucchō*, 34.
193. Bao, "Fushun de shiming"; Fushun kuangwu ju meitan zhi bianzuan weiyuanhui, *Fushun kuangqu dashiji*, 84–88.
194. "Fushun kuangwu ju dashiji," 326.
195. "Fukuang yijia."
196. Yamazaki, *Bujun tankō shucchō*, 30.
197. Yamazaki, 15–16, 36.

Beyond heightened hazards, workers faced everyday hardships as the colliery struggled to get back on its feet. Xie Shuying's call for "striving in the midst of adversity" in his anniversary address to the colliery's employees was a refrain repeated by other commission leaders during the Nationalists' short and troubled tenure in Fushun. Specifically, these hardships stemmed mainly from the workers not receiving their wages from the colliery on time or at all. During their inspection of the Fushun coal mines, the team from the Pauley mission spotted "a group of men around a large blackboard notice" by the open pit. When the visitors asked, the men shared that the notice concerned the colliery's "efforts to obtain funds to pay them for the month the mine was under Soviet management," as they had "received no pay that month."[198] The Nationalists were scouring for funds to remunerate these workers for the labor their predecessors had failed to compensate. But problems over meeting wages would continue to dog Fushun's management. And because these wages were paid in a mixture of cash and grain, delays in their dispensation quickly became for these workers a question of subsistence.

Underlying the National Resources Commission's consistent inability to pay Fushun's workers on time or, in fact, sufficiently was a larger crisis of capital. The challenge was twofold. For one thing, the country was in the throes of debilitating inflation triggered by the government constantly printing money to close the gap between its income and expenditure.[199] Food prices were both an index of this issue and themselves a source of concern. For instance, the price of sorghum per catty was 700 to 750 yuan at the beginning of November 1946, 1,000 yuan by the end of the month, and 1,500 yuan by the middle of December. By January, it hit 3,500 to 3,600 yuan.[200] Apart from confronting the prospect of raising cash wages to keep pace with such increases, the commission needed to spend more capital on purchasing the grain with which it also compensated its workers. This then related to a second problem.

In line with carbon technocracy's ideal of having cheap energy flow abundantly into all other industries, the Nationalist regime tightly controlled the price of coal and kept it low. As such, extractive sites like Fushun soon found themselves not bringing in enough from coal sales to recoup operating costs, including cash wages and grain purchases. In December

198. Pauley, *Report on Japanese Assets*, app. 4, Plant Inspection Report 1-B-4, 6.

199. On inflation during the Chinese Civil War, see Pepper, "KMT-CCP Conflict," 741–46.

200. Xie Shuying to Weng Wenhao (January 20, 1947), file no. 24-12-05-04-01, Institute of Modern History, Academia Sinica, Taipei.

1946, the cost of producing a ton of coal in Fushun was 280,530 yuan, but each ton was then sold for considerably less, at 86,000 to 88,000 yuan. In a message to commission chairman Weng Wenhao in early 1947, Xie explained that while the colliery was initially able to "use stored materials or approach the banks for loans," these options had since been "exhausted," and "production volume is falling and commodity prices rising by the day." Given widespread inflation, coal's price inertia was jarring: "The price of food and consumables can change several times in one day, while coal prices are recalibrated at most once a month." In exasperation, Xie concluded that "the reason why coal prices have not been able to support production costs should be obvious."[201] The statist imperative to suppress the price of this energy resource to more readily fuel the nation's larger industrial apparatus proved unsustainable for those involved in its extraction.

The Nationalists grappled with this problem throughout their time in Fushun. In terms of food, Xie suggested to Weng that the commission set a fixed exchange rate for coal to grain (以煤易糧) so as to ensure a steady supply of food even if its price increases outpaced those of coal.[202] The commission tried this measure in the coming months. However, once the fighting between Nationalist and Communist forces began in earnest in Manchuria in the spring of 1946, transportation was hampered, and the inflow of grain became sporadic.[203] In September 1947, the colliery purchased twenty thousand sacks of wheat flour from Shanghai to give out to its workers. During distribution, accusations of unfair rationing arose at the Longfeng mine, and more than three hundred workers armed themselves and rioted, cutting power lines and beating up the manager. Having endured hunger, these workers were unwilling to stomach inequity. Xie himself had to go and diffuse the situation.[204] As hardships continued, many workers, rather than waiting for outstanding wages without assurance of payment, just left, further jeopardizing production.[205] The troubles Fushun faced with food extended to other essentials for its operations, exacerbating shortages in equipment and materials.[206] The commission had to

201. Xie Shuying to Weng Wenhao (January 20, 1947), file no. 24-12-05-04-01, Institute of Modern History, Academia Sinica, Taipei.

202. Xie Shuying to Weng Wenhao (January 20, 1947), file no. 24-12-05-04-01, Institute of Modern History, Academia Sinica, Taipei.

203. "Fushun meikuang muqian kunnan," 366. On the fighting in Manchuria and its significance for the larger civil war, see Tanner, *Battle for Manchuria*.

204. Xie Shuying to Wang Wenhao (October 13, 1947), file no. 003-010309-0550, Academia Historica, Taipei.

205. *FKZ*, 54.

206. "Fushun meikuang muqian kunnan," 367.

borrow heavily from the central government to keep the colliery running. By October 1947, Sun Yueqi reported that Fushun was twenty billion yuan in debt.[207]

Ultimately, Fushun was unable to claw its way out from under the weight of the civil war, in which Manchuria was the central theater. Marginal improvements made in the recovery of the colliery were soon undermined by the fighting taking place around it.[208] Xie tendered his resignation in July 1947, citing an "extreme exhaustion of both vitality and strength."[209] It was six months before he was relieved. The next colliery manager, Wei Huakun, who came from overseeing the Beipiao coal mines to the west, only took office in February 1948.[210] When Sun Yueqi offered him the position, Wei "gave it much thought" before deciding that his coming to Fushun would be "akin to saving a sick person . . . no matter what, [he was] determined to not let it pass away."[211] He spent most of his time shuttling between Fushun, Mukden, and areas afar to raise funds and acquire grain to try and sustain the colliery's operations.[212] Wei was the last National Resources Commission representative to lead Fushun. Just over half a year after he arrived, the Chinese Communists succeeded in seizing the Coal Capital, concluding the Nationalist chapter in the history of this energy enterprise.

Conclusion

The story of coal and carbon technocracy in China from the 1920s to the 1940s speaks to a central tension in the history of the Nationalist regime: strides in state building on the one hand and ultimate collapse on the other. Chiang Kai-shek's "loss of China" (as anticommunist American rhetoric defined it) arose as a topic of interest right after the Nationalist defeat. Although some cited exogenous factors, including Soviet assistance to the Communists in Manchuria, many explanations centered on the failings of the Nationalist state. As a dominant argument went, venality and incompetence rendered its inception in the 1920s an "abortive revolution" and ensured its due demise through the wars of the 1930s and 1940s.[213] Shift-

207. Sun Yueqi keji jiaoyu jijin, *Sun Yueqi zhuan*, 197–98.
208. Gong, "Fushun meikuang zhi yan'ge," 26.
209. Xie Shuying to Weng Wenhao and Sun Yueqi (July 22, 1947), file no. 24-12-05-04-1, Institute of Modern History, Academia Sinica, Taipei.
210. "Wei Huakun jizhang Fukuang."
211. "Juzhang xunci," 433.
212. *FKZ*, 54.
213. Eastman, *The Abortive Revolution*; Eastman, *Seeds of Destruction*.

ing the focus away from the question of failure, studies in recent decades have paid more attention to how the Nationalist government worked, particularly through the institutions it built to collect taxes, rationalize administration, police smuggling, and fulfill other functions of state with a measure of success. Many of these institutions outlived the regime in the mainland or survived with its transplanted instantiation in Taiwan.[214]

In the case of coal, the Nationalist state, represented predominantly by the National Resources Commission, managed to tighten control over the industry and encourage its expansion in the Chinese interior during the Second Sino-Japanese War. It accomplished these goals by transferring mining machinery inland, establishing or expanding collieries, carrying out research into coal-washing and coking technologies, facilitating technical assistance from large mechanized state-operated mines to the scores of smaller privately run ones, and introducing measures to secure capital, equipment, and labor for enterprises across the industry. In so doing, the Nationalist state was able to obtain the coal it needed to jump-start wartime industrialization, allowing it to begin to realize in the interior its long-standing vision of national reconstruction—if on a much-diminished scale.

At the same time, the troubles the National Resources Commission had with reviving production at Fushun after the war point to the limits of statist control. Certainly the commission faced highly unfavorable circumstances, from the Soviet looting to the burgeoning civil war. Yet the decision to hold down the price of coal amid the inflationary spiral also undermined the enterprise from the start. Perpetually unable to recover its operating costs, the cash-strapped colliery fell behind in purchasing essentials and paying workers. That the National Resources Commission subjected workers here to suffering was not, however, just an inadvertent outgrowth of wartime constraints. It was just as much a natural progression of a technocratic vision that (departing from Sun's, which at least gestured to people's livelihood) prioritized short-term industrial output over the human beings who at the end of the day determined the long-term legitimacy of the regime. The sacrifice that Fushun's management asked of its workers may have been a common call in national mobilization during hard times, but there was a particular cruelty in undersupplying calories to those whose labor made possible the extraction of calories for the industrial economy and the war effort.

Looking at the workings of carbon technocracy in Nationalist China

214. See, for example, Kirby, *Germany and Republican China*; Strauss, *Strong Institutions in Weak Polities*; Bian, "Building State Structure"; and Thai, *China's War on Smuggling*.

alongside the same in 1930s and 1940s Japan, we may be struck by how, if we were to use the textbook definition of "technocracy" as "a government of engineers," the Chinese state actually appears to have come closer to that ideal than its Japanese counterpart. Although wartime Japan had its Cabinet Planning Board—which, like China's National Resources Commission, oversaw economic planning at the highest level of state—it was not staffed by the likes of scientists and engineers such as Weng Wenhao or Sun Yueqi, who were formally trained scientific and technical experts firmly ensconced in the corridors of power (Weng would even, if only for half a year in 1948, serve as the president of the Executive Yuan).[215] Still, the similarities between the two energy regimes exceed the differences.

Carbon technocracy as state-driven planned development of coal and other fossil fuel resources emerged in both China and Japan out of crises that prompted leaders to embrace energy self-sufficiency and pursue intensive extraction. Both appealed to the authority of science and technology and mobilized engineers as they embarked on these endeavors. For one as it would be for the other, war conditions first led to the consolidation of technocratic initiative around the management of the coal sector and subsequently to the coming apart of that energy regime. That the two looked so alike in spite of China having more engineers at the helm suggests that it may be helpful to think of technocracy less in terms of the number of engineering experts holding high office and more in terms of the extent to which a narrowly defined technoscientific rationality wedded to a proclivity for planning directed the practice of statecraft, regardless of who filled the upper echelons of government. The Communists who came after were certainly fond of lumping the Nationalists and the Japanese together as abusive, extractive regimes that they had vanquished. Their similar pursuit of fossil-fueled development would, however, take them down some of their predecessors' most well-trodden and twisted paths. It is to that last chapter that we will now turn.

215. As historians Hiromi Mizuno, Janis Mimura, and Aaron Stephen Moore have shown, Japanese engineers were able to carve a space for themselves in government in the 1930s, with one important example being Miyamoto Takenosuke and the Asia Development Technology Committee he helped found. As they note, however, this committee played at most an advisory role to the Cabinet Planning Board and did not have the power to make policy. See Moore, *Constructing East Asia*, 90–91.

SIX

Socialist Industrialization

In 1954, the intractable proletarian writer Xiao Jun 蕭軍 (1907–1988) published a socialist realist novel titled *Coal Mines in May* (五月的礦山). Set in the fictional coal-mining town of Wujin (烏金; literally, "black gold"), the novel centers on a production competition during the "red month" of May. The year is 1949, and this month follows Wujin's first Labor Day after its "liberation" (解放)—the emancipatory term the Chinese Communists used to refer to areas that had come under their control. While the Communists' national victory is still several months away, the People's Liberation Army, having already taken the Northeast, where Wujin is supposedly located, is making swift and significant gains in its southward advance. In an effort to "celebrate their enthusiasm" and to "show appreciation for their armed brethren at the frontlines," Wujin's miners decide to "dedicate labor" through a competition in which various units within the colliery vie to see who can produce the most coal.[1]

The night before the competition is scheduled to begin, however, a sudden blast rocks the explosives storehouse at the open pit. According to rumors, this was the work of "hidden reactionary Nationalist special agents" who had plotted to disrupt the competition by sabotaging the mine—an act aimed at "not just scattering sand into [the people's] rice bowls but also smashing those rice bowls to smithereens." While the incident casts a shadow of suspicion across the colliery, the competition proceeds as planned. Luckily, essential materials like explosives and detonator caps had been distributed to work teams a few days in advance. The explosion thus does not end up holding operations back. In fact, it only serves to fuel

---

1. Xiao, *Wuyue de kuangshan*, 26–27.

a sense of outrage among the miners that, in turn, boosts their morale "to the point of bursting."[2]

Wujin was a carbon copy of Fushun. Its backstory, for starters, matched Coal Capital's exactly: mined by Koryŏ Koreans a thousand years earlier, closed by the Manchu Qing rulers for fear of damaging the "dragon veins" linked to their imperial progenitor's grave, reopened by Chinese and Russian business interests at the turn of the twentieth century, seized by Japanese forces during the Russo-Japanese War, and kept under Japanese control for forty years thereafter.[3] This similarity was no coincidence. Xiao Jun was in Fushun between 1949 and 1951 and based the novel on his experiences there.

Born in Manchuria, Xiao Jun had, as a young man, dreamt of a life in the saddle, riding from one adventure to another. The closest he got to becoming a "mounted bandit hero" was joining the local military. After a few years, his penchant for questioning authority and distaste for the less than romantic realities of army life prompted him to consider other options. Channeling his energies into writing, Xiao Jun made his name with his 1935 debut novel *Village in August*, a gripping account of anti-Japanese warfare waged by a motley crew of guerilla fighters.[4] Soon after, he fell in with the Communists and traveled to their base in Yan'an, where he remained for the duration of the war with Japan. Returning to Manchuria with the People's Liberation Army in 1946, Xiao Jun was tasked by the party to run the new weekly *Cultural Gazette*, a journal that broadly covered all things literary and cultural. In keeping with his inability to unquestioningly toe the party line, he published in it articles that criticized the Communists for their alliance with the Soviet Union—which he saw as accommodating Russian imperialists—and their land reform policy—which he regarded as "an unprecedented act of robbery." The party denounced him as "a talent but a hooligan" and sent him to Fushun, ostensibly to be reeducated through hard labor. There, he ended up working in the colliery's documents office.[5] Drawn in by what he observed on the ground, Xiao Jun decided from early in his time in Fushun that the site—and its open-pit

2. Xiao, 167–68.

3. Xiao, *Wuyue de kuangshan*, 4.

4. *Village in August* has the distinction of being the first contemporary Chinese novel translated into English.

5. Lee, *Romantic Generation*, 222–28, 240–44; Goldman, *Literary Dissent*, 70–86. The quoted phrase about land reform policy is from Lee, *Romantic Generation*, 241. The quoted phrase about Xiao Jun being dismissed as "a talent but a hooligan" is from Xiao, *Dongbei riji*, 619 (April 2, 1949).

mine in particular—would make a good "literary backdrop" for a piece of writing.[6]

Xiao Jun started work on *Coal Mines in May* not long after leaving Fushun for Beijing in 1951.[7] If we read the novel alongside the diary he kept over that period, it quickly becomes apparent how heavily he derived his fiction from fact. The main episodes in the novel correspond closely to highlights of his stay in Fushun during those transitional years as the new socialist state worked to revive and increase production at this war-ravaged site of energy extraction. The novel's protagonists, whose selfless diligence Xiao Jun went to great lengths to portray, are the laboring masses and their "model workers" (勞模)—paragons of the socialist proletariat singled out for recognition by the party. As it was in reality then and over the decade that followed, these workers gave much of themselves, often at great cost, to meet ever-escalating output goals and to feed an industrializing state ravenous for carbon resources.

This chapter traces that history in Fushun and across the Chinese coal-mining sector more broadly, from the colliery's liberation in 1948 to the end of the 1950s. It begins with the first few years of recovery, during which the restoration of operations at Fushun benefited not only from Soviet assistance but also from retained Japanese technical expertise. If carbon technocracy became a framework under which the Communists, too, pursued their vision of progress, then the legacy of imperial Japan was essential in furnishing many of the techniques and technologies that underlay the establishment of this framework.

Coal came to occupy a key position in the socialist state's industrialization drive—as best exemplified by the popular slogan "coal is the grain of industry" (煤是工業的食量)—and the state took measures in Fushun and elsewhere to more intensively mine this fossil fuel. These measures ranged from mobilizing workers through campaigns similar to the fictional production competition at Wujin to introducing new technologies of extraction. Although output increased, the speed and scale at which it did so were made possible only by sacrificing quality for quantity and, most glaringly, by compromising sustainability and safety. Here was one of the main paradoxes of the socialist state. In coal mining, as in other areas, the Communists had set out to break with the Nationalist and Japanese pasts they repeatedly disparaged. In the end, however, they wound up perpetuating some of the very worst of former excesses: the wasteful extraction of

6. Xiao, *Dongbei riji*, 662 (May 17, 1949).
7. Xiao, *Wuyue de kuangshan*, 458.

resources, the ruination of the landscape, and the exploitation of workers whose labor sustained the enduring system of carbon technocracy.

"Liberation" and Recovery

Fushun's liberation was hard fought. The Communists had begun staging their offensive soon after the Nationalist takeover of Fushun in March 1946. In the mining town itself, underground Communist operatives mobilized workers in strikes against the food shortages and layoffs that arose under the National Resources Commission's management. In its outskirts, Communist guerilla fighters launched repeated onslaughts against Nationalist troops, with moderate success. A strike near Jiubingtai village to the northwest in November 1946, for example, took out half of a Fushun-based Nationalist unit. In response to these assaults, the Nationalists conducted a series of elimination drives purportedly modeled after Japanese wartime scorched-earth tactics—torching villages suspected of lending aid to the Communists. The Communists then fled to Sankuaishi, a mountain to the southeast, where they continued their targeted attacks from that harsh but defensible terrain.[8]

In September 1948, the People's Liberation Army's Northeast Field Army, under the command of military maverick Lin Biao 林彪 (1907–1971), began the Liao-Shen campaign. The Communists, who had already established a firm foothold in the countryside of southern Manchuria, moved to encircle and seize control of the region's cities, Fushun included.[9] Right around the stroke of midnight on October 31, 1948, the Communist Tenth Independent Squad advanced on Fushun. By dawn, following hours of fighting that saw both sides of the Huai River turn into a "sea of flame" and the Yongan bridge that spanned it "wildly swept by machine gun fire," Communist forces overpowered Nationalist troops and declared Fushun "liberated."[10] In part to mark their triumph, the Communists renamed the Shenfu mine, which the Nationalists had only recently named after the martyred National Resources Commission engineer. They would christen it Shengli (勝利, or "victory") instead.

The preceding years of chaos and war had not been kind to Fushun's coal-mining enterprise. As Wang Xinsan, who had led the earlier Communist takeover of Fushun and now returned to head its Mining Affairs Bureau

8. Meidu Fushun bianxie xiaozu, *Meidu Fushun*, 50–51.
9. Tanner, *Where Chiang Kai-shek Lost China.*
10. Fu, "Duli shi shi," 251.

(撫順礦務局), described it, "Three of the pits were completely flooded, and the remaining pits had largely halted operations because of the inability to carry out excavation work, and so we could only mine and collect coal fragments in shallow portions."[11] By one count, there were almost four hundred miles of damaged tunnels, 2.05 billion cubic feet of accumulated water in workings, more than five hundred sites sealed off because of fire, thirty-seven miles of destroyed railway tracks, and 2.3 billion cubic feet of debris at the open pit. Annual production had fallen to 1.07 million tons, about a tenth of its 1936 prewar peak of 9.59 million tons.[12]

Official pronouncements blamed the Nationalists for wanton and wasteful extraction, likened to "draining the pond to get the fish" (竭澤而漁) and "killing the hen to get the eggs" (殺雞取卵). At the open pit, for instance, the ideal stripping ratio was 3:1, that is, three tons of overburden removed to mine one ton of coal. Failure to observe this balance ran the risk of the sides collapsing. The Nationalists ("the scoundrels") purportedly extracted at a ratio of 1:0.8, triggering rockfalls and creating mountains of debris that needed clearing in order to resume production.[13] It is uncertain how much this debris really was the fault of the Nationalists, for the Japanese in the final squeeze of World War II and the Soviets in the short period after were, as we have seen, guilty of unsustainable extraction too. Regardless, the amount of debris was staggering. To make matters worse, many of the piles were of shale rock, which if left on site could catch fire, as Xiao Jun observed, further hampering operations.[14]

Fushun's incapacitated state at its liberation was by no means an anomaly. The new Communist government soon found itself presiding over a country stricken by poverty, rocked by inflation, and ravaged by war. Across city and countryside it put in place measures from regulated state capitalism to moderate land reform in order to restore social order, make good on selected political promises, and resume economic production.[15] In Fushun, the Mining Affairs Bureau, which was set up to run the colliery, directed efforts at reopening sealed pits, pumping flooded tunnels, and putting out raging fires.

Recovery depended on resolving a shortage in equipment. In the pandemonium of war and regime change, a large quantity of equipment from Fushun's various industries not already taken by the Soviets was looted by

11. Wang, "Di er ci lai Fushun," 280.
12. Fushun shi shehui kexueyuan, *Zhonggong Fushun difang shi*, 228.
13. Fushun kuangwu ju, *Canguan shouce*, 37–39.
14. Xiao, *Dongbei riji*, 628 (April 15, 1949).
15. Meisner, *Mao's China and After*, 75–102.

the local populace. One official account from 1959, describing the situation a decade prior, claimed that this was largely an act of resistance by "workers who wanted to show the [Japanese and Nationalist] oppressors their opposition." As this account went, these workers "hid a large volume of equipment in their homes, and when they found out that the restoration of production lacked equipment, they enthusiastically responded to the party and government's call and voluntarily took the equipment they had hidden away in their homes and donated it to the state."[16] It is more likely that the workers, in those times of uncertainty and disorder, seized tools and technologies that they had rightly discerned as valuable. In the "equipment donation campaign" (獻納器材運動) launched on December 17, 1948, and again on March 7, 1949, the Mining Affairs Bureau offered cash rewards to those who surrendered equipment. During the course of the second campaign, over thirty thousand workers and local residents came forward and turned in more than ninety thousand pieces of equipment.[17] These included invaluable motors and tools such as micrometers and hard alloy cutters.[18] In total, the bureau paid more than 470 million yuan for all these "donations."[19] The retrieval of locally looted equipment, along with the arrival of other vital machinery, resuscitated the colliery. As Wang Xinsan later recalled, "the power plants went online again, the dead machines came back to life—production at the factories and mines was restored."[20]

But Fushun's recovery needed more than machines and electrical power. It also required labor and expertise. A major obstacle the Mining Affairs Bureau faced in the beginning was a depleted workforce. According to the Department of Industry under the Northeast People's Government, there were, in 1948, around thirty-seven thousand people working at the Fushun coal mines, a considerable drop from the more than ninety-five thousand workers who labored there at the Pacific War's end. Fushun needed more hands for the restorative work ahead. The bureau thus collaborated with other local bodies and dispatched three groups to recruit workers from outside the community. These groups went not only to nearby Liaoxi, Xin-

---

16. Meidu Fushun bianxie xiaozu, *Meidu Fushun*, 55.

17. *FKZ*, 134.

18. Meidu Fushun bianxie xiaozu, *Meidu Fushun*, 55.

19. Fushun shi renmin zhengfu difang zhi bangongshi and Fushun shi shehui kexue yanjiusuo, *Fushun dashiji*, 14. The monetary amounts mentioned in this source are in dongbeibi, a local currency in Communist-occupied areas of the Northeast valued in 1951 at roughly a tenth of the yuan.

20. Wang, "Di er ci lai Fushun," 281.

min, and Panshan, but, like labor recruiters in earlier times, also to cities "within the pass," such as Beijing, Tianjin, Tangshan, Ji'nan, and Shanghai. In the first two months, they brought in 5,714 workers and 2,637 apprentices to assist in restarting Fushun's engines of production.[21]

Just as transnational technoscientific expertise had been essential to raising the infrastructure of carbon technocracy in Fushun, so, too, was it necessary for the restoration of that wrecked assemblage. Local sources often pointed to the contributions of Soviet technicians and engineers who descended on Fushun and other parts of the Northeast in that period. Soon after the People's Republic was established, Mao Zedong traveled to Moscow to seek Soviet military support and economic aid. Through the Sino-Soviet Treaty of Friendship, Alliance, and Mutual Assistance that he and Joseph Stalin signed on February 14, 1950, and in agreements that came after, the Soviet Union pledged to provide not only money and machines for China's industrial reconstruction but also expertise in the form of technical advisers dispatched to China to consult on all manner of work.[22]

In Fushun, Soviet experts were credited with speeding up the rate of recovery. For instance, the Shengli mine had been so flooded that the Mining Affairs Bureau estimated that it would take two years to completely drain. However, Soviet engineers drafted a plan that required just two more pumps than budgeted in the original plan and managed to drain the mine and restart its production in but half a year. Soviet experts were also extolled for introducing new machinery and methods that increased the efficiency of various mining operations. One notable example was at the thirty-sixth work face of the Laohutai mine. Here, Soviet technical advisers "scientifically experimented" with hydraulic cutting machines that could powerfully tear into the coal seam and with improvements to mechanized transport that doubled coal-carrying capacity. Successes at this site were replicated in other mines across the colliery.[23] A 1954 *Fushun Daily* article, taking account of Soviet contributions to coal mining in Fushun over the preceding years, noted that Soviet experts had made as many as 295 "valuable suggestions" regarding such issues as "methods of coal extraction, mechanical and electrical furnishings, transport safety, coal dressing, and mine shaft construction."[24]

Expressions of gratitude for the "meticulous help" of these "sincere and

21. Fushun shi shehui kexueyuan and Fushun shi renmin zhengfu difang zhi bangongshi, *Fushun shi zhi, gongye juan*, 71.

22. Peng, "Zhong-Su youhao tongmeng huzhu."

23. Lü, "Sulian zhuanjia gei Fushun meikuang."

24. "Sulian zhuanjia zai wu nian lai."

selfless friends" proliferated in the local press.[25] They oftentimes took a personal dimension. An Qifu, a worker at the Laohutai mine, spoke about the "care and concern" of the "Soviet elder brothers" with much fondness. To him the thing that left the greatest impression was the help they gave in redesigning the ventilation system to reduce the otherwise oppressive heat of his subterranean work: "This was most effective. There was enough air coming in, the temperature dropped, and we no longer had to strip off our clothes when working."[26] Similar accounts brimming with appreciation were told and retold in the local press then, as was consistent with dominant discourses in those heady days of Sino-Soviet cooperation.[27]

There was, of course, some irony to this lauding of Soviet technical aid. After all, had the Soviet troops not made off with a good deal of Fushun's machinery, the colliery may not have slid so deeply into disrepair in the first place. Certain sites long remained ruins. Walking through what was left of the coal hydrogenation plant, which had been completely ransacked by the Soviets, Xiao Jun noticed that "small willow trees, littleleaf boxes, reeds, and other wild grass had already started growing amid the retorts and mixers, demonstrating the degree to which this factory had become derelict, while the few mechanical parts and metallic objects there were completely covered with a layer of deep red rust."[28] Still, if the contradiction between Soviet looting and Soviet assistance was noticed (and it is difficult to imagine that it would not have been), it was not recorded in local sources at the time.

By 1952, the Fushun colliery was well on its way to recovery. Coal output that year was 5.3 million tons, almost five times that of 1948, when the Communists seized control of the mines. Old tunnels were repaired and new ones built. Facilities from workshops to dormitories were restored and expanded.[29] While the coal hydrogenation plant lay in shambles, the shale-oil plant was refurbished. In 1952, it produced 225,000 tons of shale oil, which was more than half of domestic crude output that year.[30]

Fushun's rehabilitation seemed so successful that the Communist leaders chose to showcase this colliery to a delegation of international labor union representatives visiting China that May. An internally circulated

25. "Sulian zhuanjia zai wu nian lai."

26. An, "Sulian zhuanjia bangzhu zamen."

27. For an account of Sino-Soviet tensions that ran counter to official narratives of friendship and solidarity, see Hess, "Big Brother Is Watching."

28. Xiao, *Dongbei riji*, 741 (July 18, 1949).

29. *FKZ*, 63–64.

30. *FGY*, 108.

party report described this trip with candor. After surveying the site, the German representatives likened Northeast China to their Ruhr region as a bastion of industrial modernity, declaring that "there is truly a future for industry here." The representatives from England, many of whom were themselves miners, went down to observe conditions in the pits. What they witnessed there in terms of cleanliness and safety impressed them. "China's mines are better than England's," one exclaimed. Another, who was apparently a member of the British Communist Party, raised a clenched fist and said, "Once I get back, I will definitely strive to spread the word. We absolutely need to advance toward socialism." As the group departed, the English representatives "enthusiastically burst into song, singing 'The Internationale.'"[31] Overly dramatic as it may seem, this account, given its source, was relatively reliable.[32] The scale of operations at rehabilitated Fushun was impressive enough to really inspire song. But what these visitors could not have known was how this massive, modern colliery—taken to be a symbol of socialism's greatness as it had been of imperialism's before— would consume many who worked in its depths.

Japanese Engineers in Red China

In the Mining Affairs Bureau's attempts to rebuild the material foundations of carbon technocracy and ensure the continuity of the energy regime in Fushun, it drew heavily on the expertise of Japanese engineers who ended up staying on through liberation. Celebrated as the Soviet advisers may have been, they were not the only foreign experts who assisted in the recovery here. Fushun's experience was, in this regard, not exceptional. Between 1949 and 1953, the Communists, like their Nationalist predecessors, enlisted the help of Japanese technical experts in restoring the decimated industrial landscape of the Northeast.[33] Appealing to the fact that many of them had previously worked in the region for twenty or thirty years, the party's Commission for the Management of Japanese in the Northeast, formed in October 1948, argued from early on that "today

31. "Ge guo gonghui daibiao tuan."

32. *Neibu cankao* (internal reference), the source for this account, was a bulletin consisting of investigative reports that circulated among party officials at the ministerial level and higher. Because it was not widely distributed and was meant for policy makers, it is considered trustworthy in comparison to many other centrally published materials from the Mao era. See Schoenhals, "Elite Information in China."

33. On Japanese technical experts in the early People's Republic, see King, "Reconstructing China."

the economic recovery of the Northeast region really requires the help of Japanese technicians."[34] According to one count, there were, in September 1949, almost eighteen thousand Japanese in the entire Northeast, ninety-nine of whom were based in Fushun.[35] Among them were engineers and technicians who hailed from different parts of Japan, trained at some of the top institutions in the home islands and Manchuria, and cut their teeth in Manchukuo's mining industry.[36] They would lend their technical knowledge to rehabilitating the colliery and expanding its operations.

The recruitment of Japanese engineers may have come from a central directive, but at Fushun, it was something that Mining Affairs Bureau head Wang Xinsan took personal interest in. Between his two stints in Fushun, Wang was active in northern Manchuria, where he served as party secretary and mining bureau chief in the colliery town of Hegang on the Sino-Soviet border. There, he met a Japanese coal-mining engineer by the name of Kitamura Yoshio 北村義夫 (1907–1961). A graduate of Kyoto Imperial University, Kitamura had been working in Manchurian mines since 1933 and had, in fact, even spent some time in Fushun. When Wang first arrived in Hegang, he was struck by how Japanese experts like Kitamura were marginalized by local authorities, who had them while their days away in "reeducation classes." He objected immediately. "Although these Japanese engineers and technicians may have indeed come to China during the period of Japanese invasion, some of them were forced here by the Japanese military, others were compelled to come because of hardships back home. The pain the Chinese people suffered in the war was the fault of Japanese militarists and not the Japanese people. We need to make a distinction between the two," Wang argued. "That we have Japanese technical personnel remaining behind," he continued, "is in itself something that will support and help China's revolutionary enterprise. . . . We must regard them as our teachers and friends and let them contribute to expanding production and construction in the liberated zones!"[37]

Although they had different mother tongues, Wang and Kitamura spoke a similar language of carbon technocracy. Wang regarded Kitamura first and foremost as a technical expert whose knowledge and experience could be mobilized for productivist pursuits. Kitamura, for his part, applied himself to the technical tasks he was assigned and, later, even recruited other

34. Cited in King, 150.
35. King, 149.
36. Nishikawa, "Bujun tankō no genjō," 701.
37. Cited in Ji, "Zhongguo kuangzhang he Riben gongchengshi," 39.

Japanese experts for the work of socialist industrialization. Wang made Kitamura Hegang's head engineer, and the latter oversaw increases in the colliery's coal output. When Wang was ordered by party leaders to return to Fushun after its liberation, he brought Kitamura along with him. Beholding the "ruins" of his "old haunt," Kitamura was most taken aback by how the Nationalist troops had turned the district of Yongantai—where Mantetsu had its "splendid" Colliery Club and housed its top employees—into an area for rearing horses.[38] For Wang, what surprised him was that while there were Japanese all over Fushun when he first came to take it over in 1946, there were none in sight this time round.[39]

On the eve of the Communist seizure of Fushun, only eight Japanese mining engineers and technicians remained, the rest having been repatriated over the course of the civil war. As the Communist forces drew near, these eight fled to nearby Shenyang, where they had hoped to catch a flight to Tianjin and then board a boat back to Japan.[40] Shenyang soon fell into Communist hands, however, and these men found themselves grounded. Kitamura came for them a few days later. Hokimoto Hiromi, a former machinist at Fushun's machine works who numbered among the eight, recalled how "pleased and surprised" he was to see Kitamura, especially when he learned that he, too, had previously worked at Fushun. Kitamura convinced the men to follow him back to the mining town. Upon their arrival, they were "greeted with a beaming smile" by Wang himself, who then proceeded to treat them to a "massive feast"—a seemingly happy start to the work they would undertake toward Fushun's recovery and reconstruction.[41]

According to Nishikawa Tadashi—a retained engineer who had headed the colliery's planning branch during Mantetsu times—it was difficult to discern what his compatriots really made of their work. In front of others, they often gave responses that invoked their identities both as Japanese and as engineers and technicians, saying that they were "maintaining the integrity of Japanese technology, reducing Japan's reparations for war damages, improving their own technical skills." Some claimed "to have had their thinking reformed." At least one seemed taken by the Communist cause. A Japanese accountant told Xiao Jun about an occasion where Kitamura, after a few drinks, burst into tears and said that he hoped "China could help

38. Kitamura, "Bujun tankō kara kaette," 1.
39. Ji, "Zhongguo kuangzhang he Riben gongchengshi," 39.
40. After 1949, Shenyang is much less commonly referred to as "Mukden" in English, and so I will refer to it as "Shenyang" in this chapter and in the epilogue that follows.
41. Hokimoto, "Bujun saishū zanryū ki," 675–77.

Japan launch its own revolution."[42] Nishikawa suspected, though, that most were "just trying to survive and to go back to Japan as soon as possible."[43]

Whatever their reason for working for the Communists, these men were treated particularly well by the colliery's management. This was most evident in remuneration. In order to ensure that people could obtain necessities amid the instability of ongoing inflationary cycles, the Communists introduced a system of "wage points" (工薪分) to the Northeast in which commodities like grain, salt, cloth, and coal could be redeemed with points, the number of which per good was set monthly.[44] Top engineers like Tsubota Yūichiro, who once directed excavation works at the open pit, received 1,200 points a month. Hokimoto, as a technician further down the hierarchy of expertise, received 850 points. These sums were "especially high," Hokimoto recognized, considering that skilled Chinese machine operators only got 400 to 600 points.[45] These wage points attested to just how much the new socialist regime valued the Japanese experts on whom the realization of its industrial aspirations so relied, enough to sustain asymmetries of treatment that the revolution in theory was supposed to do away with.

However the Japanese engineers may have felt about working for the Communists, their dedication to restoring and maintaining the energy enterprise in Fushun was difficult to dispute. Their first task was to help determine the colliery's route to recovery. A day after their return, these men set off by jeep for their former units—the open pit, the explosives factory, the shale-oil plant, the ironworks, and the machine works—where they surveyed conditions. After a week, they reported back to Wang Xinsan with recommendations on matters from personnel to materials, which Wang then factored into the recovery plan he drew up.[46] Over the next few years, these Japanese experts continued to work for Fushun's Mining Affairs Bureau, contributing to the expansion and maintenance of the colliery's productive capabilities. When the Korean War (1950–1953) broke out, the Northeast, which bordered the peninsula, was put in a precarious position. One concern was the potential loss of power—if, perhaps, a plant were damaged or destroyed by aerial bombardment—which would, as before, spell disaster for the entire enterprise.[47] Diesel generators seemed

42. Xiao, *Dongbei riji*, 703 (June 25, 1949).
43. Nishikawa, "Bujun tankō no genjō," 700.
44. Xiao, *Wuyue de kuangshan*, 23.
45. Hokimoto, "Bujun saishū zanryū ki," 679.
46. Hokimoto, 678.
47. The fear of aerial bombardment was not unwarranted. American planes had flown beyond the China–North Korea border to parts of the Northeast. One report claimed that these

a viable backup. Hokimoto, fellow Japanese expert Katō Sakusaburō, and two Chinese colleagues traveled to Shanghai, where they purchased from a textile mill a two-thousand-horsepower generator, which they hauled back to Fushun and installed and tested until it worked.[48]

In local memory, the most unforgettable contribution by the Japanese engineers was Kitamura's pet project: mitigating the mines' deadly methane gas. Released from coal seams and surrounding rock through mining, methane, which is flammable at certain concentrations, was typically controlled through ventilation systems that could either dilute the gas with air or carry it out of the mine. These systems did not always work, though, and methane explosions often took place, claiming multiple lives in one fiery burst. One such explosion occurred in the Longfeng mine in March 1950. A newly arrived worker who was down in a ventilation shaft conducting repairs had decided to take a smoke break and lit up. The resulting explosion killed eleven men.[49] This incident prompted Kitamura to begin researching ways of dealing with this gas. He pointed out that there tended to be considerable concentrations of methane in Fushun's subsurface workings and that this was largely a function of its age. After half a century of mining at this now "elderly" colliery, "the easily mined areas have all been mined, leaving behind only the deep, hard-to-mine areas with much gas, many explosions, and high pressure," he observed.[50]

Kitamura worked together with Longfeng's deputy head, Chinese engineer Fei Guangtai, and the two men searched for a workaround in both transnational science and local knowledge. While they consulted the international scientific and technical journals they could get their hands on, they also interviewed experienced "old miners" to better understand the specific properties of the gas at this site. The two engineers then drafted a design in which, before mining, colliers excavated a borehole adjacent to the seam being worked and drilled openings in that borehole to lay pipes for draining methane from the coal face.[51] They successfully tested this design at one part of the Longfeng mine where the methane concentration was particularly high, marking the first case of methane drainage for the Chinese coal-mining industry.[52]

planes were spotted as close to Fushun as twelve miles to the south. See, for example, "US Planes over New China."

48. Hokimoto, "Bujun saishū zanryū ki," 680.

49. Ji, "Zhongguo kuangzhang he Riben gongchengshi," 41.

50. Kitamura, "Bujun tankō kara kaette," 6.

51. For more details on this process, see Miao, "Woguo meikuang he wasi."

52. Ji, "Zhongguo kuangzhang he Riben gongchengshi," 41–42.

Kitamura and Fei subsequently introduced this technique throughout the colliery. They then followed up this achievement with further research to find a use for the drained methane. At first, they had the gas piped into homes within the mining district for household use. Later, they extended this setup into the town beyond. The pair also managed to use the methane to make carbon black—typically used as a reinforcing filler or a color pigment—and directed a portion of the gas to a facility they set up to manufacture this material. For their accomplishments, the two engineers received a commendation from the Northeast People's Government in 1952 and a prize from the Ministry of Fuel Industry in 1953.[53] Kitamura departed Fushun and returned to Japan in the fall of 1954—having chosen to stay on for over a year after all the other Japanese experts had been repatriated.[54] He left behind a lasting legacy, or at least one that was recognized over those of his peers. Half a century later, Fan Chuanxin, Kitamura's former assistant, who went on to become Fushun's deputy chief engineer, remembered him as "a Japanese of modern thinking [現代思維] . . . who dedicated all his energy to building new China."[55]

Aside from applying their expertise to various projects in concert with their Chinese colleagues, Japanese engineers like Kitamura also helped create the conditions for carbon technocracy by teaching technical skills to Chinese technicians. Many technicians from other parts of China would actually travel to Fushun specifically to receive instruction from these Japanese experts. Training mostly took place on the job, and Hokimoto remembered these technicians as "very enthusiastically throwing themselves into their work." He noted that by the time he and his Japanese compatriots went back to Japan, these Chinese technicians "were being posted all around the country."[56] If Fushun was a model of modern mining technology, this cultivation of personnel also facilitated the transfer of this technology to sites of extraction beyond, making it a hub for building and sustaining socialist China's energy regime.

The mining expertise that these Japanese engineers passed on in person was further transmitted in print. Starting in 1949, Fushun's Mining Affairs Bureau arranged for selective translations of Japanese manuals on mine work, mine safety, and mine technology. The booklets produced, expressly "presented to our worker comrades," included drawings of the various pro-

---

53. Ji, "Zhongguo kuangzhang he Riben gongchengshi," 41–42.
54. Hokimoto, "Bujun saishū zanryū ki," 681; Kitamura, "Bujun tankō kara kaette," 1.
55. Cited in Ji, "Zhongguo kuangzhang he Riben gongchengshi," 42.
56. Hokimoto, "Saishū zanryūsha no kiroku," 263–64.

cesses outlined within, presumably to enable even those with limited literacy to better understand their content. The first of these, *An Introduction to Coal-Mining Technology* (採煤技術入門), began by showing the proper posture for wielding a pickaxe so that one may "fully exert [one's] strength in mining coal" and, in so doing, "become an outstanding labor hero." By the end of this booklet, when the reader got to the dangers of gas in excavated work areas, they would have already been introduced to topics from pit props to ventilation.[57] Details within the text—such as the presumed high level of mechanization and the use of hydraulic stowage—point to it being based largely on coal mining in Fushun. The booklet's aim, though, was to convey general principles to miners everywhere who "were getting ready to welcome the revolution with production competitions" and who "urgently need to seek technological knowledge" (迫切要求學技術).[58] The Communist revolution aimed to be a technological one as well.

But this attempt to generate widely applicable knowledge from Fushun's experience was not so straightforward. As he prepared this booklet, translator Zhang Dounan became aware of one persistent issue. In Fushun and other parts of the Northeast, the Japanese colonial legacy left an imprint on work terms and machinery names. For instance, even in Chinese, "mine cars" were referred to as *gu-lu-ma* (谷爐馬), a phonetic transliteration of the Japanese *kuruma* (車), for "car" or "vehicle." Coming up with "a pure Chinese nomenclature for mining terms" was, he argued, a "fundamental task."[59] For as much as the infrastructure of carbon technocracy remained linked by techniques and technologies to the Japanese past, many Chinese in new China were also invested in papering over those connections.

"Coal Is the Grain of Industry"

The new socialist state, like the regimes that preceded it, took coal as a resource of paramount industrial importance. A popular slogan promulgated by the state in those years pithily summed up the black rock's significance: "coal is the grain of industry," a "sinicized" variant of a characterization

57. Itō, *Caitan jishu rumen*, 6. I have not been able to find the original text on which this booklet was based or any information on Itō Kobun, who is listed as the original author. There is a possibility that this persona was fabricated and the booklet's content cobbled together from multiple sources. If so, it is striking that those who prepared the booklet found it necessary to attribute authorship to a foreigner (and a Japanese one at that), presumably as a mark of this text's authority.

58. Itō, 3.

59. Itō, 95.

Vladimir Lenin had used years before. Addressing the All-Russia Congress of Mineworkers in 1920, Lenin stressed the importance of coal to modern industrial society: "Coal is the veritable bread of industry; without it industry comes to a standstill; without it railways are in a sorry state and can never be restored; without it the large-scale industry of all countries would collapse, fall to pieces and revert to primitive barbarity."[60] Similarly, the Chinese appropriation of the metaphor suggested just how fundamental that energy source was to feeding and nourishing the socialist industrialization with which the regime was so preoccupied. "Coal is the most important power source in the progress of production," Xiao Jun would remark to a local party secretary, raising this as one reason why Fushun, as a site of its extraction, "might be used as a banner for all of China's industrialization [全中國工業化的旗幟]."[61]

The socialist state's fixation with coal-fired industrialization stemmed from a vision for China's future that converged with previous expressions of carbon technocracy while also seeing this system as a means to realizing socialism. "If the Communists shared with the Guomindang the eminently nationalist goal of achieving 'wealth and power' in the modern world," historian Maurice Meisner has contended, "they differed from their vanquished predecessors in that they viewed the wealth and power of a nation not as the ultimate end but rather as the means to attain Marxian socialist ends."[62] According to their understanding of Marxist doctrine, socialism needed to be built on the foundation of material abundance laid by modern capitalistic development. The challenge for China was that it lacked that foundation. This was, however, not the first time that a revolutionary regime had failed to meet that precondition. The Soviet Union once occupied a similar position. Economically backward at the point of revolution, it nevertheless managed to industrialize not through the mechanisms of a capitalist market but through the machinations of a socialist state.[63] For China's leaders, this was an encouraging precedent. Broadly, the "noncapitalist road to socialism" the Soviet Union took involved "the combination of rapid economic development with the existence of socialist state

60. Lenin, *Collected Works*, 495. The harnessing of inanimate energy was foundational to the project of the Soviet state. Lenin's other famous slogan in this regard was "Communism is Soviet power plus the electrification of the whole country." On early Soviet electrification, which relied heavily on coal and oil, see Coopersmith, *The Electrification of Russia*, 151–257.

61. Xiao, *Dongbei riji*, 802 (October 20, 1949).

62. Meisner, *Mao's China and After*, 103–4.

63. For a useful synthesis of prewar Soviet industrialization, including some of the main debates and contestations around the means by which the industrial economy would be brought into being, see Freeman, *Behemoth*, 169–225.

6.1. "Coal Is the Grain of Industry." This poster, from 1956, states the First Five-Year Plan's goals for coal production, with output slated for an increase from 635 million tons in 1952 to 1.1 billion tons in 1957. In line with carbon technocracy's fixation on the machine, it calls for the "implementation of coal-mining mechanization" (实行采煤机械化) in the text above the circle with the image of the miner in it, an image that is flanked by depictions of mechanized shaft and open-pit mining. The arrows extending from "the uses of coal" (煤的用途) near the center of the poster point to the widespread utility of this resource. From left to right, the applications represented here are (engine) fueling, electricity generation, iron smelting, synthetic oil production, dye and explosive making, fertilizer and pesticide manufacturing, and home use. (Image courtesy of Stefan R. Landsberger Collection, International Institute of Social History, chineseposters.net.)

power and the nationalization of key means of production."[64] Moreover, the Soviet model seemed especially compelling because of recent Soviet successes in postwar reconstruction. Particularly hard hit by World War II, the Soviet economy recovered with astonishing speed, purportedly through the mobilization of society behind its Fourth Five-Year Plan.[65] What they had was an example that seemed to work, and Chinese planners readily drew on the Soviet experience in charting their own developmentalist course.

In Fushun, as elsewhere, the phrase "the Soviet Union's present is our tomorrow" (蘇聯的今天就是我們的明天) became a much-repeated rallying cry, reflecting a view of development that was as teleological as it was temporal. In the early 1950s, the *Fushun Daily* ran a number of articles that together painted a vivid picture of the Soviet Union's socialist utopia toward which China was supposed to aspire. The image captured the industrial modern ideal, with production processes endlessly expanding through the application of the latest science and technology.[66] To realize this ideal, China's leaders insisted that it was necessary to privilege the development of heavy industry: steel and iron, machinery, chemicals, oil, electricity, and the coal that was the driving force of this assemblage.[67] In a famous episode from 1950, Mao, standing atop Tiananmen, remarked that he longed for the day when he could see hundreds of smokestacks all around.[68] Carbon technocracy foreclosed other models of development. Mahatma Gandhi's vision for an industrial India constituted by a multitude of individually industrious Indians rather than a handful of mammoth industries, while ultimately upended, seemed not to have found parallels here.[69] If the Chinese socialist state took an interest in the exertions of individuals, it was mainly to direct those efforts toward big, coal-dependent dreams of an industrial tomorrow.

At the same time, as much as coal and other products of industry were to serve as the building blocks of China's imagined material future, so, too, did their output, rendered into numbers on paper, become an abstract measure for its state of development, especially when put next to similar figures from other countries. In this way, Communist planners were simi-

64. Meisner, *Mao's China and After*, 104–6.
65. Kaple, *Dream of a Red Factory*, 9–10.
66. See, for example, "Xuexi Sulian xianjin jingyan."
67. Li, "Guanyu fazhan guomin jingji."
68. Wu, "Tiananmen Square," 98.
69. For an account of Gandhi's industrial vision, see Bassett, *The Technological Indian*, 79–105.

lar to imperial Japanese and Nationalist technocrats, who likewise saw a positive and quantifiable link between coal and development. What distinguished them was their almost exclusive attention to production. While their counterparts in imperial Japan and Nationalist China fixated on the production and consumption of coal, both in aggregate and per capita, the Communists, when looking at coal, focused largely on just output.[70] The comparison with other countries that accompanied this quantification was inexorably connected with competition. In 1955, Mao declared, "We have to within roughly several decades overtake or catch up with the world's largest capitalist countries." This charge intensified in the next few years, giving rise to the notion of "overtaking England and catching up with America" (超英趕美) and, by 1957, to the specific goal of outpacing Britain in steel production within fifteen years.[71]

In Mao-era China, carbon technocracy reached unprecedented heights of productivity. The State Statistical Bureau reported that between 1950 and 1958, China's average growth rate for coal output was 26.6 percent, while Britain's was only 0.03 percent and the United States' was in decline. The focus on relative rather than absolute growth was intentional. It cast China in a particularly favorable light, making every increase appear more significant since its starting point was much lower than those of its British and American counterparts. The gains were, nevertheless, real. China moved from ninth in world coal production to third in the decade after 1949.[72] But China's leaders saw the country as competing not only with "the world's largest capitalist countries" in the present but also with its past selves. Even as the socialist regime sought to criticize its predecessors for their abuses, it also set out to surpass their accomplishments. The State Statistical Bureau kept track of the extent to which China's production of key industrial goods surpassed preliberation heights. Coal exceeded its prior record of 62 million tons in 1952. By 1957, it was 210.1 percent that of its pre-1949 peak output. By 1958, just one year later, production shot up to more than double that at 436.3 percent.[73] Quantified comparison was not

70. Karl Gerth has highlighted ways in which consumerism and concerns about consumption flourished in Mao-era China. See Gerth, *Unending Capitalism*. While his account centers mainly on the persistent desire for and use of everyday commodities, extricating a history buried under the more abstemious rhetoric of the time, it is interesting to note how the consumption of industrial products, too, received far less coverage in official statistics than their production. See, for example, Zhonghua renmin gongheguo Guojia tongji si, *Woguo gangtie, dianli, meitan, jiqi.*

71. Qi and Wang, "Guanyu Mao Zedong 'chao Ying gan Mei,'" 66–67.

72. Guojia tongji ju, *Weida de shi nian*, 95–96.

73. Guojia tongji ju, 92.

new, as earlier examples attest. What was new was the rapidity with which escalating production targets were set and seemed to be met under the centralized plans of the socialist state.

With coal, at least, the potential for growth seemed present from the start. Perceptions of China's wealth in coal had persisted past liberation. A 1949 article in the *People's Daily*, the mouthpiece of the party, celebrated the richness of this resource: "Beneath the beautiful expanse of our great homeland—from the rolling waves of the Amur river to the undulating ridges and peaks of the Hengduan mountains, from the vast Shandong plains to the open country of faraway Xinjiang—lie untapped inexhaustible natural resources. Reserves of coal, described by folks as black gold or ink jade, are particularly plentiful."[74] Prospecting work conducted under state auspices through the 1950s enabled the verification of more coal resources in many existing mining districts and the discovery of a considerable number of new coalfields.[75] In Fushun, engineers and technicians from the Mining Affairs Bureau, assisted by Soviet geological experts, uncovered more deposits to the north of the area being mined, bringing its estimated holdings to 1.5 billion tons.[76] Nationally, coal reserves were calculated to be as much as 1.5 trillion tons, placing China third in the world behind the Soviet Union and the United States.[77] The challenge for the socialist state, again as for those who came before it, was accessing this abundance.

The Communist government oversaw coal mining in the 1950s initially through its Ministry of Fuel and, when that dissolved in 1955, through the newly established Ministry of Coal Industry. Ownership and operation of coal mines were divided into four categories: (1) state enterprises, which included large enterprises run by the ministry, medium ones run by provincial authorities, and small ones run by local authorities; (2) cooperative enterprises run by agricultural and handicraft cooperatives (later, people's communes); (3) joint public-private enterprises, which many private mining operations became in the mid-1950s during the campaign for the "socialist transformation" of private enterprises; and (4) private enterprises. The trend across the decade, in this industry as in others, was the enlargement of the state sector. In 1950, state enterprises accounted for 66 percent of coal output and private enterprises 32 percent; by 1956, the proportion

74. Gao, "Woguo meikuang gongye."
75. Dangdai Zhongguo congshu bianji bu, *Dangdai Zhongguo de meitan gongye*, 257–58.
76. *FKZ*, 65.
77. State Statistical Bureau, *Major Aspects of the Chinese Economy*, 130; Guojia tongji ju, *Weida de shi nian*, 12.

had shifted to 90 percent and about 2.5 percent respectively.[78] In 1957, almost three-quarters of the total coal produced in China came from mines under the Ministry of Coal Industry, of which Fushun was to be one of the five largest—the others being Datong, Fuxin, Huainan, and Kailuan. Aside from managing mines, the ministry was also responsible for drafting plans for coal output, the dispensation of capital investments and material supplies to mines, and the distribution of coal to consumers—plans that it sent to the State Planning Commission, the central organ in charge of directing China's economic and social development, for approval.[79]

China's increases in coal production in the 1950s occurred in the context of the First Five-Year Plan, the massive state-led industrialization initiative that dominated the middle of the decade. This plan was officially announced at the second session of China's legislative First National People's Congress in July 1955 by Li Fuchun 李富春 (1900–1975), chairman of the State Planning Commission and architect of the Northeast's postliberation industrial economy. It had, in fact, been in operation since 1953. At its core, the plan embodied the state's foremost commitment to developing heavy industry. "Big industry is the material base on which a socialist society can be built. Only by building a big and powerful heavy industrial sector . . . can we manufacture various modernized industrial equipment and allow heavy industry itself and light industry to undergo technological transformations," Li asserted. "Only then can we provide for the agricultural sector tractors and other modernized agricultural machines, provide for the agricultural sector enough fertilizer, and allow agriculture to undergo technological transformations. Only then can we produce modernized means of transport, such as locomotives, automobiles, steamships, and airplanes, and allow transportation to undergo technological transformations. Only then can we manufacture modernized weaponry to equip the soldiers who protect our country and allow our national defense to be further strengthened."[80] Li and other like-minded leaders imagined a modern socialist nation as a mechanized one. The benefits for the individual socialist subject were implied: "technological transformations" in agriculture and transportation put more food on the table and provided more personal mobility. These benefits were not their focus, however. Their vision was explicitly concerned with buttressing statist military power.

78. Ikonnikov, *Coal Industry of China*, 30–34.

79. Bazhenov, Leonenko, and Kharohenko, "Organization of Production, Labor Productivity and Costs," 1–2.

80. Li, "Guanyu fazhan guomin jingji," 163–64.

Coal was central to the First Five-Year Plan as the fuel for all its targeted processes. The plan aimed for a national output of 113 million tons and the production of sufficient coking coal to meet the needs of the steel industry by 1957. Furthermore, it also sought to "rationalize" the industry's geographic inefficiencies where coal producers were often located great distances from coal consumers. To these ends, the government opened new mines, some closer to the sites of industry they supplied, even as it expanded existing ones. The plan privileged scale, setting out to ensure that by the end of the five-year period there would be thirty-one coal-mining enterprises with a capacity of more than one million tons each.[81] As many as 24 of the 156 large industrial projects at the heart of the plan—to which the Soviet Union committed considerable amounts of aid—were coal projects, 4 of them based in Fushun.[82] The intensification of mechanization, too, was regarded as imperative. Reflecting on the plan in its immediate aftermath, Minister of Coal Industry Zhang Linzhi stressed how there had been a concerted effort "to push the existing mines from the facade of extreme technological backwardness onto the proper course of technological modernity."[83] These moves on various fronts appeared to reap results. By the end of the First Five-Year Plan, China's coal output had almost doubled from 67 million tons to 130 million tons, exceeding the target by more than 10 percent.[84]

In Fushun, the First Five-Year Plan called for a massive increase in production. According to the State Planning Commission's directive, output should reach about 9.3 million tons by 1957, just a little less than double that of 1952. In order to not only meet that target but to "resolve the long-term development and construction issues of all the shaft mines and the open pit" and to "resolve the long-term planning issues of the entire coalfield" that impinged on productivity, it was necessary, the directive suggested, for the Fushun colliery to undertake "wholesale reform" (總體改造).[85] This process involved extensively deepening operations at the existing mines and excavating three new ones—the Eastern open pit, a vertical shaft mine in northern Laohutai, and another vertical shaft mine in northern Longfeng. The state poured much capital into this enterprise to facilitate its expansion. At the Western open pit alone, eighty-seven million yuan of central government funds were spent on more machines

81. Li, 173; Dangdai Zhongguo congshu bianji bu, *Dangdai Zhongguo de meitan gongye*, 24.
82. Thomson, *The Chinese Coal Industry*, 36.
83. Zhang, "Wei gao sudu fazhan," 5.
84. Kharohenko, "Development of the Coal Industry," 4.
85. *FKZ*, 65.

and supporting facilities before the end of 1958. Not everything went as planned, however. Because of insufficient materials and manpower at the commencement of work on the Eastern open pit, there were delays that resulted in eight million yuan in losses. Still, by 1956, Fushun's total output was already 9.04 million tons, well within range of the target.[86]

One particular measure from the period credited for the rise in production at Fushun was the adoption of the so-called Pang-Zhou coal-mining method (龐周採煤法). Proposed by worker Pang Guanxiang and technician Zhou Guangrui, both from the Laohutai mine, this method had three main components: (1) synthesizing various advanced techniques gleaned from the experiences of the masses, one example being the use of wider drills with interchangeable heads to bore deeper shot holes for explosives; (2) changing the basic unit of mining operations from squads working eighty-meter faces without much division of labor to teams working 160-meter faces in the manner of an assembly line; and, most notably, (3) exceeding previous parameters in terms of tunnel height and width, depth of shot holes, amount of explosives, and length of work face. The result was more output at a faster pace.[87] That such a method, devised in part by a worker, would be put into practice and applauded for its success was consistent with the party state's recognition of ordinary laborers as bearers of knowledge and skill. We will discuss this phenomenon in the next section. Here, it is sufficient to point out how carbon technocracy in the People's Republic made room for such forms of expertise, so long as these served to propel the productivist impulses that first drove and then defined the project of socialist industrialization.

Workers in a Workers' Society

For all the novel methods and new machines, Fushun's expanding output in the first decade of the People's Republic depended heavily on its workers. In addition to actively recruiting more labor, as mentioned earlier, the colliery began mobilizing its workforce to generate higher yields through production campaigns. The Communists had adopted these campaigns from Soviet managerial practice, introducing them earlier in Yan'an.[88] In Fushun, these campaigns proved vital for directing human energy toward

86. *FKZ*, 65–67.
87. Liu, "Fushun meikuang kaicai fangfa," 71–72.
88. Kelkar, "Chinese Experience of Political Campaigns," 50–51; Kaple, *Dream of a Red Factory*, 8.

the extraction of carbon energy. The competition Xiao Jun described in *Coal Mines in May* was based on one such campaign that took place in Fushun in May 1949. The morning of the competition, he headed down into one of Laohutai's pits, where he observed workers at various coal faces "working quietly . . . working excitedly . . . all seeking to outperform other squads."[89] This was the first of many such competitions held across Fushun in the years that followed.

The premise behind production campaigns seemed straightforward enough. Different units within the colliery competed with one another to see who could either meet a target first or complete the most work within a fixed amount of time. While workers typically vied over who mined more coal, campaigns also saw them battle it out in other tasks like tunnel clearing and equipment repair. Prized in all of these processes were "setting new records" (創造新紀錄), "surmounting conditions" (突破條件), and "exceeding plans" (超過計劃), with rates often recorded to technocratic precision.[90] For instance, in November 1950, workers in the rockfall section at the open-pit mine were singled out for success in an ongoing production competition, having completed a month of work in just half the time. The local paper would report that this was a result of them raising the daily amount of debris removed from 1,500 cubic meters to 1,900 cubic meters, an accomplishment to which one particular group contributed greatly with 100 percent work attendance and each worker increasing their rate of clearance a day from 4.4 cubic meters to more than 6 cubic meters.[91] While there were occasional monetary rewards for units who performed well, in Fushun, as elsewhere, remuneration most often came solely in the form of public acclaim.[92] Often launched in conjunction with May Day celebrations or other national movements—such as the one to "resist America and assist Korea" during the Korean War—these campaigns established and reinforced a quantifiable connection for workers between their industrial production on the one hand and their revolutionary fervor and patriotic sentiment on the other.[93]

The Communists' success in mobilizing workers in this initial phase through production campaigns rested not only on appealing to competitiveness or the desire for glory. It also relied on winning workers over by

89. Xiao, *Dongbei riji*, 649 (May 7, 1949).

90. For the importance of quantification to the early Chinese socialist state, see Ghosh, *Making It Count*.

91. "Lutiankuang bengyan gu."

92. See, for instance, Frazier, *Making of the Chinese Industrial Workplace*, 227.

93. Qi, "Yingjie hong wuyue."

promising them better lives. In the beginning, it was a matter of tending to basic needs. Because of the runaway inflation and grain shortages that had marked the Nationalists' short tenure, many of Fushun's residents went unpaid and allegedly had to fill their bellies with leaves and bean cakes—what is left behind after soybeans are pressed for oil.[94] Paying overdue wages was a priority. "I want to announce to all current staff and worker friends," Wang Xinsan proclaimed in his first public address in liberated Fushun, "from hereafter, wages will be paid according to your original positions. . . . When I arrived, I already knew well of the hardship that this colliery's employees have been facing. Our superiors have already approved . . . the payment of some of your wages in advance so that you may have enough to support yourselves."[95] Moreover, the newly installed socialist government quickly distributed relief rations and requested more grain to be sent from earlier liberated areas in order to alleviate the dire food situation.[96] It was not uncommon to associate the coming of the Communists, as one worker from the open-pit mine did, with "eating one's fill" (吃飽了飯).[97]

The Communist revolution was, however, committed to delivering more than just subsistence. If the Soviet Union provided a model of how industrial production should be organized, it also furnished an ideal of how workers' lives were to be transformed for the better. In the "Soviet Union under the great comrade Stalin," a *Fushun Daily* article began, "workers not only are not oppressed nor exploited but receive the most respect. They all have the highest knowledge and skill, and they all lead extremely happy lives. And each day is just getting better than the one before. This country really deserves to be called the great paradise of the working people."[98] To the Chinese socialist state, 1949 marked a point of rupture, in which everything that came before was to be dismissed as old and everything that came after declared new, inching toward the Soviet ideal. The local press frequently ran accounts of how workers' lives had supposedly improved across that transition. "Before liberation, we workers labored like cows and horses; after liberation, we took charge of our own affairs," one account went. It proceeded to detail the new creature comforts workers enjoyed, from "eating white flour" (instead of "acorn flour") to "wearing comfortable cadre uniforms and leather shoes" (instead of "having elbows stick out above and toes stick out below" in their worn-out clothes

94. Meidu Fushun bianxie xiaozu, *Meidu Fushun*, 53.
95. Wang, "Zai di yi ci jun, kuang."
96. Meidu Fushun bianxie xiaozu, *Meidu Fushun*, 53.
97. Zhang, "Jiefang liao de caimei gongren," 287.
98. "Sulian renmin de xingfu shenghuo."

and footwear), and the range of leisure activities for which they now had the time and the opportunity to partake in. "Before liberation, workers were trampled on into trash; after liberation, we are like scrap metal that has been smelted into steel," the account concluded—and with an appropriately industrial metaphor no less.[99] Although such characterizations carried more than a hint of hyperbole, there is little reason to suspect, especially given the troubled times that came before, that the day-to-day living standards of Fushun's workers saw marked improvements in the early People's Republic.

Designated by the new socialist state as "masters of the country" and the "leading class," workers were also promised more dignity in labor.[100] One important aspect of this was acknowledging the expertise of workers. In Mao-era China, the question of expertise was a question of class politics, and the state repeatedly sought to dislodge entrenched associations between knowledge and the traditional elite. This question came to the fore nationally in 1957, when party leaders enjoined intellectuals to be "both red and expert," which is to say they should meld revolutionary fervor ostensibly of and for the masses with technoscientific ability.[101] At the Fushun colliery, the issue of expertise, which had arisen by the early 1950s, seemed mainly about leveling the hierarchy between its many technicians and the multitude of workers who were previously regarded as beneath them—the two groups being divided by technical skill, which the former were said to have and the latter not. To begin with, the colliery openly endorsed worker expertise, particularly of "old workers" (老工人), who had labored in the coal-mining industry for decades. At a colliery-wide "elderly appreciation" celebration in 1951, Fushun's general union chairman urged old workers to "redouble efforts to impart work experience and technical skill to young apprentice workers, developing them into talents for building the nation."[102]

Official discourse also promoted the idea that workers had things to teach not only one another but technicians as well. Accordingly, technicians were encouraged to learn from workers through active interactions at work sites and informal discussions. An example among many that made the local paper was that of Wang Xi. Wang, a technician at the Shengli mine, was highly praised for curbing his "purely technological viewpoint"

99. Wu and Gu, "Gongren de shenghuo."

100. For a discussion of how these promises quickly proved hollow in the case of trade unionism in Shanghai, see Perry, "Masters of the Country?"

101. On this issue of "red and expert," see Schmalzer, "Red and Expert."

102. "Zonggonghui zhaokai chunjie jinglao dahui."

(單純技術觀點) and going down to the coal face on multiple occasions to talk to workers about production problems they might resolve together. One of these issues was recovering pit props for reuse once their support was no longer needed. Wang sketched the design of a device for this purpose, discussed this design with workers, and made adjustments according to their input. The pit-prop removal machine that resulted from these exchanges (a contraption consisting of an engine with several chains extending from it mounted onto an adjustable tripod) turned out to be a success, allowing for the recovery of forty-five cubic meters of pit props at each coal face per month.[103] In her study of agricultural science in Mao-era China, historian of science Sigrid Schmalzer shows how *yang* experts associated with foreign knowledge, such as University of Minnesota–trained entomologist Pu Zhilong, often relied on the native expertise and the inventiveness of *tu* experts to develop and introduce new farming technologies.[104] In Fushun, collaborations between workers and technicians similarly resulted in fixes and modifications that furthered the socialist state's nagging pursuit of increasing coal production.

Moreover, like Schmalzer's *tu* experts, workers in Fushun were also celebrated for innovations of their own. One such instance involved Wei Xijiu, a worker at the Laohutai mine's electrical machinery plant. Wei had found several old rock drills from the Japanese era that had been sitting in the mine's warehouse and fixed them up for use. When these refurbished drills were switched on, however, they gave out a *"da da* sound . . . as earsplitting as machine-gun fire," preventing workers from hearing one another, which was not only an annoyance but a safety hazard and, perhaps more importantly, an impediment to smooth production. Wei took up the problem immediately. After "concentrating all attention" on it for several days, he decided that the issue was in the narrowness of the drills' vents, which he proceeded to widen. Later, inspired by the mufflers on automobiles, he designed a similar noise-deadening device for the drills. Together, these two measures mitigated the problem, prompting workers who operated the newly modified drills to, in one account, jump up and declare, "Our ears have been liberated!"[105]

Circulated as propaganda, such episodes of instruction and innovation by workers were almost certainly more aspirational than descriptive of a reality in which worker expertise received widespread recognition. Still, it is

103. "Jishi Wang Xi zao gongren bangzhu."
104. Schmalzer, *Red Revolution, Green Revolution*, 47–72.
105. Liu, "Wei Xijiu chuangzao xiaoyinji"; Zhang and Lin, "Wei Xijiu chuangzao xiaoyinqi."

notable that there were these attempts to acknowledge, even if just rhetorically, worker knowledge and skill.[106] In the socialist period, carbon technocracy drew from a less elitist pool of expertise than under past regimes. At the same time, it bears repeating that while a broader cast of characters might be considered experts, their expertise still served narrow statist ends. Useful knowledge, be it from formally trained engineers or experienced workers, was that which helped further production for the advancement of the state.

If Fushun's workers remained unconvinced of the official line that they stood on equal ground with technicians, they could, technically, also become technicians. Over the years, the colliery set up several training institutes, beginning in 1949 with a professional night school that aimed to "lift employees' levels of culture and technology." By taking classes in mechanical, electrical, and chemical engineering as well as in mining and machinery, workers who finished courses of study could rise in occupational rank. Workers became machinists; skilled workers became technicians; some workers and machinists became workshop heads or section chiefs; some skilled workers and technicians became assistant engineers; several trainee machinists became machinists; and several workers became technical team leaders. However, not many of Fushun's workers took advantage of this educational opportunity: in the first three years, only 361 workers completed courses, a tiny fraction of the mining town's more than 100,000 employees. This was perhaps because of the school's capacity (it had only seventeen classes in 1953) or because workers were busy with other matters outside of work.[107] Or it could even be because many workers had in fact accepted their essential equivalence with technicians or, conversely, because they were resigned to the distinctions that divided them.

In mobilizing workers for production, the colliery also targeted women who, under the new regime, entered the industrial workforce in significant numbers, liberated, as they were, for wage labor. In Fushun, women remained out of the mines and sites of direct extraction, but they did go into processing facilities, such as coal-washing plants, and the factories of auxiliary industries. The colliery began employing women soon after the Com-

106. For a helpful set of reflections on how to read reports of scientific accomplishment in socialist China (especially by ordinary workers) that can seem far too rosy and recover from them a promise of progressive politics, see Schmalzer, "On the Appropriate Use of Rose-Colored Glasses."

107. "Fushun meikuang gongye zhuanke yexiao"; "Fushun jishu yexiao"; Fushun shi renmin zhengfu difang zhi bangongshi and Fushun shi shehui kexue yanjiusuo, *Fushun dashiji*, 74.

munist takeover. "Most of them worked desk jobs; a few of them worked in electrical equipment or mining lamp factories; not one of them had a grasp of technology," the local women's federation chair He Zhi claimed in a 1952 address, "Later, under our leaders' directions . . . a portion of women were able to get a grasp of technology. . . . Now the number of women workers in state and private enterprises is twenty-two times more than in 1949."[108] To facilitate women with children taking up this kind of work outside the home, units within the colliery started their own nurseries. The one at the open-pit mine was a "light gray and exquisite Western-style house" in which there were cots surrounded by "walls scrubbed white" with posters on them showing how much formula the children should take, when they should feed, and when they should sleep—a sterile and structured environment designed so that mothers could "go to work with their minds at ease."[109] Like their male counterparts, these women became essential to the mines' operations. For "enthusiastically finding solutions to problems" and for "demonstrating a high degree of zeal for production," women workers at the electrical machinery factory would, for instance, be characterized as "an indispensable great force in the battlefront of production."[110]

However, even as women now participated actively in industrial work, in many ways, the revolution brought little change to earlier understandings of their role in society, tethered as those understandings often were to family and the reproductive labor of the household. In the discourse of Fushun's women's federation, the old ideal of the "virtuous wife and good mother" (賢妻良母) was upheld. For example, in 1952, a group of workers' wives at the colliery's steel mill noticed that their husbands seemed to be performing badly at work. The men were churning out an inordinate volume of scrap metal, which could result from inefficiencies in handling the material. After looking into this matter, the women concluded that the problem was that "many wives among them frequently asked their husbands to carry children, fold bedding, and chop firewood, preventing the men from properly resting." They resolved to "no longer ask the men to do these things," and their husbands' section purportedly "never again put out

108. He, "San nian lai Fushun."

109. "Lutiankuang de tuoersuo." On the party's creation of nurseries across China so that it could benefit from the labor of women "liberated" from childcare and on the role of cow's milk in nourishing the children in these nurseries under the new regime of work, see Braden, "Serve the People," 59–66.

110. "Jidian chang nügong."

scrap metal."[111] Similarly, Wang Xueqin, a miner's wife who was designated a "model woman" (婦女模範), won that accolade for work she did organizing other wives behind a "household plan." As one account went, "she knew that only if the housework were done well would the men be able to focus on production at the mine; only if more were produced would their days improve."[112] Historian Gail Hershatter, in her research on women's lives in rural Shaanxi just before and after the revolution, noted the "unprecedented intensity" with which the "campaign time" of the state and the "domestic time" of the home came together in the massive production drive of the late 1950s.[113] Fushun, as one of socialist China's towering heights of industry, experienced this conflation of times from early in the decade. No one felt this more acutely than its women workers.[114]

As part of its effort to encourage increased output, the state provided workers models to emulate. This model worker system, which went hand in hand with production campaigns, also had Soviet origins, based as it was on the Stakhanovite movement. Named after Aleksei Stakhanov, a miner who in 1935 was acclaimed for purportedly mining 102 tons of coal in just six hours (fourteen times the regular rate), this movement called on workers to follow his example and exceed productivity standards.[115] Fushun's most famous model worker was one of its first. Zhang Zifu 張子富 (1915–1990) was a worker at the open pit who made his name by organizing a highly successful "shock team" (突擊組) that went around the mine setting new production records. For "his great strength for manual labor, his outstanding ability at labor mobilization, and his optimistic and steadfast work ethic," the colliery bestowed on him the title of "all-round labor hero" (全面勞動英雄).[116] During his time in Fushun, Xiao Jun befriended Zhang, whom he once described as "representing the spirit of a certain revolutionary romanticism, with zeal, bravery, boldness, responsibility, creativity, resourcefulness, and the ability to be agitated."[117] He thought Zhang would make a good main character for a novel and ended up modeling the protagonist in *Coal Mines in May*, the fearless Lu Dongshan, after him.[118]

111. Yu, "Feiyue qianjin."

112. "Funü mofan Wang Xueqin."

113. Hershatter, *The Gender of Memory*, 236.

114. For an account of how the demands of the socialist state together with those of local social convention amplified costs borne by women in rural settings, see Eyferth, "Women's Work."

115. On the model worker system, see Yu, "'Labor Is Glorious.'"

116. Fushun kuangwu ju, *Canguan shouce*, 45.

117. Xiao, *Dongbei riji*, 771 (August 16, 1949).

118. Xiao, 662 (May 17, 1949).

Over the years, the colliery named many more model workers, all of whom similarly distinguished themselves by their rapidity or volume of production.[119] At the end of the day, for all the attention the socialist state gave to their living standards and social status, the workers at Fushun, as elsewhere in the country, were to be defined by their work, subject as this work was to carbon technocracy's ever-growing demands. The lofty coal-centered industrial schemes of statist planners did, after all, depend dearly on the exertions of the masses who labored on the ground.

The Problem of Thermal Inefficiency

Despite the energy poured into coal production, the expanding supply did not seem able to catch up with the growing demand. By the late 1950s, China faced a coal shortage. Premier Zhou Enlai 周恩來 (1898–1976), writing in the *People's Daily* in November 1957, spoke of an "extreme scarcity" that emerged because of this gap between supply and demand, in which "individual work units' stockpiles of coal were more depleted than before, while the demand for coal in various sectors such as manufacturing, transportation, and households had vastly increased." The State Council, which Zhou headed, called for the "concurrent increasing of coal production and saving on coal consumption" (增產煤炭和解約用煤同時並據), which he claimed would be essential to easing the tense situation.[120]

The demand for coal continued to rise with the onset of the Great Leap Forward, the disastrous production drive from 1958 to 1961 that precipitated a famine that killed tens of millions. In the second half of 1958, Mao, convinced that steel would serve as the "key link" for a great leap in industry, called for doubling the production of this most industrial of alloys from 5.4 million tons to 10.7 million tons. In the steel campaign that followed, people across the country were mobilized to set up in their backyards and beyond homemade blast furnaces in which they would not only smelt ore but, infamously, also melt down scrap metal and otherwise useful tools from cooking vessels to farming implements. This endeavor proved fruitless, memorably likened by war hero Peng Dehuai 彭德怀 (1898–1974) to "beating a gong with a cucumber." The party purged him for raising concerns. A good deal of the supposed steel churned out was actually pig iron of low quality, in part a product of the inferiority of the inputs, which included cook ware and farm implements. As one observer

119. See, for example Wei, "Chang mofan Li Yuming."
120. Zhou, "Zengchan meitan he jieyue yongmei."

at that time remarked, "to me the stuff that was being poured out at the bottom of the furnaces looked exactly like the stuff being poured in at the top."[121] But insofar as the backyard furnace represented a waste of time, labor, and iron resources (at an ultimately calamitous cost), so too did it signal a certain squandering of fuel. This wastefulness was not just because the furnaces ended up producing so much worthless iron. It was also because these furnaces consumed more coal than deemed normal for smelting iron to begin with. This was a problem of thermal inefficiency.

The Ministry of Coal Industry had its finger on this problem from early on. In November 1958, its coal preparation division noted that in order to smelt one ton of iron, many backyard furnaces burned as much as three tons of coke (or five to six tons of mine run coal—coal straight out of the ground—which was processed to make the coke).[122] This was more than four times the consumption rate at China's modern steel mills, which was already higher than the average in the Soviet Union and the United States.[123] In its assessment, the division blamed this high coal consumption on the "unfamiliarity [of those working the furnaces] with the operations, the low grade of the iron ore used, and the high ash content of the coke."[124] But the problem of thermal inefficiency extended beyond backyard furnaces to the rest of China's industrial sector. In 1963, economist Yuan-li Wu wrote that one of the main reasons this problem was so pervasive across industries was the widespread use of low-quality coal, including, but not limited to, coke of high ash content.[125]

The question of quality had in fact been on the minds of state planners and coal producers since the beginning of socialist industrialization. The quality of coal, which determines how much heat or thermal energy it releases when burned, may be undermined by the presence of ash and other impurities within it. These impurities can occur in the formation of the coal itself or be introduced in the process of its extraction and transportation. Impurities that occur during coal formation are typically inorganic minerals mixed up with the plant matter before burial or precipitated into fissures during coalification. Other impurities that can be chemically reduced and trapped in coal at that time include nitrogen and sulfur. Impurities introduced during the mining and movement of coal largely consist of

121. MacFarquhar, *Origins of the Cultural Revolution*, 88–90, 113–16, 193–200, quotation on 115.

122. "Strengthen Coal Washing Work," 1.

123. Wu, *Economic Development and the Use of Energy*, 198.

124. "Strengthen Coal Washing Work," 1.

125. Wu, *Economic Development and the Use of Energy*, 199.

surrounding rocks and debris or even parts of machines and tools used in extractive operations.[126]

If impurities occurring in coal's formation were an accident of nature, then those introduced during production often were, official pronouncements suggest, the result of human error. In July 1951, the *Fushun Daily* published a piece written from the point of view of anthropomorphized coal that presented that position in almost endearing detail. The coal introduces itself as "an important player in developing national defense and the economy" who, after Fushun's liberation, had been "emancipated" (翻身) alongside its "master" (the man who mined it) and who has been "working in the service of the people's homeland." However, the coal complains, its "neighbor waste rock liked to mix with it and make mischief, causing its quality to deteriorate." The fault lay with the miner: "My master does not care whether the elements around me are good or bad. He just sets off explosives and then goes at it with the electric shovel and does not thereafter actively pick out [the coal from the waste]. He is also careless about how he loads me, merely chasing quantity [over quality]." Thermal efficiency was being compromised, much to this coal's chagrin. The tale does, however, end on a happy note. Following an inspection that brought the poor quality to light, the miner changes his ways, and our carbon protagonist then boasts of its "quality improving" and its "strength for economic and national defense development increasing," allowing it to "join the might of the masses in 'resisting America and assisting Korea.'"[127]

But compromised coal quality did not just result from human carelessness. It could also have been an unintended outcome of technological advancements. In recent years, historians writing about science and technology in Mao-era China have begun to overturn prior assumptions about the supposed deficiency or failure of the technoscientific enterprise under socialism.[128] Many have demonstrated how, often through approaches foreign to capitalist societies such as "mass science," science and technology in China actually achieved desired results. But coal-mining mechanization is one case in which the technology worked perhaps a little too well. The Ministry of Coal Industry had for years seen in mechanization the possibility of greatly increasing production. While mechanizing extraction did in-

126. Raask, *Mineral Impurities in Coal Combustion*, 29–30.

127. Fan, "Lutiankuang 'mei' de zishu."

128. The literature is now sizable and growing. Some examples include Neushul and Wang, "Between the Devil and the Deep Sea"; Schmalzer, *The People's Peking Man*; Fan, "'Collective Monitoring, Collective Defense'"; Mullaney, "The Moveable Typewriter"; Schmalzer, *Red Revolution, Green Revolution*; and Jiang, "Socialist Origins of Artificial Carp Reproduction."

deed bring about greater and faster output, it also tended to churn out coal mixed with greater amounts of impurities. This was particularly so with longwall mining, in which colliers employed big combined cutter-loaders to shear elongated coal faces. The machines' rotating teeth broke off not only chunks of coal but often also inadvertently portions of the surrounding debris, which were mixed into the coal pieces as conveyors carried the material out from the work site. Between 1952 and 1956, the number of coal combines used in mines under the Ministry of Coal Industry went up from just four to eighty-eight.[129] As production expanded to meet the needs of the socialist energy regime, technological developments here exacted costs in other areas—in this case, thermal efficiency.

The main way of dealing with impurities was to wash the coal. This process of coal washing takes advantage of the different densities of the coal and its impurities to sort and separate—through water, screens, and jigs—coking coal, coal with higher amounts of impurities, and waste.[130] According to one report by the Ministry of Coal Industry, washing was a means of conserving coal. By its estimate, one hundred tons of raw (unwashed) coal, converted into coke, could go toward producing twenty tons of pig iron. If the coal was first washed, then only about ninety tons of the raw coal would be needed to make the coke necessary for the same output of pig iron. Furthermore, washing also generated over twenty tons of "middling coal" per hundred tons of raw coal that, while unsuitable for smelting, could be burned in homes or power plants. "Great savings can be made on a nationwide scale if more coal washing is done," the report declared.[131] Linking this practice to the priorities of the present, a *People's Daily* article proclaimed, "Coal washing is an important connector for increasing steel output."[132] Long in practice at larger mines like Fushun, coal washing became by 1959 much championed among state planners, who pushed for its widespread adoption amid the escalation of steel and iron production.[133]

However, although there was much attention given to this particular solution to thermal inefficiency, the problem still persisted. One factor was the multitude of small mines that opened in this period. Under plans for expanding production in the Great Leap Forward, the Ministry of Coal Industry had raised the possibility of "soldier [corresponding to] soldier,

129. State Statistical Bureau, *Major Aspects of the Chinese Economy*, 147, 158.
130. Huagong, "Ximei."
131. "Strengthen Coal Washing Work," 2.
132. "Ximei shi zengchan gangtie."
133. Zhou, "Zengchan meitan he jieyue yongmei."

general [corresponding to] general," in which large mines would supply large steelworks and small mines would support the backyard furnaces that Chinese citizens were erecting across the countryside.[134] Most of these small mines did not practice coal washing, though, and the coal they put out suffered for it. Another factor was that even with the large mines that had coal-washing facilities, the rate at which coal was mined in the period exceeded their coal processing capabilities. These plants were unable to handle the increase in coal production during the Great Leap Forward. As a result, while total coal output increased significantly from 130 million tons in 1957 to 425 million tons in 1960, the average quality of coal actually dropped.[135]

Moreover, although Chinese engineers and technicians worked on improving the efficiency of furnaces, boilers, and engines through the 1950s, feeding these devices large amounts of inferior coal essentially canceled out any gains.[136] Between the inferiority of the coal and the wastefulness of industrial consumers, aggregate coal consumption had reached a point at which the entire output of small mines—about one-fifth of total production—went toward making up the energy lost through such inefficiencies.[137]

Speeding up production not only compromised coal quality but also further endangered the lives of miners, subject as they already were to the common perils of a hostile environment. Between January and May 1958, the number of mining accidents in larger state mines rose by 53 percent and those in smaller local mines by 10 to 30 percent. Apart from gas explosions and mine cart collisions, many of these accidents occurred because of overextraction, as workers mined beyond recommended recovered-to-retained coal ratios, triggering deadly tunnel and (at open-pit mines) slope collapses.[138] Troubles with thermal inefficiency due to declining coal quality were, in this sense, but a symptom of a deeper malady. Insofar as these troubles were an outcome of increased mechanization and production, they also pointed to yet another contradiction within the energy regime of carbon technocracy—the costs were arguably greater than the benefits.

134. "Strengthen Coal Washing Work," 1.
135. Wu, *Economic Development and the Use of Energy*, 40 and 125.
136. Wu, 106–8, 125.
137. Wu, 199.
138. "Quanguo meikuang yi zhi wu yue fen."

"The Coal Is Paid for in Human Lives"

The socialist state sought to distinguish itself from preceding powers by its safety record.[139] In Fushun, local authorities often cited earlier accident statistics as evidence of just how oppressive the Japanese and Nationalists had been. A popular ditty that circulated among miners during Mantetsu times painted a picture of indifference to the loss of life: "In this old society, miners sing many bitter songs / Miners' blood and tears flow and form a river / On this run mine cars day after day / Who has seen how many miners are still alive?"[140] One miner, speaking in 1952, put it more plainly: "The [Japanese] devils took the lives of us workers and traded them for coal."[141]

In postliberation Fushun, the local newspaper frequently reported on safety issues. Articles often celebrated extended periods of time in which particular mines or units within them managed to go about their day-to-day work without accidents occurring. For instance, when Kong Xiangrui's coal extraction team from the Shengli mine achieved over three years of safe production, the paper interviewed Kong, as team leader, for tips on how to ensure safety in operations. In that case, it seemed that vigilance was the answer: he did not tolerate even the slightest safety infraction from his team members, nor did he let the smallest indication of potential danger, such as a fragment of coal falling from above, go uninvestigated.[142]

Even more common, however, was coverage of safety problems, with most of the reported breaches in safety arising during production competitions. Although these competitions may have been central to the colliery's productivity, there were numerous instances in which colliers cut corners and compromised safety on account of them. For example, in November 1950, the Shengli mine was singled out for sacrificing safety in its bid to win a competition. In the preceding month, it experienced two major accidents that claimed the lives of three workers. Bureaucratic thinking was blamed, and more attention to safety urged, but this was a pattern that nevertheless repeated itself over the next few years.[143] Perhaps more egregious was another incident, also at Shengli, that happened just about a year and a half later. Li Qiliang, who headed the eastern pit's third mining

---

139. On state responses to accidents in Mao-era China and their legacies in the reform era, see Brown, "When Things Go Wrong."

140. Fushun shi shehui kexueyuan and Fushun shi renmin zhengfu difang zhi bangongshi, *Fushun shi zhi, gongye juan*, 112.

141. Luo, "Kong Xiangrui xiaozu."

142. Luo.

143. "Shengli kuang jingsai zhong hulue anquan."

zone, covered up an accident that happened under his watch, allegedly to win an award and a monetary prize for exceeding production quotas in an ongoing "red flag competition."[144] While it would still be a few more years before the Great Leap Forward, in which systemic misreporting of grain production figures exacerbated the nationwide famine, instances like these foreshadowed the disaster that was to come.

Many of these accidents were avoidable. For example, in three incidents within a week at the Laohutai mine in January 1951, workers heated dressed coal along with shale rock (which should have been treated separately), overloaded a cart, and operated a vehicle without proper supervision. All resulted in accidents and injuries.[145] There were thus numerous calls for mine leaders to educate their workers about safety through propaganda or instruction and to raise their level of self-awareness when it came to how they carried out their work.[146] In a letter to the local newspaper in 1954, Jiang Xing of the National Safety Techniques Supervision Bureau reported that his investigations revealed an increasing number of workers bringing matches and cigarettes into Laohutai's pits, which he interpreted as the mine's management not providing sufficient safety education.[147] While this admittedly marked a difference from Mantetsu times, in which Japanese colonial authorities blamed workers and not managers for such mishaps, it did not stop the frequency of accidents.

The trend of accidents due to speeding up production continued through the decade. In the first quarter of 1954, for instance, the Fushun colliery not only met the centrally set target but exceeded it by more than 80,000 tons. At the same time, however, there were as many as 823 accidents that injured 821 miners and killed 10.[148] Despite the attention these accidents drew, the situation failed to improve and actually worsened. In 1956, the Fushun municipal government reported the worst year for production-related accidents since liberation. Between January and October, there were over 2,100 accidents, claiming 36 lives (30 of them in the mines).[149] Some workers, echoing all too closely critiques of the former Japanese management, would complain that "the coal is paid for in human lives" (煤是用人命換來的).[150]

144. "Shengli kuang dong keng."
145. "Laohutai kuang fangsong anquan gongzuo."
146. "Jieshou Shengli kuang Laohutai kuang."
147. Jiang, "Laohutai kuang fangsong baoan jiaoyu."
148. "Fushun meikuang shigu."
149. "Fushun shi ge shengchan changkuang."
150. "Fushun meikuang shigu."

The government's greatest concern, though, was that these accidents would hamstring efforts to increase output and "bring massive losses to the country." In the case of 1956 Fushun, these losses, calculated as they were with great precision, centered not only on repair costs but on forgone labor time and coal unmined: "From January to August, 53,694 workdays were lost because of accidents. On average, every day 268 workers were unable to participate in production. Because of the damage in equipment that resulted in work stoppages, coal-mining operations lost 15,377 hours, and 67,980 fewer tons of coal were excavated. In order to repair equipment damaged by the accidents, as much as 116,329 yuan had to be spent."[151] In the project of socialist industrialization, as it had been with extractive efforts under the earlier regimes from which the Communists sought to break, the priorities were clear: production took precedence over people.

Conclusion

One blustery February morning in 1958, Chairman Mao came to Fushun, arriving, as he often did, at the shortest of notices. As part of his visit, Mao, like countless others before him, was given a view of the open pit. As he peered down into its depths, he noticed plumes of smoke rising from the darkness. When he asked what those were, his handlers from the colliery quickly responded that those were old fires from Japanese and Nationalist times. Mao replied: "We have to take good care of our resources. Do your best to extinguish these fires as soon as possible. We cannot let our resources just burn away for nothing."[152] There is some irony to Mao's injunction. The carbon regime that the Communist state brought into being in the 1950s perpetuated much waste by privileging production. At the same time, Mao's focus on the material damage—the loss of these precious energy resources—in not taking consideration of the loss of life that often accompanied such fires, seemed all too consistent with the years that were to follow in which the regime sacrificed more lives in the pursuit of extensive energy extraction and the false promises of industrial modernity.

The Communists built their energy regime in Fushun and in Manchuria more generally on the ruins of the Japanese empire. In this endeavor, they tapped into transnational technical expertise, not only from Soviet advisers who arrived in the early 1950s, but also from Japanese engineers and technicians who stayed on postliberation. With an economic infrastructure

151. "Fushun shi ge shengchan changkuang."
152. Shen, "Nanwang '2.13,'" 157.

毛 主 席 视 察 抚 顺

6.2. "Chairman Mao Inspects Fushun." Poster depicting Mao Zedong at the Western open-pit mine during his visit to Fushun on February 13, 1958. In accordance with a strong theme in socialist realist art, machines loom large in the background. (Poster from the author's collection.)

assembled over decades of Japanese rule, Manchuria stood out as China's most industrialized region at the birth of the People's Republic. This was notwithstanding the devastation of war and Soviet looting, from which the colliery and the region recovered through the help of retained Japanese experts.[153] That part of the Japanese legacy in Manchuria was industrialization and economic development might seem like an inconvenient truth for those of us opposed to the idea of imperialism, but this is by no means a reason to cherish the memory of Japan's empire.[154] Aside from the fact that Manchuria's economic infrastructure was established on the backs of countless Chinese workers, it also needs to be stressed that the Japanese architects of the region's industrial edifices almost certainly had not raised them for the benefit of their colonized subjects. Furthermore, this issue of Japanese contributions to the postcolonial Manchurian economy

153. On the contributions of retained Japanese experts to the reconstruction of industrial Manchuria, see also Matsumoto, *"Manshūkoku" kara shin Chūgoku e;* and Hirata, "From the Ashes of Empire."

154. In scholarship, as in wider public discourse, there have been frequent attempts to talk about imperialism as a good thing in retrospect, particularly in regard to the supposed long-term economic development of the formerly colonized. For one exchange around the question of whether Japanese imperialism in Manchuria led to the development or underdevelopment of the Manchurian agrarian economy, see Myers and Bix, "Economic Development in Manchuria."

may be easier to accept if we recognize that large-scale, coal-fired industrial expansion did not yield positive outcomes for all or even many sectors of the local population, to say nothing of the environment. We can, then, acknowledge Japan's "positive" economic impact on Manchuria without celebrating it.[155] A portion of the fraught inheritance imperial Japan bequeathed to its Communist successor, so visible in Fushun, was the technological infrastructure and technical expertise for the constitution of carbon technocracy.

For the Chinese socialist state, carbon technocracy presented the possibility of securing the material conditions to realize its vision of a socialist utopia. At the same time, it shared with its predecessors an understanding that industrialization would equip it with the "wealth and power" necessary for survival in a hostile world of states, many of which regarded its existence as an aberration. Also like its predecessors, it was committed to mobilizing science and technology to boost productivity for its industrial aspirations. To gauge its success on these fronts, the socialist state was obsessed with the aggregate output of resources like iron and steel, electric power, and coal, markers of national progress that had been established earlier in the age of industrial capitalism and that it did not (or perhaps could not) even begin to reimagine, in spite of all its rhetoric of revolution. The problem with those abstractions, as we have seen in this chapter, was that they tended to mask issues that arose in the process of their production: questions of quality, of sustainability, of human cost.

Not too long after *Coal Mines in May* hit the shelves, Xiao Jun found himself in trouble with the party again. He faced problems with publishing the novel, and when it became clear to him that he had been blacklisted as a writer, he sent a petition to Mao in 1953 to plead his case. All it took was a word from the chairman. *Coal Mines in May* came out in 1954, initially without much issue.[156] One year later, however, a harsh critique by two relatively unknown writers, penned purportedly with party approval, appeared in the leading literary journal.[157] "After reading this book, we could feel only extreme indignation," the piece protested, "We believe that, through his

155. For a powerful ethnography of how successive generations in China and Japan have grappled with the legacies of the Japanese empire in Manchuria, see Koga, *The Inheritance of Loss*.

156. Cohen, *Speaking to History*, 181–82.

157. Goldman, *Literary Dissent*, 221.

use of revolutionary clichés and a gaudy vocabulary and through his affectation of zeal, Hsiao Chün [Xiao Jun] has tried to camouflage his seriously intended caricature and calumny of socialist enterprises, of the working classes and their political party."[158] The novel drew particular ire by showing ordinary workers accomplishing superhuman feats independent of party direction.[159] As another critique charged, Xiao Jun was suggesting that "the working class does not need leadership, political consciousness, nor technical abilities, it only needs a strong subjective spirit."[160] Even more critiques followed. Xiao Jun would be purged by the party in the Anti-Rightist Movement (1957–1959), only reemerging in the late 1970s following the Cultural Revolution (1966–1976).

The climax of *Coal Mines in May* is an accident at the open-pit mine. Two workers, Yang Pingshan and Li Fengde, had been trying to recover an electric shovel buried under debris from an earlier collapse when another rockfall occurs, entombing both men and the machine so that observers could see only "a portion of the neck of the shovel sticking out from the pile." Atop the debris, a blaze breaks out, "a fire from which rolls forth a thick plume of black smoke." The men had, in fact, suspected that such a rockfall would occur, but when they cautioned their superiors about it in the days leading up to the accident, their warnings were largely dismissed. When their fellow workers dig them out two days later, "the faces on the corpses were no longer recognizable, and they could only be told apart from the remains of clothes and shoes that had not yet been burned."[161] Xiao Jun based this episode on an incident from his time in Fushun that claimed the lives of electric shovel operator Liu Honglin and his assistant Li Wensheng.[162] As in the novel, the two men were celebrated as "heroes" who had "sacrificed" for the revolution.[163]

It does not take a leap of imagination to read *Coal Mines in May* as a critique of the productivist tendencies of carbon technocracy. In accounting for the cause of the accident that killed the characters Yang and Li, both the Mining Affairs Bureau's party secretary and the protagonist Lu Dongshan, speaking on behalf of the colliery's workers, blamed "bureaucratism" (官僚主義): the failure of leaders to listen to the masses or to take into

---

158. Cited in Hsia, *History of Modern Chinese Fiction*, 278.
159. Goldman, *Literary Dissent*, 221.
160. Cited in Goldman, 221.
161. Xiao, *Wuyue de kuangshan*, 434–36.
162. Xiao, *Dongbeio riji*, 843 (March 25, 1950).
163. *FKZ*, 62.

consideration their well-being.[164] At the back of the book, after the story it-self ends, Xiao Jun appended three similarly fictional pieces concerning the novel's climatic accident that had supposedly been clipped from a newspaper: a portion of an article, an investigative report, and an editorial summary. Together, these pieces provide additional detail about the accident and an extended castigation of the colliery's management who were held responsible for it. Wujin's leaders were faulted for "merely blindly emphasizing increases in production and reconstruction figures." Although "this impulse, this subjective desire, may appear to be good," it was ultimately "petty and self-aggrandizing." As a result, "whether consciously or not, [these leaders] have sacrificed long-term and collective interests and even the health and lives of workers by seeking 'victory' only in form."[165] In Fushun, as in other sites of industry across socialist China, such impulses, although periodically censured, would dominate for decades.

164. In the early 1950s, bureaucratism was one of the targets of the Party Rectification campaign of 1950 and the Three Antis campaign of 1951. See Andreas, *Disenfranchised*, 32.

165. Xiao, *Wuyue de kuangshan*, 444–45.

Exhausted Limits

Sometime during my first trip to Fushun in the summer of 2011, I found myself in the middle of a construction site. I had been brought to a half-completed building on the edge of the Western open pit by the colliery representative who had earlier that afternoon taken me down into its depths. The exterior of the main structure, still encased in scaffolding, was white and red with thin, tall windows of reflective blue. As we trudged through the dusty corridors and chambers within, my guide gestured to empty spaces and unadorned walls, giving me a quick tour of yet-to-be-installed exhibits that I strained to see with my mind's eye. This was a museum in the making. The colliery's management, reorganized in 2001 as the Fushun Mining Group (抚顺矿业集团), had decided to build it right at this spot, where the pit's viewing platform once stood and countless visitors before me surveyed its workings. Several months later, the Fushun Coal Mine Museum (抚顺煤矿博物馆) opened its doors to the public.

I was excited to see the finished structure when I returned to Fushun the following summer. In front of the building now sat, upon a dais, a bronze statue of Chairman Mao in an armchair, left leg crossed over right, a piece of coal in hand. The statue was cast and put up here beside the pit several years earlier to commemorate the fiftieth anniversary of Mao's visit to Fushun but had been temporarily removed when construction was underway. I did not know whether this was the case before, but the statue now faced Beijing to the southwest, which, a plaque explained, "alludes to the fact that the centuries of great causes initiated by the Chinese Communist Party and the revolutionaries who came before will, under the core leadership of Party Central Committee from generation to generation, be as enduring as heaven and earth, as eternal as the sun and the

E.1. The Fushun Coal Mine Museum with a bronze statue of Mao Zedong in front of it. The wording by his feet reads, in larger characters, "Showing favor on the Coal Capital" and, in slightly smaller characters, "Standing up and remembering Mao Zedong / happily giving thanks to the Communist Party." (Photograph by the author, August 2012.)

moon."[1] Here was a vision of the future. Behind it, in the museum and the pit beyond, lay the past.

The museum tells the story of coal in Fushun from prehistoric times to the present. In their variety and vividness, the exhibits capture the richness of that history. Dioramas depict scenes from earlier eras, from the primeval vegetation from which Fushun's coal seam formed to men in traditional tunics and topknots working kilns fired by Fushun coal. The museum's focus, though, is on the period starting from the turn of the twentieth century, when, as a preface plaque declares, "national capitalist Wang Chengyao resolutely mined the 'dragon veins' and, in so doing, raised the curtain on Fushun's industrial revolution."[2] In a display recounting the beginnings of modern mining in Fushun, a mannequin in the dark blue robes of a Qing official represents Wang, while a reproduction of Zeng Qi's memorial requesting that the court permit coal mining in Fushun shows the Guangxu emperor's approval in vermillion ink.

The presentation here of Fushun's subsequent history, like local narratives advanced since the 1950s and most official tellings of the past in the People's Republic in general, is divided by the Communist revolution. Here, the account of the period before the revolution is predominantly one of Japanese exploitation. Photographs in black and white and sometimes sepia depict the growth of the mines and the surrounding town, while artifacts from flimsy old mine carts to the solid wooden furniture in a replica of Kubo Tōru's office give visitors a glimpse into the material culture of life and labor at the colliery over those decades. The exhibits emphasize that the industrial transformation of this site, considerable as it may have been, was little more than an expression of "crazed plunder" (疯狂掠夺), with escalating figures of worker injury and death presented on charts as indices of the human cost underlying Japan's intensifying extraction. By contrast, in the museum's rendering of the period after the revolution, the narrative shifts to one of unbroken progress, gauged largely in terms of coal output. "Even under the unceasing impact of the 'Cultural Revolution's powerful currents,' the Fushun colliery never halted coal production," one write-up stresses. Visitors are called to appreciate how such progress was made possible through the guidance of the party, the grit of Fushun's workers, and the greatness of technological developments, the last of which is

1. "Mao Zedong shicha Fushun Xi lutiankuang jinian tongxiang jieshi" [An explanation of the bronze statue commemorating Mao Zedong's inspection of Fushun's Western open-pit mine], Fushun Coal Mine Museum, Fushun, China.

2. "Qianyan" [Preface], Fushun Coal Mine Museum, Fushun, China.

showcased in the multitude of mining machines and equipment displayed throughout the premises.

The museum's main attraction, however, is perhaps not Fushun's coal-mining history, engagingly told as that history is through the carefully curated collection of images and objects. Rather, it is the ten-story observation tower at the back of the building. From the top, one has a good view of the pit below. Although fog often masks much of the man-made chasm, when the weather is clear, the sight is stunning. In a 2018 blog post, one visitor described the scene as "mystical, profound, ethereal, and enchanting . . . the layered winding paths, with their graceful and round curves and rich display of lights and colors, like the trajectories of planets in the firmament." As he watched the pit's operations, from oil-shale-bearing electric trains weaving east and west to coal-laden trucks that "crawled up their specially designated route like a long snake," he professed this a sight that "struck my vision and shook my soul."[3] He was consumed by the spectacle of the technological sublime, as Yosano Akiko had been over ninety years ago.

The museum's designers evidently intended for viewers like this man to be awed by the wonder of the mechanized pit in full swing—of carbon technocracy in motion. And yet, there was soon little activity left to see. In 2019, after almost a century of extraction, the Western open-pit mine ceased operations completely, the latest in Fushun's mine closures that began over four decades back with Shengli's shuttering in 1976.[4] Countless museums have sprung up across China since the 1980s, a good number of these focusing, as this museum does, on an industry of deep local significance.[5] In the case of the Fushun Coal Mine Museum, it may have been established to celebrate the industry that made this city, but its construction also sounded a death knell for coal mining here—a search for remembrance in the midst of ongoing loss.

Fushun's decline, beginning as it did a few decades back, sits between two seeming contradictions. First, this mining metropolis, in what was once China's most industrialized region, fell into recession at the same time that the national economy was making its spectacular ascent. Second, coal mining in Fushun is coming to an end even though there are still considerable deposits in the ground and coal production and consumption

3. Wang, "Fushun meikuang bowuguan."

4. "Fushun Transforms Century-Old Coal Mine."

5. On museums in postsocialist China, see Denton, *Exhibiting the Past*. Fushun is, incidentally, home to three other museums of note: the Pingdingshan Massacre Memorial Hall, the Lei Feng Memorial Hall, and the Fushun War Criminals Management Center.

globally remains stubbornly high. This epilogue explores how these contradictions emerged from the limits of carbon technocracy. The path we will take goes outward, as through concentric circles, the passage from one section to the next moving up in scale from region to nation (both China and Japan) to the world, with Fushun at the center.

Of "Never-Rusting Screws" and Rustbelts

The intercity highway from Shenyang to Fushun was wide and, as highways in China go, not particularly busy. Looking out of the bus window, I initially found my surroundings unremarkable, a flat expanse of familiar greens, yellows, and browns, the monotony occasionally broken by a building or a transmission tower in the distance. About midway through the hour-and-a-half journey, though, we started passing through clusters of unfinished high-rises, the outlines of their lofty forms blurred in the haze of the summer heat. This construction was part of the Shenfu New Town project, designed to house excess businesses and residents from Shenyang and Fushun, which were anticipated to grow following their integration under the framework of a recently established regional economic zone. Northeast China, the country's industrial heartland, had been in steady decline since the start of the reform era in the mid-1970s. Accounting for as much as a quarter of China's industrial production in the 1950s, the region, by the 2000s, generated less than 10 percent of national output.[6] Shenfu represented larger efforts the central and local governments had been undertaking to revive the region by attracting capital from beyond, as the Qing court did a century earlier.[7] Yet the buildings went up, but the people barely came, and Shenfu joined the ranks of China's overbuilt and underoccupied "ghost cities."[8]

The bus I was riding across this haunting landscape had been named after Lei Feng 雷锋 (1940–1962), a young People's Liberation Army sol-

6. Rithmire, *Land Bargains*, 69–70.

7. On the regional economic zone centered on Shenyang, see Wang et al., *Old Industrial Cities*, 181–203.

8. Denyer, "In China, a Ghost Town"; Wu and Liu, "Eight Cities Partner." In a bid to place Shenfu on the map, local officials approved the construction of a massive ring at the center of the new town, which stands over five hundred feet tall, was made from about three thousand tons of steel, and cost around ten million yuan. Because of how expensive it was and how useless it seems, this structure has become an object of much critique and ridicule in China and abroad. When photographs of this ring appeared online, some Chinese netizens joked, for instance, that it was a "cosmic portal" or an "entrance for aliens to invade the Earth." See "Landmark Building."

dier who met his end in Fushun. There was something slightly ironic about naming a transport service after someone killed in a traffic accident: Lei Feng had been struck down by a telephone pole knocked over when the truck he had been directing reversed into it. In death, he became larger than life. The Chinese Communist Party promoted "Comrade Lei Feng" as the model socialist citizen, a selfless do-gooder earnestly devoted to the cause of Chairman Mao. In his posthumously published diary, Lei Feng famously wrote that he wanted nothing more than to be a "screw" that "never rusts" in the revolutionary "machine."[9] The socialist state may have depended on coal from Fushun and other collieries across China to drive its industrial apparatus, but it also relied on countless workers whom it mobilized as "never-rusting screws" to meet its unconscionable production targets and hold together that unwieldy assemblage.

In Fushun, as in most of the country, the 1960s began in readjustment and ended in revolution. It is indicative of Fushun's importance that it was shielded from the worst of the famine that resulted from the Great Leap Forward. As food shortages arose, workers began experiencing high instances of edema from malnutrition, and production suffered. In response, the central government sent over 190,000 pounds of supplementary grain, while the local government distributed food substitutes and increased rations of nonstaple foods, measures that together kept starvation at bay.[10] The colliery announced massive increases in coal output over the Great Leap, which reached an all-time peak of 18.6 million tons in 1960, more than double that produced in 1957, at the end of the First Five-Year Plan.[11] Although overreporting was prevalent during this period, there is at least some reason to regard this figure as accurate.[12] When the colliery cited the same figure in official accounts later, it did so not to boast. Rather, it ruefully acknowledged that this massive jump in production was only possible because of an obsession with output to the exclusion of all other necessary considerations, from maintaining the ideal mining ratio to fixing broken or worn equipment, creating problems for subsequent operations.[13] Following the Great Leap's unceremonious end in 1961, the colliery under-

9. Mao zhuxi de hao zhanshi Lei Feng jinianguan, *Lei Feng riji*, 94. For more on the history and historicity of Lei Feng's diary, see Tian, "Making of a Hero."

10. Fushun's experience cleaves closely to the urban bias in food politics during the Great Leap famine. See Brown, *City versus Countryside*, 53–76; and Wemheuer, *Famine Politics*, 62–74.

11. *FKZ*, 71.

12. For one account of how overreporting contributed to food shortages during the Great Leap Forward, see Li, *Village China*, 87–92.

13. *FKZ*, 72.

took a series of measures to readjust the mines' workings, resuming essential tasks it had forgone in preceding years, even as output plummeted from its overheated heights.[14]

When the Cultural Revolution erupted in 1966, Fushun was swept by its revolutionary winds. Workers wrote big character posters and held struggle sessions in which those who had the misfortune of being branded class enemies were subjected to public violence and humiliation. Mines took on new revolutionary names: the Western open pit became "Red Guard" (红卫); Laohutai became "Red Star" (红星); Longfeng became "Red Flag" (红旗). Miners carried with them Mao Zedong's codified wisdom in the *Little Red Book* when they descended into the pits and recited lines from it before commencing work. As elsewhere, two rival factions (loosely characterized as radical and conservative) emerged and clashed violently. In total, these altercations claimed the lives of 119 and gravely injured 72 before the factions were dissolved in the fall of 1968. A revolutionary committee organized by the Mining Affairs Bureau and headed by a representative from the People's Liberation Army ran the colliery until the Cultural Revolution ended in 1976.[15] Through this all, Fushun, as the colliery museum exhibit states, "never halted coal production." Output may have dipped to a low of 5.3 million tons in 1967, but it rose again after, staying above 10 million tons yearly from 1969 to 1979.[16]

For Fushun's miners, whose labor bolstered this extraction, work in these times of turmoil was often treacherous. In 1970, the productivist impulses that had culminated in the Great Leap Forward resurfaced. The colliery's revolutionary committee, seized by a fanatical fervor, began making remarks such as "17 million tons a year is short, 20 million tons is safe, 25 million tons is splendid," essentially calling for a quick doubling of output, which they then formalized as part of the enterprise's production plan. Those who raised objections were labeled "old rightists" or "old conservatives" and criticized and denounced at public meetings. At

14. *FKZ*, 76–78. On the post–Great Leap Forward readjustment more broadly, see Meisner, *Mao's China and After*, 260–72.

15. *FKZ*, 79–82.

16. *FKZ*, 143–48. Nationally, coal production increased over the course of the Cultural Revolution. This was to the point that output far exceeded the transportation capacity to move the coal to sites of consumption. See Thomson, *The Chinese Coal Industry*, 53. One engine of coal-mining expansion in this period was the massive military-industrial Third Front project through which the socialist state sought to build up a strategic base in China's interior in preparation for potential war with the United States or the Soviet Union. Between 1964 and 1980, the Third Front region's share of coal production increased from about 40 percent to just under 50 percent. See Meyskens, *Mao's Third Front*, 212.

the Western open pit, colliers once again abandoned the work of stripping overburden and attended solely to extracting coal. Multiple rockfalls resulted, to the point that the large volume of coal being mined could not be transported out of the pit. Left to pile up, the coal caught fire and burned to ash and cinder. The metaphor is hard to resist: the accomplishments gleaned through the preceding years of readjustment similarly went up in smoke.[17] Just as the working environment turned increasingly dangerous, safety standards were relaxed. Under prevailing discourses of the Cultural Revolution, expressing concerns about matters of safety could be taken as a sign of bourgeois survivalism and imposing safety regulations evidence of a desire to discipline or control workers.[18] This spelled disaster.

I first heard about Laohutai's massive explosion of 1976 from Mr. Wang, a retired miner I got to know on my second trip to Fushun. We were deep into our conversation about his work at the colliery in the 1970s and 1980s when he brought up this disaster. He had, as it turned out, been down in the pit when it happened. Gesticulating excitedly as he spoke, he described the sudden force that threw him to the ground and being in disarray until the wafting odor of charred flesh shook him to his senses. Ordered by his supervisor to flee immediately, he was fortunate to have made his way out. He recalled that when he first joined the colliery, a superior told him that "under the evil old society, the Japanese devils only cared about plundering coal and absolutely did not bother about the safety of Chinese workers, and gas explosions frequently occurred." Sometime later, an older worker mentioned that "even after liberation, there was a gas explosion of some scale at B mine that killed quite a number of people." Back then, he found it hard to believe. Little had he expected to experience a disaster like this firsthand.[19] In the official record, ninety-two workers perished in the 1976 explosion.[20]

Despite disasters such as this one, Fushun's leaders held out much hope for the colliery's industrial future as the reform era dawned. Mao died in September 1976. Deng Xiaoping 邓小平 (1904–1997), his unchosen successor who consolidated power by December 1978, soon oversaw the introduction of market mechanisms aimed at boosting China's productive forces that had been flagging under the strictures of central planning. In this

17. *FKZ*, 83.

18. Wright, *Political Economy of the Chinese Coal Industry*, 183.

19. Mr. Wang later also kindly shared his personal writings with me. One of these was an unpublished essay titled "That Unforgettable April 14" (难忘的四一四), in which he further detailed what he had told me about this disaster. Quotations are drawn from this piece.

20. *FKZ*, 84–85.

period, Fushun bore the distinction of being labeled a "Daqing-style enterprise" (大庆式企业), which meant that the central government deemed it important for the planned expansion of the Chinese economy.[21] There were, however, challenges from the get-go. By 1976, the Shengli mine's subsurface workings extended under Fushun's urban area, causing subsidence in the vicinity of the amusement district and the power plant. The Ministry of Coal Industry sanctioned its closure in 1979.[22] Although colliery leaders initiated reforms in following years that, for instance, made safety and efficiency priorities, Fushun's coal-mining industry languished, like many large state-owned enterprises amid the throes of market transition.[23]

Its workers would suffer for it. In 1994, one of the mines announced that it was going to lay off 20,000 workers, triggering labor unrest that prompted provincial authorities to intervene and call off the dismissals. But problems persisted. In 1999, Longfeng declared bankruptcy, and almost 100,000 workers found themselves unemployed. The next year, Laohutai discharged 24,000 workers, or 80 percent of its workforce. "Protests of desperation," which sociologist Ching Kwan Lee identifies as characteristic of collective action here in China's rustbelt, ensued.[24] In the spring of 2002 about 10,000 laid-off miners and their counterparts from other industries staged sit-downs against their paltry severance payments. They blocked the railway line and the main road into Fushun, only dispersing when local officials gave each a small sum of seventy-five yuan. Those who refused to budge were forcibly removed by the People's Armed Police.[25]

Fushun's unarrested decline is part of the story of how Northeast China slid into recession over the preceding few decades as the state-owned enterprises that dominated the region's economy registered continuous losses, the "iron rice bowls" they proffered rusting and falling apart. It is also part of a global story of coal-mine closure. The World Bank estimates that industrial restructuring in Europe, the United States, and China over the past

21. The Daqing oilfield, discovered in 1959 and developed into a colossal petroleum-producing enterprise within a few years, became a model of socialist industrialization that privileged self-sufficiency and the revolutionary conquest of nature. In the late 1970s, the Daqing model referred more generally to enterprises that were efficiently managed and, consequently, highly productive. On Daqing's development, see Hou, *Building for Oil*. On the Daqing model in the late 1970s, see Hama, "The Daqing Oil Field."

22. *FKZ*, 88.

23. *FKZ*, 88–96. On problems that state-owned enterprises began facing in China's market transition, see Naughton, *Growing Out of the Plan*, 284–88.

24. Lee, *Against the Law*, 11–12, 69–153.

25. Human Rights Watch, *Paying the Price*, 34–35. On problems of unemployment and reemployment in Fushun, see Smyth, Zhai, and Wang, "Labour Market Reform."

half century has put about four million coal miners out of work.[26] From Appalachia to the Kuzbass, coal-mining towns like Fushun have become economically blighted by the slowdown or stoppage of extraction. In China, the late 1990s and early 2000s saw a slew of mines shutting down. One driver for closure has been environmental concerns. In a bid to curb carbon emissions, many governments have introduced policies to regulate or restrict coal burning, turning to energy sources that pollute the air less and leave a lighter carbon footprint in their production and consumption.[27]

Herein lies a point of tension. If the manifold disruptions that closure has brought to mining communities were taken as something of a necessary evil, it seems as though the greater good this evil was expected to serve is not being realized in the aggregate. The use of coal worldwide may have taken a hit in 2020 with the COVID-19 crisis, but it was, until the onset of the pandemic, resolutely high and as of early 2021 has just begun its rebound.[28] In China, coal consumption reached its peak in 2013, a decade or so after mine closures commenced in earnest. While this figure proceeded to fall between 2014 and 2016, it started climbing again from 2017, when the Chinese government, which for those few years in the mid-2010s oversaw cuts in fossil fuel consumption in a battle against air pollution, began loosening restrictions on the expansion of coal power as part of efforts to boost a slowing national economy.[29] More recently, in September 2020, Chinese president Xi Jinping pledged before the United Nations that China would go carbon neutral by 2060, a move that has excited many environmental analysts.[30] We are, however, still at the very early stages of seeing how that target might be reached. For now and certainly from a global perspective, coal seems here to stay.

East Asia in the Great Acceleration

Fossil fuels are the stuff of miracles. The second half of the twentieth century witnessed the rise of East Asian economic powers, beginning with Ja-

---

26. World Bank Group, *Managing Coal Mine Closure*, 9.

27. World Bank Group, 20.

28. Alvarez, "Global Coal Demand Surpassed Pre-Covid Levels."

29. Feng and Baxter, "China's Coal Consumption on the Rise"; Gao, "China Relaxes Restrictions on Coal Power Expansion." In another instance worth highlighting, the Chinese government had banned the use of coal in the winter of 2017 but had to reverse this ban for a number of cities in the Northeast when it became apparent that many people had not managed to switch to alternative fuels, especially with gas supply unable to meet the surge in demand, and were left without proper heating. See "China Does U-Turn."

30. Braun, "Is China's Five Year Plan a Decarbonization Blueprint?"

pan, whose rate of growth from the 1950s to the 1980s seemed nothing short of miraculous. By the turn of the new millennium, it was China that was the miracle, its economy from the reform era onward growing annually by double digits on average. Multiple factors may help account for each of these two instances of extraordinary growth, from specific governmental policies to the particular international political economic contexts in which they unfolded. Yet both share one feature: explosive increases in fossil fuel use, which environmental historians John McNeill and Peter Engelke, among others, identify as a main characteristic of the "Great Acceleration" in earth-changing human activity under the Anthropocene since 1945.[31]

Between 1953 and 1990, coal's share of Japan's primary energy supply fell from just under half to around a sixth, but the annual energy derived from coal went up almost threefold. Oil, which made up under a tenth of domestic primary energy supply in 1953, rapidly increased in relative importance, outpacing coal by 1962.[32] In 1973, the year of the global oil crisis, Japan consumed in one week, on average, the same volume of petroleum it did in the entirety of 1941.[33] Japan used, in 1990, over thirty-five times more oil than it did in 1953, and oil accounted for just below half of its total available energy that year.[34] Meanwhile, in China, the three decades from 1980 to 2010 witnessed a rise in the consumption of coal from 715 million to 3.2 billion tons annually and of petroleum from 1.9 million to 9.3 million barrels a day.[35] That many might think of these incredible increases in energy use as but part and parcel of economic expansion speaks to just how naturalized the otherwise contingent relationship between carbon and the idea of growth has become.

Japan's thorough devastation by the end of World War II made the already speedy postwar growth appear even more miraculous. Following its surrender, Japan was for seven years occupied by the Allied powers under the leadership of the United States, and American geopolitical interests had much bearing on the fossil-fueled recovery of the postwar Japanese economy. American occupying authorities were initially committed to dismantling the political and economic apparatuses of the Japanese imperial state. Among other measures, they expelled from government and business over 200,000 men accused of setting in motion Japan's war machine and began

31. McNeill and Engelke, *The Great Acceleration*, 7–27.
32. Japan Statistical Association, "Ichiji enerugii kokunai kyōkū."
33. Samuels, "Sources and Uses of Energy," 49.
34. Japan Statistical Association, "Ichiji enerugii kokunai kyōkū."
35. "China," U.S. Energy Information Administration; National Bureau of Statistics of China, "7-5 Coal Balance Sheet," *China Statistical Yearbook 2012*.

taking apart the zaibatsu that had helped fuel it with capital. However, with the advent of the Cold War amid the harsh socioeconomic realities still facing millions of ordinary Japanese, the occupiers pivoted, thinking it most prudent to privilege "stability" instead, lest Japan "fall" (as China soon would) to communism and its particular appeal to the disaffected. As one defining feature of what was soon called the "reverse course," the American authorities returned to power purged bureaucrats and capitalists whom they believed would best facilitate the country's economic rehabilitation, a notable example being Kishi Nobusuke, who had been imprisoned but not indicted as a "Class A" war criminal. The outbreak of the Korean War in 1950 proved to be a key catalyst in that rehabilitation. The Japanese industrial sector, ravaged as it had remained since the earlier war, greatly profited from American military procurement and started to turn a corner.[36]

Japan's swift economic ascent followed. The rising Japanese economy was dominated, as before, by large conglomerates and directed by a developmental state that found its clearest expression in the technocratic Ministry of International Trade and Industry (通商産業省 or MITI).[37] In successive decades, this Japanese model served as a source of inspiration for the "Four Little Dragons" of Hong Kong, Singapore, South Korea, and Taiwan, which underwent rapid industrialization along a broadly similar trajectory.[38] At the same time, Japan, which was expected to pay considerable reparations to its Asian neighbors for its acts of aggression in World War II, benefited from having brokered agreements with individual recipient governments to compensate them in the form of goods and services furnished by Japanese firms. Through these arrangements, which MITI and other bureaucratic bodies coordinated with private sector interests, Japan was able to secure access to both export markets and raw materials essential for its economic rise.[39]

36. The classic work on the American occupation of Japan is Dower, *Embracing Defeat*. For a revisionist take that attributes the reverse course not so much to shifting American interests but to the agitation of conservative elements within Japan, see Masuda, *Cold War Crucible*, 31–37.

37. MITI was created in 1949 out of the Ministry of Commerce and Industry, which, during the tail end of World War II, had been briefly reorganized as the Ministry of Munitions. On MITI and the key role it played in making the Japanese miracle, see Johnson, *MITI and the Japanese Miracle*. For a series of highly productive engagements with the developmental state idea and its iterations beyond Japan, see Woo-Cumings, *The Developmental State*.

38. For a comparative analysis of the "Four Little Dragons," see Vogel, *The Four Little Dragons*; and Castells, "Four Asian Tigers."

39. Arase, "Public-Private Sector Interest Coordination," 173–77. Hiromi Mizuno, Aaron Stephen Moore, and others have stressed the important connections between the postwar dispensation of aid, under which these reparations fell, and Japan's colonial development in Asia. See Mizuno, "Introduction"; and Moore, "From 'Constructing' to 'Developing' Asia."

Cheap energy fueled the Japanese economy's postwar takeoff. Right after World War II, Japan had, in fact, confronted a coal crisis, with output at but a fraction of the wartime average.[40] This was a product both of persistent problems that had arisen out of Japan's warscape of intensification (chapter 4) and, more immediately, of the American authorities' decision to repatriate about 9,000 Chinese and 145,000 Korean miners, a large proportion of whom had been brought to Japan as forced labor during Kishi's tenure as commerce and industry minister following his time in Manchuria. Japanese officials acted quickly to address this energy shortage. They moved workers from other kinds of mines to coal mines and gave them sizable food-ration allotments—calories neither wartime slave labor nor the postwar general public enjoyed.[41] When those measures proved insufficient, they introduced "priority production" (傾斜生産), a system in which the coal industry was accorded significant public financing (second only to what was given to the occupying authorities) and its output allocated first to the steel industry on which coal mining, in return, heavily depended.[42] Although this met with success, and coal production almost reached its targeted output of thirty million tons in 1947, the coming decade saw Japanese coal, with its relatively high cost of extraction, struggle against cheaper oil imports.[43]

The Japanese government and MITI tried to shore up the domestic coal industry with assistance packages while placing restrictions on oil imports. In this, they were motivated both by older anxieties around energy self-sufficiency and by ongoing concerns about safeguarding influential coal interests and minimizing anticipated social disorder should coal mines close and coal miners lose their livelihoods.[44] Yet the demand for cheap energy among industrial consumers eventually won out, and state planners gradually withdrew the support that kept many coal mines afloat. Between 1961 and 1975, the number of mines dropped from 570 to 35, the number of miners from 210,000 to 20,000, and annual coal production from 55

40. For an account that places this coal crisis within the context of a larger energy crisis in the opening years of the occupation, see Metzler, "Japan's Postwar Social Metabolic Crisis."

41. Johnson, *MITI and the Japanese Miracle*, 179–80.

42. Johnson, 181–84; Samuels, *Business of the Japanese State*, 92–95; and Hein, *Fueling Growth*, 124–28.

43. For more on the Japan's postwar "energy revolution" that witnessed coal being overshadowed by oil, see Samuels, *Business of the Japanese State*, 108–32; and Hein, *Fueling Growth*, 316–28.

44. Samuels, *Business of the Japanese State*, 108.

to 19 million tons.[45] In contrast, oil imports surged, while liquefied natural gas, liquefied petroleum gas, and nuclear power—with the first reactor coming online in 1966—enlarged their share in Japan's energy portfolio.[46] By the 1973 oil crisis, oil, having overtaken coal a decade earlier and having widened the distance in the intervening years, accounted for almost three-quarters of Japan's primary energy supply.[47] Nevertheless, Japan continued to rely much on coal, though mostly through imports, which were cheaper and generally of a higher quality than the local varieties they began to displace by the early 1970s.[48]

Over the last quarter of the twentieth century, Japan came to be widely associated with a certain degree of efficiency, including in energy. Japan boasts one of the lowest energy intensities in the industrial world, meaning that less energy is spent generating each unit of gross domestic product. This is attributable in large part to efforts by MITI and large corporations to increase energy efficiency after 1973, which built on earlier attempts Japanese industry took to conserve fuel (chapter 3).[49] Along with measures that MITI and other government agencies put in place to mitigate environmental issues like air pollution—such as mandating the installation of desulfurization equipment—this high level of energy efficiency contributes to an image of Japan as "an environmental technology and policy leader."[50]

Japan is, however, still a fossil-fueled power. As of 2017, it is the world's third-largest importer of coal after India and China, the third-largest consumer and net importer of oil after the United States and China, and the

45. Samuels, 131. On the local effects of mine closures in Japan, see Allen, *Undermining the Japanese Miracle*; and Culter, *Managing Decline*.

46. On the beginnings of nuclear power in Japan, see Yoshimi, "Radioactive Rain."

47. Japan Statistical Association, "Ichiji enerugii kokunai kyōkū."

48. Yoshida, "Japan's Energy Conundrum," 3.

49. Samuels, "Sources and Uses of Energy," 50. For an account of Japan's transwar search for energy efficiency, particularly in the energy-intensive iron and steel industry, see Kobori, *Nihon no enerugii kakumei*. It was also at this time that MITI launched the Sunshine Project, through which it sought to spur the further development of solar and other forms of energy. One unplanned outcome of this initiative was the extensive integration of photovoltaic technologies into consumer electronics such as calculators. See Nemet, *How Solar Energy Became Cheap*, 90–96.

50. Imura and Schreurs, "Learning from Japanese Environmental Management," 2. For a somewhat triumphalist take on Japanese air pollution management, see Hashimoto, "History of Air Pollution Control," 83–90. On the flip side, one study estimates that in 2015 Japan still experienced over 60,000 deaths due to air pollution. See Cohen et al., "Estimates and 25-Year Trends," 1914. Environmental historians of Japan have both underscored the environmental devastation that accompanied Japan's industrial rise in the modern era and pointed to ways in which that devastation left legacies that have yet to be fully reckoned with. See, in particular, Walker, *Toxic Archipelago*; and Miller, Thomas, and Walker, *Japan at Nature's Edge*.

largest importer of liquefied natural gas.[51] That Japanese industries burn coal and other fossil fuels more efficiently and cleanly should not over-shadow the fact that they still burn huge amounts of these carbon re-sources. Furthermore, Japan's environmental footprint in terms of fos-sil fuels extends beyond the archipelago. Japan has moved much of its carbon-intensive industry offshore, particularly to China.[52] It has also been involved in overseas petroleum development (to secure for itself oil sup-plies) in Indonesia and—achieving what interwar Japanese fuel experts could only admire in the examples of the American and British empires—the Middle East.[53] In the wake of the Fukushima Daiichi nuclear disaster of 2011, Japan took offline the nuclear reactors that up to that point had supplied about a third of the country's electricity (as of early 2021, nine of Japan's thirty-three operable reactors have been put back online and sixteen are awaiting restart approval). To compensate for the shortfall in power generation, the government intensified its ongoing pursuit of renew-able energy but, with seemingly greater enthusiasm, concurrently pushed for increases in the use of coal, which, as one analysis noted, "was seen as the fastest, cheapest and most reliable way [to] keep the lights on."[54] In 2020, Japan announced plans to build twenty-two new coal-fired power plants, of which fifteen have already commenced construction. If all are completed, Japan, which is already the world's fifth-largest greenhouse gas emitter, will be adding the capacity to emit another eighty-three million tons of carbon dioxide a year, a potential increase that alone dwarfs the an-nual emissions of most countries.[55]

In the case of China, observers may have expected increases in fossil fuel use to proceed in tandem with post-1978 economic growth, but what most found surprising was the acceleration of carbon consumption after the turn of the century. From 1978 to 2000, China's economy grew 9 per-cent on average, and its demand for energy 4 percent. However, from 2001 to 2006, the economic growth rate edged up to 10 percent, but the rate of growth for energy demand almost trebled to 11 percent.[56] Over the first two decades of the reform era, coal production had doubled, reaching a high of

51. "Japan," U.S. Energy Information Administration.

52. Samuels, "Sources and Uses of Energy," 50; Cole, Elliott, and Okubo, "International Environmental Outsourcing."

53. On Japan and Indonesian oil, see Tanaka, *Post-War Japanese Resource Policies*, 84–91; and Dinmore, "The Hydrocarbon Ring." On Japan and Middle Eastern oil, see Hamauzu, "Changing Structure of Oil Connections"; Takahashi, "Iran-Japan Petrochemical Project."

54. Bochove, Takada, and Clark, "Nine Years after Fukushima."

55. Irfan, "Why the World's Third-Largest Economy."

56. Bergsten et al., *China's Rise*, 137.

1.4 billion tons in 1996 before declining to a billion by the century's end.[57] Much of this increase was carried by small, local mines (so-called township and village mines [乡镇煤矿]), rural enterprises that could be collectively or privately owned but that were, regardless, oriented toward the market.[58] These mines multiplied quickly during the first decade and a half of economic transition. In 1995, they accounted for as much as 47 percent of total output.[59] Scholars have attributed the relatively low energy intensity of China's pre-2000 growth to both the structure of an expanding economy dominated by labor-intensive light manufacturing and the reform of state-owned enterprises that yielded some returns on efficiency.[60]

After 2000, Chinese coal consumption rose rapidly. From 2001 to 2007, it went up twofold from 1.3 billion to 2.6 billion tons. It then continued its speedy climb, peaking in 2013 at 4.2 billion tons.[61] Among the reasons for this change were China's shift toward a development model firmly grounded in energy-intensive heavy industry and infrastructure as engines of growth and its move to substitute imports of energy-demanding basic materials like steel with domestic production.[62] Chinese coal output generally kept pace with escalating demand. This was in spite of a spike in mine closures, which the government had initiated in 1998 because of overcapacity amid the economic slowdown triggered by the Asian financial crisis and, ostensibly, increased concerns over safety, efficiency, and the environment. The number of small, local mines fell by at least half between 1997 and 2000, from over 70,000 to around 35,000.[63] Large enterprises were not entirely unscathed, as evident in the case of Fushun.[64] Many of the remaining mines received state investment for technological upgrades

57. National Bureau of Statistics of China, "7-4 Coal Balance Sheet," *Yearly Data 1999*; and National Bureau of Statistics of China, "7-5 Coal Balance Sheet," *China Statistical Yearbook 2001*.

58. These township and village mines were examples of the township and village enterprises (TVEs) often credited with China's economic rise in the reform era. On TVEs more generally, see Naughton, *The Chinese Economy*, 271–94; and Huang, *Capitalism with Chinese Characteristics*, 50–108. On the township and village coal mines specifically, see Wright, *Political Economy of the Chinese Coal Industry*, 93–137.

59. Wright, *Political Economy of the Chinese Coal Industry*, 25.

60. Bergsten et al., *China's Rise*, 138–39.

61. National Bureau of Statistics of China, "9-5 Coal Balance Sheet," *China Statistical Yearbook 2015*; National Bureau of Statistics of China, "6-5 Coal Balance Sheet," *China Statistical Yearbook 2008*; and National Bureau of Statistics of China, "7-5 Coal Balance Sheet," *China Statistical Yearbook 2003*.

62. Bergsten et al., *China's Rise*, 141–44.

63. Andrews-Speed et al., "Impact of, and Responses to," 490–91.

64. For a brief overview of changes to the coal industry wrought by the 1998 reforms, see Andrews-Speed, *Energy Policy and Regulation*, 181–83.

that boosted their productivity.[65] At the same time, China, which had been a net exporter of coal and was, in fact, the world's second-biggest exporter after Australia until 2003, became a net importer in 2009.[66] Together, output and imports fed China's growing appetite for coal. The environment has been poorer for it. Incidentally, this was not just because of more coal being combusted but because much of this coal was unwashed, an outcome of water shortages in North China, where the country's largest coalfields lie—water shortages that, in the vicious cycle of environmental destruction, have themselves been exacerbated by coal-fired anthropogenic climate change.[67]

China's fossil-fueled economic rise coincided with its government's tightened embrace of science and technology as the primary tools of governance. From Deng Xiaoping to Xi Jinping, its paramount leaders have espoused an ideology that privileges scientific and technological fixes for the myriad problems facing the ascendant superpower.[68] By the late twentieth century, the Chinese state was closest to being a technocracy in the dictionary sense of the word: "government or control by an elite of technical experts."[69] Officials from the ruling Politburo Standing Committee down to the county level were increasingly college educated, many specializing in engineering or the natural sciences.[70] If economic growth and social control were their preoccupations, then science and technology were the means toward those ends. In regard to coal mining, technoscientific improvements facilitated the scaling up of extraction. In 2001, there was only one mine that, through fully mechanized mining technologies, had an output of over ten million tons. By 2013, when China started to temporarily cut back on its dependence on coal, the country boasted fifty-three of these massive mines.[71]

China has also been harnessing its growing arsenal of science and

65. Andrews-Speed and Ma, "Energy Production and Social Marginalisation," 115.

66. Wei et al., *Energy Economics*, 181.

67. Roumasset, Burnett, and Wang, "Environmental Resources and Economic Growth," 266.

68. Greenhalgh, "Governing through Science."

69. *Oxford English Dictionary Online*, s.v. "Technocracy," http://www.oed.com.

70. On the emergence of the technocratic class in post-1978 China, see Li and White, "Thirteenth Central Committee"; Li and White, "Fifteenth Central Committee"; Li, *China's Leaders*, 25–50; and Andreas, *Rise of the Red Engineers*. The most overt expression of technocratic governance in China was probably between 2002 and 2007, when all nine members of the Sixteenth Politburo Standing Committee were trained engineers. China's top leadership has since moved away from such a configuration to include those with economics, law, and political science degrees, but technocratic thinking still prevails.

71. Wang, "Development and Prospect," 253; Hu, Liu, and Cheng, "Caimei shi shang de jishu geming."

technology to remake the world beyond its borders.[72] In 2013, Xi Jinping launched the Belt and Road Initiative, a vast infrastructural venture aimed at redrawing the map of global connections through new or improved roads, railways, ports, and powerlines financed by Chinese loans and built with Chinese expertise. Under this initiative, China has taken up hundreds of coal-fired power projects in partner countries across South Asia, Eastern Europe, and Africa.[73] The contradiction between concurrent domestic decarbonization and carbon exportation has not been lost on observers. "The result," journalist Isabel Hilton notes, "is that while China is making commendable efforts to clean up at home and to reduce its carbon emissions, the Belt and Road Initiative threatens to lock China's partners into the same high-emission development that China is now trying to exit."[74] Like Japan and, indeed, many other so-called advanced economies, China has seemed generally unperturbed by its offshoring of pollution or the potential to pollute, as though it would actually be unaffected by any resultant impacts on our shared planetary habitat.

If the Chinese and Japanese economic miracles were defined in this era of the Great Acceleration by their speed, then the large-scale combustion of fossil fuels, which enabled this speed, has involved a twofold expenditure of compressed time. On the one hand, as Matthew Huber, Jennifer Wenzel, and others have contended, to use fossil fuels is, in a sense, to spend the millions of years it took for these energy resources to form.[75] On the other hand, by consuming carbon to the extent that we have to this point, we are, within a short span of geologic time, leaving imprints on the earth system that would have otherwise taken millions of years to occur. These changes, from ocean acidification to biodiversity loss, will probably remain for millions of years to come, long after the presumed passing of the fossil fuel age. Viewed in this light, one thing is clear: industrial society has been burning up both past and future for the sake of an unsustainable present.

Fossil Fuel Fantasies of Disenchantment

When they first deployed miners and machines to work Fushun's coal seams, Japan's empire builders crafted a case for their extractive endeav-

---

72. For one account of how such efforts fit within the emergence of a spatially distributed network of infrastructures and technologies that has served to redraw the global economic geography of resource extraction in late capitalism, see Arboleda, *Planetary Mine*.

73. Inskeep and Westerman, "Why Is China Placing a Global Bet."

74. Hilton, "How China's Big Overseas Initiative."

75. Huber, *Lifeblood*, 9; Wenzel, "Introduction," 7–8.

ors predicated on an understanding of the earth as secular and its interior disenchanted. Deriding the Qing rulers for the "superstitions" that supposedly blinded them to the value of Fushun's carbon riches, they drew not upon the power of "dragon veins" but on that of modern science and technology, channeling that power toward what they presented as the right and rational exploitation of these precious fossil fuels. In the process, the site and, indeed, the region were completely transformed. The open-pit mine, as the most visible embodiment of such a venture, would represent not only Fushun but the ideal of carbon technocracy that enthralled many across East Asia at this juncture in the industrial modern age. But what the Japanese and the Chinese who came after them seemed not to have considered was that the notion of Fushun as a "limitless" "treasure house" that "cannot be depleted"—a notion that had animated the intensity of their mining activities—was itself a fantasy.

Until recently, two existential questions regarding fossil fuels dominated public discourse.[76] Both concern limits. The first is about limits to the availability of fossil fuel resources. The most prominent expression of this is "peak oil," the idea that there is a point at which global petroleum production reaches its apex, after which it invariably declines.[77] Such a scenario is a source of anxiety both for societies that have become so deeply dependent on fossil fuels as well as for societies that desire the same material comforts and security the former enjoy through their extravagant expenditure of energy. In the first half of the twentieth century, numerous Chinese and Japanese, like people in other places, harbored similar fears, which they spoke of in terms such as "coal famine" or the "fuel question." Still, in these instances, although some agonized over the finitude of the earth's carbon resources, most were more narrowly troubled by the depletion of national holdings. For Japanese imperialists, in particular, anxieties over the anticipated exhaustion of domestic reserves made the prospect of an expanding empire and further access to sites of purportedly endless energy like Fushun all the more appealing.

The second question concerns limits to the planet's ability to support human and other forms of life were we to continue our ravenous consumption of fossil fuels. In recent years, as we witness report after report of fires burning more brazenly, ice melting more quickly, and other calamities birthed by the extreme weather conditions brought about by anthropogenic climate change, there is little doubt that this latter question is the

---

76. For another take on this pair of questions, see Mitchell, *Carbon Democracy*, 231–54.

77. On the theory of peak oil, see Deffeyes, *Hubbert's Peak*; and Priest, "Hubbert's Peak."

more pressing of the two. We are much more likely to destroy ourselves before we run out of the means to do so. Some of our historical counterparts recognized the harm that burning coal in large quantities inflicted on the environment and worked toward curbing combustion. Among them were, as we have seen, those who fought the long fight for smoke abatement in Osaka (chapter 3). Yet few people at the time, these conservationists included, apprehended the sheer extent and irrevocability of the damage caused by introducing into the atmosphere as much carbon dioxide as industrial society has through fossil fuel use.[78] As such, while we can acknowledge that those who had earlier advocated for or directed efforts at expanding coal production and consumption, like many of the Chinese and Japanese figures we encountered in this book, were complicit in sparking off processes that have led to our current climate crisis, it is difficult to condemn them solely on the basis of their having promoted fossil fuels given the general ignorance of the planetary consequences.[79]

What we most certainly can fault them for, however, is disregarding human life and well-being in their pursuit of intensified carbon-energy extraction. Modern fossil fuel mining was often imperial at its core. This was most apparent in the extension of power abroad to seize foreign resources but also evident in projects of "internal colonization," such as with coal mining and energy development spearheaded by government officials and metropolitan elites on Navajo territory in the American Southwest sunbelt.[80] Aside from dispossessing land through force or ultimately unfulfilled promises of prosperity for original inhabitants, such extractive enterprises frequently visited violence on the many who labored within them. Under a framework that privileged production above all, violence was both a disciplinary tool wielded in the name of efficiency and a consequence of placing output ahead of things like safety or sustainability, as clearly was the case in Fushun. Coal mine operators often employed violence to master their workers even as they exposed them to the violence of the exploited earth.

Since World War II, the scale of such violence appears to have diminished across the globe, in part due to the labor force contracting as coal

78. On the emergence of the scientific consensus around climate change, see Weart, *Discovery of Global Warming*.

79. The extent to which past emitters may be held morally responsible or legally liable for their actions, especially if they did not know back then what others in subsequent eras discovered, is a topic of interest in the environmental justice literature. For one take, see Kingston, "Climate Justice and Temporally Remote Emissions."

80. Needham, *Power Lines*, 246–57.

mines close and many of the remaining ones are further mechanized. In China, coal mining was, however, until very recently extremely dangerous. Small local mines were infamous for abysmal safety records, with official fatality rates seven to eight times that of large state-owned ones. Yet hazards dogged the latter too. While less frequent, accidents were also more devastating when they occurred. In 2005, a gas explosion at the Sunjiawan mine in the state-run Fuxin colliery to the west of Fushun killed at least 214 miners, the highest number in one accident since the founding of the People's Republic.[81] In old northeastern mines such as Sunjiawan, colliers often mined at greater depths since the coal closer to the surface had already been extracted. The farther down they dug, the more risks they exposed themselves to, particularly from gases.[82] In 2005, there were close to 6,000 coal mine deaths. This figure has fallen since, reaching a new low of 333 in 2018.[83]

But violence also takes other forms, beginning with the precarity former miners are thrust into when coal mines shut down. In addition, the extraction of oil, which eclipsed coal globally in 1965, remains rife with violence over access to this energy resource and the wealth and power it brings.[84] Furthermore, with the acceleration of climate change, the environmental violence once mostly experienced by those living and working in or near sites of extraction or the industrial cores of urban centers now reaches far afield. Its heaviest hand falls on the world's most vulnerable. For example, in the event of flooding due to rising sea levels, the hardest hit will be disadvantaged groups who live in more flood-prone areas, reside in homes made of cheaper materials that are less likely to withstand flooding, and possess few if any means to recover following a flood.[85] Just as inequalities of power enabled the violence through which the modern fossil fuel economy was made and is maintained, so, too, has the copious consumption of carbon produced effects that are all too unequally felt.[86]

81. Weston, "Fueling China's Capitalist Transformation," 77–78.

82. Wright, *Political Economy of the Chinese Coal Industry*, 173.

83. "Deaths from Coal Mine Accidents."

84. On the violence that can seem almost endemic to petroleum production, see Watts, "Petro-Violence."

85. Islam and Winkel, "Climate Change and Social Inequality," 6.

86. My analysis here and in other parts of this book jibes with that of scholars such as Jason Moore and Andreas Malm, who see our climate crisis as more accurately designated as an outgrowth of a "Capitalocene" rather than the "Anthropocene," with the former foregrounding how the intensification of fossil fuel use proceeded in step with the rise of modern capitalism and how the inequalities engendered by the capitalistic order are reproduced in the unequal impact of the unfolding crisis. See Moore, *Capitalism in the Web of Life*, 169–92; and Malm, *Fos-*

Amid our ongoing climate crisis, China looms large. It may still be the world's foremost emitter of carbon dioxide, but it is also, paradoxically, a potential model for how we might confront this planetary problem. In 2013, the Chinese government put tackling air pollution at the top of its agenda. Winter typically brings a seasonal smog to North China as more coal is burned for heating, but that winter, the air quality in Beijing and dozens of other cities dropped to unprecedented lows. Concentrations of the deadly fine particulate matter PM2.5 were almost forty times more than what the World Health Organization deemed healthy. China, environmental policy analyst Barbara Finamore writes, "was choking on its own high-speed and unbalanced development."[87] Spurred by this crisis, which some commentators nicknamed "airpocalypse," the government introduced sweeping measures that targeted reductions in coal use. These included placing coal consumption caps, closing coal-fired power plants marked as excess capacity, introducing stringent power plant efficiency standards, experimenting with a national cap-and-trade program, and empowering the newly established Ministry of Environmental Protection to oversee efforts at addressing issues of emissions and climate change. Enforcement has been an issue, especially given how many of these measures impinge upon entrenched local and industrial interests, but for several years, some headway was being made.[88] In a somewhat happy coincidence, this development converged with and fueled China's ongoing efforts to ramp up its solar and wind power capacity and to galvanize its electric vehicle market.[89] China seemed, in that moment, truly poised to be, as President Xi Jinping put it in a 2017 address, a "torchbearer in the global endeavor for ecological civilization."[90]

Although China appears to have, until most recently, relapsed on coal power, the question remains whether, given the immensity of the climate crisis, it would take a strong state, like the Chinese one today, to muster the required response. In *The Collapse of Western Civilization*, their fable of a near future wrecked by climate change, historians of science Naomi

---

sil Capital. Where my account distinguishes itself is in stressing how historically the state was an engine of this environmentally devastating capitalism. In this, it joins Christian Parenti's contention about how we should regard states as central to the workings of the Capitalocene. See Parenti, "Environment-Making in the Capitalocene."

87. Finamore, *Will China Save the Planet?*, 25.

88. Finamore, 36–60.

89. Finamore, 61–101.

90. Cited in Finamore, 7. For an analysis of how "ecological civilization" can prompt us to reexamine our study of the environment in the social sciences, see Zinda, Li, and Liu, "China's Summons for Environmental Sociology."

Oreskes and Eric Conway invite us to imagine China as the main survivor of a "Great Collapse." Unlike the liberal democracies that proved "at first unwilling and then unable to deal with the unfolding crisis," China, their narrator—a historian from the "Second's People's Republic"—tells us, took steps to check population growth and convert its economy to run on renewables. When sea levels rose, it quickly moved most of its coastal-dwelling citizens inland. "China's ability to weather disastrous climate change vindicated the necessity of centralized government," the narrator concludes.[91] If the state, in its manifold aspirations and ambitions, paved the way for the rise of the fossil fuel regime, as the preceding pages have shown, then might the state also help redirect us from our current collision course?

While not discounting that possibility, I would like to offer some cautionary considerations that emerge out of this present study on the presumption that such a state would be technocratically inclined, as China's is.[92] These are not, I should say at the onset, critiques against scientific and technological expertise per se.[93] Rather, they underscore the limits to the idea that questions of society can and should be identified, delineated, and solved solely through the mobilization of science and technology, which I take as the central premise (and problem) of technocracy.[94]

First, technocracy does not necessarily achieve what it sets out to accomplish. Although this much might be said about all human plans whatever the system under which they are devised, it seems particularly striking here given technocracy's overwhelming confidence in delivering outcomes,

91. Oreskes and Conway, *Collapse of Western Civilization*, 6, 51–52.

92. Although, as will be evident in the following pages, I have reservations about leaving the mitigation of the environmental crisis entirely in the hands of a technocratic state, I have even deeper doubts that this crisis can be tackled through the workings of the market (insofar as we can conceive of it independent of the state), primarily because capital (like the calorie) often serves as the basis for false equivalencies between incommensurable things (in this case, through the concept of "offsets") while reinforcing the inequalities between those who have amassed more of it and those who have not. As Nancy Fraser has nicely put it, "The idea that a coal-belching factory here can be 'offset' by a tree plantation there assumes a 'nature' composed of fungible, commensurable units, whose place-specificity, qualitative traits, and experienced meanings are of little import and can be disregarded. . . . Produced by epistemic abstraction, financialized nature is at the same time an instrument of expropriation." See Fraser and Jaeggi, *Capitalism*, 100.

93. For a compelling defense of science that makes a case for the reliability of scientific findings not in spite of but because of its social construction, see Oreskes, *Why Trust Science?*

94. Yifei Li and Judith Shapiro have offered an arresting analysis of China's state-led environmentalism that acknowledges its efficacy while also pointing out the ways in which its green measures serve as means by which the party state can further consolidate its control at home and extend its influence abroad. See Li and Shapiro, *China Goes Green*.

grounded as that confidence is in both technocracy's claim to be operating in accordance with the principles of science and its belief in science's infallibility. Throughout this book, we have seen multiple instances in which technocratic plans did not bring about intended effects, whether it was Mantetsu's attempt to reduce its reliance on labor through increased mechanization, the National Resources Commission's endeavor to suppress coal prices for the sake of fueling other industries, or efforts by regime after regime to meet or maintain high levels of production. These points of slippage did not merely result from poor execution and untoward circumstances. Rather, they were often consequences of technocratic plans being shortsighted from the start, those who drew them up having placed too much faith in technoscientific factor inputs and the power of centralized planning.[95]

Second, specific to issues of energy, technocracy has typically only entertained supply-side solutions. Confronted with present or future scarcity, Chinese and Japanese planners and publics in the first half of the twentieth century, like many of their contemporaries elsewhere around the globe, sought to access more energy by discovering additional deposits, developing fuel substitutes, and taking hold of foreign resources. Although some promoted fuel efficiency, few considered the possibility of curtailing demand, keeping, instead, to patterns of consumption set in motion earlier in the carbon age, when fossil fuels were presumed essentially inexhaustible (at least at the planetary level) and the negative impact of their combustion on our biosphere was largely unknown.

The hope of finding scientific fixes that would allow us to preserve those ultimately unsustainable patterns remains something many hold on to firmly in our unravelling present. Tethered as it is to ideas of progress that conceive of the good life in terms of intensive energy use and the excesses that it has enabled, technocracy has cultivated habits of mind that have limited our ability to imagine alternatives. And yet we must. In her conceptual genealogy of energy, Cara New Daggett traces how, in the second

---

95. Political theorist Jeffrey Friedman has issued an extended critique of technocracy that centers not on its undemocratic nature, which others like Jürgen Habermas have taken issue with, but on its failure to be as efficacious as its proponents claim. He writes, "Technocracy, I will suggest, could be effective in achieving its ends—whether or not these ends are democratically determined—only if the human behavior that technocrats attempt to control can be reliably predicted. But the prediction of human behavior is an extremely difficult task, far more so than the predictive tasks at which natural science excels. An effective technocracy, therefore, may very well be out of reach." See Friedman, *Power without Knowledge*, quotation on 2–3.

half of the nineteenth century, the ascendant science of thermodynamics served as a catalyst that facilitated the conjoining of energy with work and progress with productivism.[96] Drawing from postwork traditions, she then proposes that effectively confronting the crisis of the Anthropocene requires not so much a decoupling of economic growth from carbon-fueled environmental destruction as a liberation of energy from work, what she calls "energy freedom," or "an attempt to free more energy from the strictures of waged, productive work" through measures like universal basic income and shorter working hours.[97] If this alternative seems radical, it is because it is. But considering how the industrial modern ideal of progress has brought us to the brink of planetary collapse, its radical reimagination is one of our most urgent, if admittedly Augean, tasks.[98]

Third, technocracy, in consistently underestimating or overlooking the human factor, has been at the root of many of the expressions of violence mentioned above and throughout this book. Promoters of technocratic plans frequently present them as furthering the interests of states or populations in aggregate. Yet these plans often rest on assumptions about knowledge and power that too readily marginalize the well-being and self-determination of large swaths of people, as evident not just in the emergence of the carbon-energy regime in East Asia but time and again in the workings of the wider industrial world. In the most nefarious cases, technocratic plans are willfully engineered to benefit the few at the expense of the many. Even in more benign instances, plans are frequently directed toward abstracted targets that do not actually align with the goals they were designed to meet, so that it becomes possible to have what historian of technology David Noble has referred to as "progress without people."[99] One may then be prompted to ask whether we can in fact construct a better technocracy, retaining its positive aspects (such as efficiency and recognition of expertise) and rejecting its negative ones (like authoritarianism and inattention to the human factor) while reorienting it toward a more humane and sustainable future. I do not presume to have a firm answer, but

96. On the history of thermodynamics, see also Smith, *The Science of Energy*.

97. Daggett, *The Birth of Energy*, 187–206, quotation on 204. The pioneering study of how the emergence of thermodynamic theories led to changes in conceptions of work and labor is Rabinbach, *The Human Motor*.

98. Literature may be one site for such reimagination insofar as it allows for more narratives of the future that break from fantasies of surplus and endless growth. See Szeman, "Literature and Energy Futures."

99. Noble, *Progress without People*.

thinking with the example of worker expertise in Mao-era Fushun (chapter 6), I would like to suggest that if one way of tempering the top-down tendencies of technocracy is to make room for bottom-up perspectives, then efforts at being inclusive should not only be about soliciting solutions to completely predetermined problems but should also involve inviting input from on-the-ground expertise in formulating and framing what exactly the problems are to begin with.[100]

When you step into the Fushun Coal Mine Museum, chances are the first thing you will see in the cavernous entrance hall is a large sculpture of coal miners against a backdrop of a tunnel through a coal seam. The style is decidedly socialist realist. The miners, hard hats on heads, stand tall, two with arms raised in triumph. Behind them, fellow miners toil in the excavated tunnel, one pair using an electric pick on a coal face, several others lugging coal and debris up an incline. In front of this sculpture, large characters read, "This is the treasure of the Chinese people and also the glorious symbol of human labor—it is the considerable labor, blood, sweat, and lives that have been paid." This phrase comes from Xiao Jun's *Coal Mines in May*, as part of the reader's introduction to the fictional coal-mining town of Wujin.[101] It applies certainly to Fushun, on which Wujin was based, but also, more broadly, to similar sites of extraction beyond. For although a jaded onlooker may dismiss the entire display as yet another piece of propaganda, it serves, I think, as a useful reminder of the considerable amounts of human energy that have gone into supporting and sustaining the world that carbon made. In a now-seminal essay on writing history amid the reality of climate change, Dipesh Chakrabarty notes how

---

100. In his analysis of what he and others term the "eco-developmental state," Stevan Harrell, drawing from cases across East Asia, has pointed out the limits of the Environmental Kuznets Curve (the idea that the environment's condition will worsen in early industrialization but then gradually improve with more economic growth) that may inspire optimism among those seeking statist solutions. One critique he raises is the overwhelming importance of development as a source of legitimacy for the states in question. That is, states are most likely to invest in environmental remediation only when this does not clash with the realization of their developmental visions. Still, he expresses a measure of hope, particularly when considering examples of environmentally conscious citizens who have placed pressure on the state for positive change and who have had their efforts meet with some success. See Harrell, "The Eco-developmental State." For another account that seeks to excavate from grassroots Asian traditions pathways toward a more sustainable future, see Duara, *Crisis of Global Modernity*, 18–52.

101. Xiao, *Wuyue de kuangshan*, 4.

intensive energy consumption has been baked into ideas of progress over the past two and a half centuries. "The mansion of modern freedoms," he remarks, "stands on an ever-expanding base of fossil-fuel use."[102] This book has attempted to show that there was and still is a deep seam of unfreedom—of human costs exacted by industrial modern states operating under the logics of carbon technocracy—that runs beneath the building of this mansion and the laying of its fossil fuel base.

102. Chakrabarty, "The Climate of History," 208.

ACKNOWLEDGMENTS

I learned a great deal over the course of writing this book—about mining coal, telling stories, making arguments, and other acts of creative destruction. I am so grateful to all those who taught me things along the way.

This book had its beginnings in my graduate training at Harvard, where I was fortunate to have studied under Mark Elliott. As a teacher, Mark was both devoted and demanding. He set a high bar for scholarly rigor and linguistic precision while investing much time and energy into guiding his students toward those ideals. I owe him my sincerest gratitude. I am also thankful to Ian Miller, who saw potential in my work long before I did and pushed me to realize it, as he often does, with a gentle hand. Andy Gordon was and remains a thoughtful interlocutor and a source of encouragement. Liz Perry, who probably knows more about China's long twentieth century than I ever will, helped cultivate my interest in looking beyond it. I am honored to now call you all my colleagues and friends.

At Harvard, I also had the good fortune of learning from several other scholars whose work has shaped my own. Henrietta Harrison got me thinking about the global in the local, Elisabeth Köll about corporate institutions and managerial expertise, Emma Rothschild about the scope of economic life, and Sheila Jasanoff about science and the state. I probably would not have taken away as much as I did from my time as a graduate student here were it not for the many wonderful teachers that I had before. I want to acknowledge, in particular, Maggie Kuo, Robin Yates, Griet Vankeerberghen, Brian Lewis, Elizabeth Elbourne, Fred Dickinson, Eiichiro Azuma, Siyen Fei, Ben Nathans, Eve Troutt Powell, and Sue Naquin. You have each, in your own way, contributed to my development as a teacher and a scholar. I hope that I can be as much of an inspiration to my students as you have been to me.

It is testament to Fushun's reach that investigating its history took me to many archives and libraries across China, Japan, Taiwan, and the United States. For handling my various requests with professionalism and, at times, good humor, I would like to thank the staff of the Fushun Municipal Archives, the Liaoning Provincial Archives, the Liaoning Provincial Library, the Jilin Provincial Archives, the Jilin Provincial Library, the Shanghai Municipal Archives, the Second Historical Archives, the National Diet Library, the Tōyō Bunko, the Hitotsubashi University Library, the Waseda University Library, the University of Tokyo Library, the Hokkaido University Archives, the Research Center for Coal Mining Materials at Kyushu University, the Institute of Modern History Archives at Academia Sinica, the Academia Historica Archives, the Hoover Institute, the Library of Congress, and the Wason Collection on East Asia at Cornell University. For over a decade and counting, I have been racking up debts with Harvard librarians, especially Ma Xiao-he and Kuniko Yamada McVey of the Harvard-Yenching Library, who have provided tireless assistance in locating and procuring materials for this and other research projects. Given how sources are the seeds of a historian's scholarship, I literally could not have done it without you.

In China, I benefited from the unwarranted kindness of many individuals. For their support, I am particularly grateful to Feng Xiaocai of East China Normal University, Qu Hongmei of Jilin University, and Yu Zhiwei of the Liaoning Provincial Academy of Social Sciences. Sincerest thanks are also due my friends and contacts in Fushun. They gave me materials, entertained my questions, and shared with me much of what they know about the place they call home. I especially want to recognize Fu Bo and Wang Pinglu of the Fushun Municipal Academy of Social Sciences, Jin Hui of the Pingdingshan Massacre Memorial Hall, and Wang Jie, whose help made all the difference. In Japan, Enatsu Yoshiki hosted me at Hitotsubashi University and went out of his way to ensure that I got what I needed to do my work. Linda Grove and Kubo Tōru of the Modern Chinese History Research Group and Setoguchi Akihisa and Fujihara Tatsushi of the Environmental History Research Group welcomed me into their vibrant intellectual communities. I am also thankful to Ikegami Shigeyasu of the Hokkaido University Archives for sharing copies of practicum reports he collected from other former imperial universities and to Miwa Munehiro of Kyushu University's Research Center for Coal Mining Materials for introducing a number of valuable sources, especially the papers of Matsuda Buichirō, Fushun's first colliery manager under Mantetsu. In Taiwan, Chen Tsu-yu, Lin May-li, and Chang Ning at Academia Sinica's Institute of Modern History gave me good advice just as I was embarking on this project and facilitated my research in Taipei. I truly appreciate all that each of you has done for me.

I began writing this book after joining the Department of History at Cornell. There is perhaps no group of historians as unpretentious as they are brilliant. For their friendship and collegiality, I want to thank Ernesto Bassi, Judi Byfield, Holly Case, Sherm Cochran, Ray Craib, Paul Friedland, Maria Cristina Garcia, Durba Ghosh, Sandra Greene, TJ Hinrichs, Julilly Kohler-Hausmann, Vic Koschmann, Katie Kristof, Jon Parmenter, Russell Rickford, Kay Stickane, Robert Travers, Claudia Verhoeven, and especially Larry Glickman, Itsie Hull, Tamara Loos, Aaron Sachs, and Eric Tagliacozzo. I am also grateful to the late Ann Johnson, Ron Kline, and Sara Pritchard in Science and Technology Studies; Nick Admussen, Chiara Formichi, and Suyoung Son in Asian Studies; Eli Friedman in International and Comparative Labor; Jack Zinda in Development Sociology; and Josh Young and Mai Shaikhanuar-Cota at the East Asia Program. Lindy Williams, Ravi Kanbur, and Jack Elliott, my fellow fellows at the Atkinson Center for a Sustainable Future in spring 2016, were steadfast compatriots. If I miss Ithaca, it is because of all these good people.

My new institutional home, the Department of the History of Science at Harvard, has been warm and welcoming. I have learned much from conversations and interactions with my colleagues Eram Alam, Soha Bayoumi, Allan Brandt, Janet Browne, Alex Csiszar, Peter Galison, Evelynn Hammonds, Anne Harrington, Matt Hersch, David Jones, Hisa Kuriyama, Becky Lemov, Liz Lunbeck, Hannah Marcus, Naomi Oreskes, Ahmed Ragab, Sarah Richardson, Sophia Roosth, Charles Rosenberg, Sara Schechner, Gabriela Soto Laveaga, Dave Unger, Nadine Weidman, and Ben Wilson. Hisa has been an attentive and inspiring mentor, and my chairs, Janet and Evelynn, unflagging in their support. Gabriela deserves special mention for organizing our weekly departmental Write-On-Site, where, fueled by pastries, fruit, and coffee, I drafted many of the sentences that made it into this book. It was also a treat to get to know Warwick Anderson while he was here for a year. I often walked away from our lunch meetings with new ideas and fresh ways of approaching familiar problems. I look forward to future collaborations. Allie Belser, Emily Bowman, Sarah Champlin-Scharff, Ellen Guarente, Michael Kelly, Brigid O'Connor, Linda Schneider, Nicole Terrien, Deborah Valdovinos, Karen Woodward Massey, and Robin Yun have helped make my life in the department not only easier but enjoyable.

Elsewhere on campus, I am thankful for the good company and wise counsel of many other colleagues, including Paul Chang, Carter Eckert, Arunabh Ghosh, Susan Greenhalgh, David Howell, Sun Joo Kim, Bill

Kirby, Arthur Kleinman, Daniel Koss, Ya-Wen Lei, Jie Li, the late Rod Mac-Farquhar, the late Ezra Vogel, and Alex Zahlten. Special thanks to David Atherton and Durba Mitra, who have been particularly faithful comrades and are both just exemplary human beings in my book. The Asia Center, the Fairbank Center for Chinese Studies, the Harvard-Yenching Institute, the Joint Center for History and Economics, the Reischauer Institute of Japanese Studies, and the Weatherhead Center for International Affairs have provided many precious opportunities for scholarly engagement, and I am additionally grateful to their directors, Karen Thornber, James Robson, Michael Szonyi, Liz Perry, Emma Rothschild, Sunil Amrith, Mary Brinton, and Michèle Lamont, for supporting the research and publication of this project and other endeavors.

Many of the ideas in this book were first tested out in talks. For their questions and reactions, which often took me in new and productive directions, I want to extend thanks to those who attended presentations I gave at Columbia University, Ewha Womans University, the New School, Nanyang Technological University, the National University of Singapore, the University of California, Berkeley, the University of Pennsylvania, the University of Washington, and the annual meetings of the Society for the History of Technology, the Association for Asian Studies, and the American Society for Environmental History. For invitations to speak on portions of this work and for their support overall, I am grateful to Anindita Banerjee, Tim Bunnell and Greg Clancey, Sei-Jeong Chin, Stephanie Dick, Madeleine Dong, Mark Fraizer and Manjari Mahajan, Dong-Won Kim, Micah Muscolino, Chris Nielsen, Annelise Riles, Norman Smith, Hallam Stevens, Matti Zelin, and Ling Zhang.

Fellow historians of science and technology working on East Asia have challenged and inspired me to constantly reflect on the question of how our collective enterprise and this project in particular might help rewrite narratives within global histories of science and technology. Aside from those mentioned above and below, I am thankful for dialogues around such matters with Francesca Bray, Mary Brazelton, Phil Brown, Susan Burns, BuYun Chen, Kaijun Chen, Hyungsub Choi, Jacob Eyferth, Fa-ti Fan, Yulia Frumer, Wendy Fu, Miriam Gross, Marta Hanson, Chihyung Jeon, Lijing Jiang, Tae-Ho Kim, Aleksandra Kobiljski, Wen-Hua Kuo, Jung Lee, Seung-joon Lee, Sean Hsiang-lin Lei, Angela Leung, Hiromi Mizuno, the late Aaron Stephen Moore, Tom Mullaney, Takashi Nishiyama, Lisa Onaga, Dagmar Schäfer, Wayne Soon, Soyoung Suh, Honghong Tinn, Zuoyue Wang, Shellen Wu, and Daqing Yang. I hope that you will find this book a meaningful contribution to our ongoing conversations.

It really is no exaggeration to say that it took a village to raise this book. At workshops, seminars, and other venues, or just because they were kind enough to offer or to say yes when I asked, numerous friends and colleagues have read chapters from or the entirety of the manuscript at various stages. I am particularly grateful for feedback I received from He Bian, David Biggs, Pete Braden, Kate Brown, Sakura Christmas, Alex Csiszar, Julian Gewirtz, Arunabh Ghosh, Andy Gordon, Steve Harrell, Anne Harrington, Evan Hepler-Smith, Stefan Huebner, Sophia Kalantzakos, Mikiya Koyagi, Hisa Kuriyama, Eugenia Lean, James Lin, Susan Lindee, Sidney Lu, Mark Metzler, Ian Miller, Durba Mitra, Suzanne Moon, Projit Mukharji, Emer O'Dwyer, Ken Pomeranz, Lissa Roberts, Jennifer Robertson, Ruth Rogaski, Aaron Sachs, Grace Shen, Peter Shulman, Gabriela Soto Laveaga, Eric Tagliacozzo, Ying Jia Tan, Julia Adeney Thomas, Heidi Voskuhl, Yvon Wang, Ben Wilson, Wen-hsin Yeh, and Louise Young. I also benefited from the capable research assistance of Niall Chithelen, Lawrence Gu, John Hayashi, Rui Hua, Justin Wong, and especially Jongsik Yi, who carefully read several versions of the manuscript. I count myself blessed to be able to work with and learn from these and other amazing students. Ron Suleski, Norman Smith, and Koji Hirata graciously shared sources. David Atherton and Mikiya Koyagi thought through several tricky passages of Japanese with me. Audra Wolfe and Beth Sherouse helped tame my prose and clarify my thoughts in the process. Scott Walker at the Harvard Map Collection drew the terrific map of coal mines and railway lines near the beginning of the book. Sigrid Schmalzer and Thomas Andrews reviewed the manuscript for the press, offering extensive and generous comments that greatly improved the final product. Any remaining errors of fact or interpretation are my own.

At the University of Chicago Press, Karen Merikangas Darling has been extraordinarily patient, understanding, and supportive. I am so grateful for her interest and investment in this project. I also very much appreciate all the work that Tristan Bates, Rebecca Brutus, Jenni Fry, Deirdre Kennedy, Steve LaRue, Christine Schwab, and others at the press have put into the making of *Carbon Technocracy*. At Columbia's Weatherhead East Asian Institute, in whose series this book appears, I am thankful to Ross Yelsey, Ariana King, Eugenia Lean, and the donor who made possible the First Book Award that helped defray production costs.

Generous grants and fellowships supported the research and writing of this book. For these, I am grateful to the Mrs. Giles Whiting Foundation, the D. Kim Foundation for the History of Science and Technology in East Asia, the Overseas Young Chinese Forum, the Japan Foundation, and multiple research centers and funding sources across Harvard and Cornell. Early in the process of putting together this book, I had the chance to spend a memorable summer at the Rachel Carson Center in Munich. Many thanks to the directors, Christof Mauch and Helmuth Trischler, for the fellowship that enabled my stay, to the staff who labor to make this as special a place as it is, and to the other fellows, especially Erika Bsumek, Sophia Kalantzakos, and Dan Lewis, for camaraderie and generative conversations. I finished this book while on sabbatical as a visiting scholar at the Program in Science, Technology, and Society at MIT. I had only recently gotten the key to my office and made it to one program luncheon before COVID-19 drove us all remote, but I am grateful to Jennifer Light for the opportunity and look forward to more interactions in the coming years.

Many friends took the time to discuss this project with me and lent support when I sorely needed it. The ones I made during those formative graduate school years influenced how I came to see the work of a historian. They were also great fun to be around. I want to recognize Bina Arch, He Bian, Jack Chia, Jamyung Choi, Sakura Christmas, Devon Dear, Devin Fitzgerald, Todd Foley, Chris Foster, Koji Hirata, Kuang-chi Hung, Masa Itoh, Kyle Jaros, Macabe Keliher, John Kim, Mikiya Koyagi, John Lee, Philipp Lehmann, Ren-yuan Li, James Lin, Yan Liu, Shi-Lin Loh, Sidney Lu, Ian Matthew Miller, Steffen Rimner, Jon Schlesinger, Nianshen Song, Holly Stephens, Fei-Hsien Wang, Xiaoxuan Wang, Yuanchong Wang, Yvon Wang, Jake Werner, Tim Yang, and Wen Yu. Apart from others mentioned elsewhere in these acknowledgments, I want to also express thanks to Enid Schatz, Gurpreet Singh, and Tiffany Trzebiatowski for being there.

Family means the world to me, and I am grateful for their love and support throughout the years. My sisters, Trudy and Lynette, provided the right ratio of ribbing and encouragement. I love you both and cannot wait till our next reunion. I am thankful to my brother- and sister-in-law, Jason and Goeun, for often being so thoughtful and especially for housing us between our move from Ithaca back to Boston. My parents-in-law, Kim Seung Hee and Kim Chang Gon, welcomed me into their family and have provided immeasurable help over the years, particularly after the children were born and when I traveled abroad for research. Thank you for your labor and your love.

My wife and best friend, Minju, has been here with me from the very beginning. Thank you for believing. I love you and the life we have built together. Our children, Ji-woo and Seo-yun, keep our hands and hearts full. They probably added to the time it took to complete this book, as little ones do. But their arrival into our lives also rendered its message even more personal and urgent. The world that carbon made, so deeply despoiled and unjust, is the one that they and their generation will inherit unless radical transformations take place. May I be a cog, however small, in the engines of change. Finally, my deepest thanks go to my parents, my very first teachers. I am not, I think, a confident person, but they taught me what it means to be secure. Their unfailing support has made me who I am; their example of selflessness guides who I want to be. It is to them that I dedicate this book, with love and gratitude.

BIBLIOGRAPHY

Abbreviations

BJTK   Minami Manshū tetsudō kabushiki gaisha Bujun tankō. *Bujun tankō* [The Fushun coal pits]. [Dairen]: n.p., 1909.

CMWE   United States Strategic Bombing Survey, Basic Materials Division. *Coals and
Metals in Japan's War Economy*. Washington: United States Strategic Bombing
Survey, Basic Materials Division, 1947.

DBDX   Zhongguo bianjiang shidi yanjiu zhongxin and Liaoning sheng dang'anguan,
comps. *Dongbei bianjiang dang'an xuanji: Qingdai, Minguo* [Selections of archival documents from the northeastern frontier: Qing and Republican eras]. 151
vols. Guilin: Guangxi shifan daxue chubanshe, 2007.

FGY   Fushun shi zonggonghui gongyun shi zhi yanjiushi. *Fushun gongren yundong
dashiji* [Major events in the history of Fushun's labor movement]. Fushun: Fushun shi zonggonghui gongyun shi zhi yanjiushi, 1991.

FKZ   Fushun kuangwu ju meitan zhi bianzuan weiyuanhui. *Fushun kuangqu zhi
(1901–1985)* [Gazetteer of the Fushun mining area (1901–1985)]. Shenyang:
Liaoning renmin chubanshe, 1990.

FMBN   Fushun shi zhengxie wenshi ziliao weiyuanhui and Fushun kuangye jituan
youxian zheren gongsi, eds. *Fushun meikuang bai nian, 1901–2001* [A century of
the Fushun coal mines, 1901–2001]. Shenyang: Liaoning renmin chubanshe,
2004.

ITM   Kokugakuin daigaku toshokan, comp. *Inoue Tadashirō monjo* [Inoue Tadashirō
archives]. Tokyo: Yūshōdō firumu shuppan, 1994. 168 microfilm reels.

KWD   Zhongyang yanjiuyuan Jindaishi yanjiusuo, comp. *Kuangwu dang* [Archives of
mining affairs]. Vol. 6, *Yunnan, Guizhou, Fengtian*. Taipei: Zhongyang yanjiuyuan Jindaishi yanjiusuo, 1960.

MGDA   Zhongguo di er lishi dang'anguan, comp. *Zhonghua minguo shi dang'an ziliao
huibian* [Compilation of archival materials on the history of the Republic of
China]. 5 ser. Nanjing: Jiangsu guji chubanshe, 1979–1999.

MSZ   Xie Xueshi, comp. *Mantie shi ziliao* [Documents on the history of the South
Manchuria Railway Company]. Vol. 4, nos. 1–4. Beijing: Zhonghua shuju,
1987.

MT10   Minami Manshū tetsudō kabushiki gaisha. *Minami Manshū tetsudō kabushiki
gaisha jūnen shi* [Ten-year history of Mantetsu]. Dairen: Minami Manshū tetsudō kabushiki gaisha, 1919.

NGB   Gaimushō, comp. *Dai Nihon gaikō bunsho* [Documents on Japan's foreign rela-
       tions]. 73 vols. Tokyo: Nihon kokusai rengō kyōkai, 1938–1963.
NMKY   Bujun tankō shomuka. *Nichi-Man taiyaku kōzan yōgo shū* [Japanese-Manchurian
       dictionary of mining terms]. 5 vols. Fushun: Minami Manshū tetsudō Bujun
       tankō, 1935.

Works of Unknown Authorship

"1,000 Cars Will Transport Coal for This City." *China Press*, November 20, 1931.

"1907–1945 nian Fushun meikuang shigu siwang renshu tongji biao" [Tabulation of
    number of people who died in accidents at the Fushun colliery, 1907–1945]. Fushun
    Coal Mine Museum, Fushun.

"1943-nen gensan gen'in" [Reasons for the fall in production in 1943]. Fushun Mining
    Affairs Bureau, Japanese-Language Archival Materials 8–10, no. 336. In *MSZ*, no. 2,
    418–22.

"Agreement Concerning Mines and Railways in Manchuria, September 4, 1909." In *Trea-
    ties and Agreements with and Concerning China, 1894–1912*, compiled by John V. A.
    MacMurray, 1:790–92. New York: Oxford University Press, 1921.

"Agreement Concerning the Southern Branch of the Chinese Eastern Railway, July 6,
    1898." In *Treaties and Agreements with and Concerning China, 1894–1912*, compiled by
    John V. A. MacMurray, 1:154–56. New York: Oxford University Press, 1921.

"Aiguo hu quan, gedi xian you youxing" [Patriotic protection of rights, demonstrations
    continue across the country]. *Dagongbao*, February 25, 1946.

"Anshan Pig-Iron and Fushun Oil Shale." *Manchuria Daily News*, August 13, 1925.

"Artificial Oil Made in Japan." *Oil, Paint, and Drug Reporter*, March 14, 1921.

"Basic Fuel Problem Being Investigated." *Japan Advertiser*, September 24, 1929.

"Bei gakusha dan o heishite Anzan tetsu to Bujun tan o chōsa" [Team of American scholars
    invited to survey Anshan iron and Fushun coal]. *Ōsaka asahi shinbun*, June 24, 1921.

"Beiping luotuo yinian yidu zhi tuo mei shenghuo" [The yearly lives of Beiping's coal-
    carrying camels]. *Kuangye zhoubao* 411 (December 1936): 418–19.

"Bujun ni okeru Chūgokujin kyōsan undōsha taiho ni kansuru ken" [Case regarding the
    arrest of Chinese Communist operatives in Fushun]. Mukden consul Hayashi Kyūjirō
    to Foreign Affairs Minister Shidehara Kijūrō, file no. I-4-5-2-011, Diplomatic Archives
    of the Ministry of Foreign Affairs of Japan.

"Bujun shi yūran no shiori: Yamaguchi yūran basu" [A tourist guidebook to Fushun: Ya-
    maguchi tour buses]. Harvard-Yenching Library, Manchuguo Collection, J-0744.

"Bujun tankō bakuhatsu sanjūgo shishōsu" [Thirty-five killed and injured in Fushun coal
    mine explosion]. *Yomiuri shinbun*, June 29, 1940.

"Bujun tankō sōmukyoku shomuka tei kōchō bun" [Fushun colliery general affairs bu-
    reau to colliery manager] (June 7, 1941). Fushun Mining Affairs Bureau, Japanese-
    Language Archival Materials 8-8, 226, no. 11. In *MSZ*, no. 2, 593–94.

Bujun tankō tōkei nenpō [Annual statistical report of the Fushun coal mine]. Fushun: Mi-
    nami Manshū tetsudō kabushiki gaisha Bujun tankō, 1942, 33. In *MSZ*, no. 1, 246.

"Burgulary at Fushun." *Manchuria Daily News*, June 21, 1917.

"Caizheng, shiye, waijiao san bu hui cheng gao" [Draft of a joint petition by the Minis-
    tries of Finance, Industry, and Foreign Affairs] (December 10, 1932). In *MGDA*, ser. 5,
    no. 1, vol. 4, pt. 6, 492–93.

"Canzao shahai de Zhang Shenfu xiansheng" [Mr. Zhang Shenfu, who was murdered in cold blood]. *Qianxian ribao*, February 21, 1946.

"Chiang Assures Nation on Northeast Problem." *China Weekly Review*, March 2, 1946.

"China." U.S. Energy Information Administration. https://www.eia.gov/international/overview/country/CHN.

"China, Soviet Union: Treaty of Friendship and Alliance" (August 14, 1945). *American Journal of International Law* 40, no. 2 (April 1946): 51–63.

"China Does U-Turn on Coal Ban to Avert Heating Crisis." BBC News, December 8, 2017. https://www.bbc.com/news/world-asia-42266768.

"Chinese Civil War Deals Blow to Coal Mines: Suffers More Acutely Than Any Other Industry." *China Press*, February 7, 1929.

"Chinese Miners of Fushun Collieries: Payment in Gold Preferred." *Manchuria Daily News*, May 10, 1927.

"Chōsen shi dai kōgyō" [The four big mining industries of Chosōn]. *Keijō nippō*, May 14, 1923.

"Chungking Students Strike." *China Critic*, February 28, 1946.

"Coal and the Manchu's Ghost." *Wall Street Journal*, March 11, 1922.

"Coal Liquefaction and the South Manchuria Railway Company." *Contemporary Manchuria* 4, no. 1 (January 1940): 17–27.

"Coal Shortage Threat." *North-China Herald and Supreme Court and Consular Gazette*, April 7, 1931.

"Colliery Disaster: Explosion of Coal Dust." *Shanghai Times*, January 22, 1917.

"Colliery Explosion: The Fushun Disaster, Theories as to the Cause." *Shanghai Times*, January 23, 1917.

"Commercial and Industrial." *Japan Times*, April 28, 1897.

"Dark Future for Coal." *Japan Times*, April 4, 1923.

"Deaths from Coal Mine Accidents in China Fall to New Low of 333 in 2018." *China Labour Bulletin*, January 24, 2019. https://clb.org.hk/content/deaths-coal-mine-accidents-china-fall-new-low-333-2018.

"Detailed Regulations for Fushun and Yentai Mines, May 12, 1911." In *Treaties and Agreements with and Concerning China, 1894–1912*, compiled by John V. A. MacMurray, 1:792–93. New York: Oxford University Press, 1921.

"Dōin hōan iinkai" [Commission for the mobilization law]. *Yomiuri shinbun*, March 25–26, 1918.

"Early Estimates on China's Coal Supply Exaggerated: New Survey of Various Fields Now Complete." *China Press*, January 19, 1928.

"Eruptive Typhus at Fushun." *Manchuria Daily News*, April 16, 1919.

"Establishment and Persons Engaged in Mining by Industry (1893–2003)." Statistics Bureau, Ministry of Internal Affairs and Communications, Japan. Last updated May 1, 2018. https://warp.da.ndl.go.jp/info:ndljp/pid/11095459/www.stat.go.jp/english/data/chouki/08.html.

"Explanation Given of Disaster at the Fushun Mine in Manchuria Which Cost Lives of 470 Workers." *China Press*, May 1, 1928.

"Fengtian jiaoshe shu gei Fushun xian zhishi de xunling" [Instructions from the Fengtian negotiations office to the Fushun magistrate] (May 19, 1926). Fushun County Archives, no. 1-12860. In *MSZ*, no. 1, 221–22.

"Frank Hutchinson Entertains with Instructive Story of Manchuria." *Skillings' Mining Review* 10, no. 45 (March 25, 1922): 4.

"Fukuang yijia" [The Fushun colliery as one family]. *Fukuang xunkan* 2, no. 5 (November 1947): 277.

"Funü mofan Wang Xueqin" [Model woman Wang Xueqin]. *Fushun ribao*, September 20, 1952.

"Fushun Coal." *Far Eastern Review* 23, no. 12 (December 1927): 58.

"Fushun Coal Exports to Japan May Be Limited." *Japan Times and Mail*, April 27, 1935.

"Fushun Coal Fight Settled Amicably." *Japan Advertiser*, July 17, 1932.

"Fushun Coal Mines." *Far Eastern Review* 12, no. 5 (May 1909): 438–44.

"The Fushun Coal Mines." *Japan Times*, October 9, 1906.

"Fushun Collieries." *Shanghai Times*, June 15, 1917.

"Fushun Colliery Development: American Experts Arrive." *Far Eastern Review* 17, no. 8 (August 1921): 513–17.

"The Fushun Disturbance." *Japan Weekly Chronicle*, March 30, 1911.

"Fushun jishu yexiao (yuan kuang zhuan gongren yexiao) xuzhao bianjisheng" [Fushun technical night school (originally, the mining specialty workers' night school) continues to accept students]. *Fushun ribao*, July 23, 1951.

"Fushun jumin Chen Rong deng wushiba ren cheng Fengtian jiaoshe si wen" [Text of the petition from Fushun resident Chen Rong and fifty-seven others to the Fengtian negotiations office]. Fushun County Archives, bag 24, bundle 2, no. 15561. In *MSZ*, no. 1, 138.

"Fushun kuangwu ju dashiji" [Chronology of the Fushun colliery]. *Fukuang xunkan* 2, no. 9 (December 1947): 326–28.

"Fushun meikuang gongye zhuanke yexiao san nian lai peiyang le dapi jishu rencai" [Fushun colliery professional night school cultivates much technical talent in three years]. *Fushun ribao*, January 30, 1953.

"Fushun meikuang muqian kunnan qingxing" [The Fushun colliery's current troubled situation]. *Fukuang xunkan* 2, no. 12 (January 1948): 366–67.

"Fushun meikuang shigu ji wei yanzhong" [Situation with accidents at the Fushun colliery extremely serious]. *Neibu cankao*, April 26, 1954, 259.

"Fushun meikuang wanhu Huaren shengming caichan san ze" [Fushun colliery trifles with the lives and properties of Chinese, three instances]. *Kuangye zhoubao* 139 (April 1931): 6.

"Fushun meikuang zhiwen guanli guicheng zhaiyi" [Selected translations from relations on fingerprint management at the Fushun colliery]. In *MSZ*, no. 1, 315–16.

"Fushun Miners All Not Like Lambs." *Manchuria Daily News*, September 7, 1927.

"Fushun Shale Oil Experiments: All Hopes Fulfilled." *Manchuria Daily News Monthly Supplement*, January 1, 1927, 8.

"Fushun Shale Oil Industry: Almost Ready to Be Submitted to Central Government," *Manchuria Daily News Monthly Supplement*, June 1, 1927, 8.

"Fushun Shale Oil Industry Decided to Be Started Next Year." *Manchuria Daily News*, September 3, 1925.

"Fushun shi ge shengchan changkuang jinnian shigu bi jiefang yihou linian yanzhong" [Situation with accidents at the various factories and mines in Fushun the most severe since liberation]. *Neibu cankao*, December 6, 1956, 116–17.

"Fushun Transforms Century-Old Coal Mine into Tourist Site." *China Global Television Network*. July 13, 2020. https://news.cgtn.com/news/2020-07-13/Fushun-transforms -century-old-coal-mine-into-tourist-site-S5NhJE7oo8/index.html.

"Gas Companies in Japan." *Far Eastern Review* 24, no. 9 (September 1928): 425, 427.

"Ge guo gonghui daibiao tuan zai Dongbei Fushun canguan shi de fanying he women gongzuo Zhong de wenti" [Responses from the group of labor union representatives from various countries to the tour of Fushun in the Northeast and problems with our work]. *Neibu cankao,* May 12, 1952, 89–92.

"God-Send to Japan to Head Off Fuel Famine." *Manchuria Daily News Monthly Supplement,* May 1, 1927, 7.

"Guomei jiuji jian you banfa: fanmai Rimei zhe zhi chufa" [Increasingly more plans for national coal relief: Penalties for those who traffic in Japanese coal]. *Kuangye zhoubao* 216 (November 1932): 761–62.

"Guomei jiuji weiyuanhui cheng" [A petition by the National Coal Relief Commission] (August 25, 1932). In *MGDA,* ser. 5, no. 1, vol. 4, pt. 6, 485–86.

"Guomei jiuji weiyuanhui wei bao song zhangcheng beian zhi Shiye bu cheng" [A petition by the National Coal Relief Commission to the Ministry of Industry to report its constitution for the record]. In *MGDA,* ser. 5, no. 1, vol. 4, pt. 6, 471–72.

"How the Chinese Coal-Mining Industry Is Throttled." *North-China Herald and Supreme Court and Consular Gazette,* June 30, 1928.

"Implementation of Soviet Objectives in China" (September 15, 1947). In CREST: General CIA Records, document no. CIA-RDP78-01617A003000080001-9.

"Japan." U.S. Energy Information Administration. https://www.eia.gov/international/analysis/country/JPN.

"Japan Industries Retarded by High Production Costs." *Weekly Commercial News,* March 17, 1923.

"Japan Moves Town to Reach Coal Vein." *New York Times,* May 19, 1929.

"Jidian chang nügong jiji zao qiaomen wa qianli zai shengchan zhanxian shang chuangzao bushao chengji" [Electrical machinery factory women workers enthusiastically finding solutions to problems and tapping potential, making many achievements in the battlefront of production]. *Fushun ribao,* July 11, 1952.

"Jieshou Shengli kuang Laohutai kuang lianxu fasheng shigu de jiaoxun" [Learning from the series of accidents that had occurred at the Shengli and Laohutai mines]. *Fushun ribao,* January 12, 1951.

"Jingji bu Ranliao guanlichu zuzhi zhangcheng" [Regulations on the establishment of the Ministry of Economic Affairs' Fuel Management Administration]. *Jingji bu gongbao* 1, no. 2 (March 1938): 65–66.

"Jingji bu Zhiyuan weiyuanhui zuzhi tiaoli" [Ordinances on the establishment of the Ministry of Economic Affairs' National Resources Commission]. *Jingji bu gongbao* 1, no. 3 (March 1938): 96–97.

"Jinian Zhang Shenfu xiansheng" [Remembering Mr. Zhang Shenfu]. *Fukuang xunkan* 2, no. 12 (January 1948): 361–62.

"Jishi Wang Xi zao gongren bangzhu cai jiang bamu yanjiu chenggong" [Technician Wang Xi succeeds in pit-prop removal research only after seeking help from workers]. *Fushun ribao,* July 13, 1952.

"Jiuji meihuang zuori xu kai huiyi" [Yesterday's meeting for coal famine relief continues]. *Shenbao,* November 19, 1931.

"Jumin Liu Baoyu cheng Fushun xian jiandu wen" [Text of the petition from resident Liu Baoyu to the Fushun magistrate] (May 17, 1927). Fushun County Archives, bag 16, bundle 2, no. 12916. In *MSZ,* no. 1, 222–23.

"Juzhang xunci" [Colliery manager's instructions]. *Fukuang xunkan* 2, no. 18 (March 1948): 433–34.

"Kashiwaba Yuichi." *Riben zhanfan de qin Hua zuixing zigong* [The confessions of the Japanese war criminals]. National Archives Administration of China. https://www.saac .gov.cn/zt/2014a/rbzf/rbzf/36by.htm.

"Kensetsu tojō no Manshū keizai o kataru" [On the Manchurian economy in the midst of construction]. *Ōsaka asahi shinbun*, May 10, 1938.

"Ketsuganyu no jitsuyōka ni Mantetsu yoyaku seikōsu" [Mantetsu plan to commercialize shale oil succeeds]. *Jiji shinpō*, April 14, 1934.

"Kōfu sū hyaku mei taikyo Fukuoka ni mukau" [Hundreds of miners head to Fukuoka]. *Asahi shinbun*, June 27, 1932.

"Kongo no Mantetsu" [Mantetsu henceforth]. *Ōsaka asahi shinbun*, May 31, 1914.

"Korea Starts Industry on Large Scale: Rivals with Manchuria in Japanese Capital." *China Press*, November 12, 1938.

"Labor Management at the Fushun Coal Mines." *Far Eastern Review* 34, no. 10 (October 1938): 378–82.

"Landmark Building Should Respect the People's Feeling." *People's Daily Online*, November 22, 2012. http://en.people.cn/90882/8029356.html.

"Laohutai Fire Scented by Rats." *Manchuria Daily News*, April 8, 1915.

"Laohutai kuang fangsong anquan gongzuo, qi tian nei jing fasheng san ci shigu" [Laohutai mine slackened safe work practices, three accidents have occurred in a week]. *Fushun ribao*, January 11, 1951.

"Loading Black Powder into 'Gopher Holes.'" *Engineering and Mining Journal* 111, no. 2 (January 8, 1921): 62.

"Local Industrialists Are Alarmed by Coal Shortage." *China Press*, May 8, 1931.

"Long Live May Day—Down with War—Long Live the United Socialist Soviet Republic!" (1929). In *Investigation of Communist Propaganda*, pt. 5, vol. 4, compiled by Special Committee to Investigate Communist Propaganda in the United States, 389–93. Washington, DC: United States Government Printing Office, 1930.

"Lutiankuang bengyan gu he dong caimei duan shengli tupo shengchan jingsai tiaojian: chanliang, chuqin lü buduan shangshen" [The rockfall section and the eastern coal-mining section of the open-pit mine have successfully surmounted the conditions of the production competition: Production volume and work attendance rates are going up without stopping]. *Fushun ribao*, November 21, 1950.

"Lutiankuang de tuoersuo" [The nursery at the open-pit mine]. *Fushun ribao*, July 7, 1951.

"Manchuria Mine Survey Completed." *Trans-Pacific* 5, no. 4 (October 1921): 39.

"Manshū ni okeru Mantetsu no san jigyō" [Mantetsu's three enterprises in Manchuria]. *Ōsaka asahi shinbun*, December 25, 1923.

"Manshū sekitan zōsan daini ji keikaku" [Increases in Manchurian coal production under the second plan]. *Yomiuri shinbun*, February 16, 1942.

"Mantetsu chōsa bu shiryō kachō chōsa hōkoku" [Report by the documents section chief of Mantetsu's Research Bureau] (December 19, 1940). In *MSZ*, no. 2, 403–6.

"Mantetsu no keizaiteki ichi" [Mantetsu's economic position]. *Yomiuri shinbun*, January 12, 1910.

"Manzhou shengwei yi Fushun wei zhongxin de difang baodong de juti jihua de baogao" [Report by the Manchurian committee on the concrete plan concerning the local insurrection centered on Fushun]. In *FGY*, 30.

"Meeting Discusses Shortage of Coal." *Japan Times and Mail*, April 24, 1938.

"Meitan ye shangxie fenhui Shanghai meitan gonghui jiji tonggao" [Urgent notice from the local branch of the Coal Merchants Society and the Shanghai Coal Trade Association]. *Shenbao*, May 17, 1928.

"Mining Industry of Manchukuo." *Japan Times and Mail,* January 11, 1938.

"Mining in the Far East." *The North-China Herald and Supreme Court and Consular Gazette,* September 21, 1906.

"Mo Desheng huiyi cailiao" [Recollections of Mo Desheng]. Archives of the Pingding-shan Massacre Memorial Hall 27-18. In *Zuixing, zuizheng, zuize,* vol. 2, *Riben qinlüezhe zhizao Pingdingshan can'an zhuanti,* compiled by Fu Bo and Xiao Jingquan, 228-30. Shenyang: Liaoning renmin chubanshe, 1998.

"Mokutan ya sekiyu no jinzō jidai" [The man-made age of charcoal and oil]. *Chūgai shōgyō shinpō,* October 3, 1930.

"More Driving Power." *Japan Times and Advertiser,* March 28, 1942.

"More on Fushun Oil Shale Question." *Manchuria Daily News,* September 1, 1925.

"Moukden: Official Sanction of Mining." *North-China Herald and Supreme Court and Consular Gazette,* July 10, 1896.

"Nenryō kyōkai sōritsu sanshūnen kinen taikai" [Commemorative symposium for the third anniversary of the Fuel Society]. *Nihon nenryō kyōkai shi* 4, no. 7 (July 1925): 427-28.

"Nenryō mondai no jūyo naru koto o ippan kokumin ni shirashimuru ronbun" [Essay to inform the general public of important issues concerning the fuel question]. *Nenryō kyōkai shi* 5, no. 7 (July 1926): 736-49.

"The New Cabinet in Peking." *China Review* 2, no. 1 (January 1922): 4.

"New Fushun Shale Oil Plant." *Far Eastern Review* 25, no. 2 (February 1929): 58-62.

"A New Oil Extracting Contrivance Invented by Fushun Colliery Experts." *Manchuria Daily News Monthly Supplement,* October 1, 1925, 7.

"New Oil Shale Plant, Fushun: Axle of Japan's Fuel Supply." *Manchuria Daily News Monthly Supplement,* February 1, 1930, 10.

"Ogawa gishi no Bujun tankō dan" [Engineer Ogawa on the Fushun colliery]. *Tokyo asahi shinbun,* July 23, 1906.

"Oong Ordered to Reorganize Honan Mines: Generalissimo Appoints Geologist to Run Joint British Coal Mines." *China Press,* December 1, 1934.

"The 'Proto' Breathing Apparatus." *State Safety News* 32 (June 1, 1918): 2-6.

"Public Bath Rate-Cutting." *Japan Times and Mail,* February 21, 1938.

"Quanguo meikuang yi zhi wu yue fen shangwang shigu yanzhong" [Rates of accidents resulting in injury and death in mines across the country from January to May severe]. *Neibu cankao,* June 27, 1958, 16-17.

"Railway Electrification in Japan." *Far Eastern Review* 19, no. 8 (August 1923): 511-16.

"Record of a Meeting between T. V. Soong and Stalin" (August 10, 1945). History and Public Policy Program Digital Archive, Victor Hoo Collection, box 6, folder 9, Hoover Institution Archives, contributed by David Wolff. http://digitalarchive.wilsoncenter .org/document/134355.

"Rotenbori zenkōan no naiyō ni chōsa iinkai no shinsa hōshin" [The contents of the provisional plans for the open-pit mine and the review plan of the investigative committee]. *Manshū nichinichi shinbun,* December 25, 1923.

"The Ryuho Coal Mine: Additional Pride of Fushun." *Contemporary Manchuria* 1, no. 4 (November 1937): 55-65.

"Sangyō kaihatsu gonen keikaku daisan-nendo seisan jisseki hōkoku" [Report on the actual production in the third year of the five-year plan] (April 1940). Fushun Mining Affairs Bureau, Japanese-Language Archival Materials 8-7, no. 140. In *MSZ,* no. 2, 401-3.

"Science and Life in War-Time China" (Broadcast from London, December, 1944). In *Sci-*

ence Outpost: Papers of the Sino-British Science Co-operation Office (British Council Scientific Office in China), 1942–1946, edited by Joseph Needham and Dorothy Needham, 50–55. London: Pilot, 1948.

"Shanghai Faces Coal Shortage." *China Press*, October 28, 1931.

"Shanghai shi meiye tongye gonghui kang Ri weiyuanhui tonggao" [A notice from the Shanghai Coal Trade Association and the Resist Japan Association]. *Shenbao*, September 26, 1931.

"Shengli kuang dong keng di san caimei qu zhang Li Qiliang yinman shigu pianqu jiangjin" [Li Qiliang, chief of the third mining area of Shengli mine's eastern pit, obscuring the occurrence of accidents]. *Fushun ribao*, March 22, 1953.

"Shengli kuang jingsai zhong hulue anquan shengchan, shiyuefen fasheng liang ci zhongda shigu" [Shengli mine compromised safe production in the midst of competition, two major accidents occurred in October]. *Fushun ribao*, November 12, 1950.

"Shenyang gejie zhuidao Zhang Shenfu" [Various sectors across Mukden mourn the death of Zhang Shenfu]. *Qiaosheng bao*, May 22, 1946.

"Shi bu zhaoji jiuji meihuang huiyi taolun ge an you xiangdang jueding" [Considerable resolution regarding the various items discussed at the Ministry of Industry's meeting for coal famine relief]. *Shenbao*, November 18, 1931.

"Shishaku giin hoketsu senkyo wa Inoue Tadashirō ga dōsen" [In special election for viscount peer, Inoue Tadashirō elected]. *Yomiuri shinbun*, October 2, 1910.

"Shi shanghui jiuji meihuang huiyi" [Shanghai Chamber of Commerce meeting on relief for coal famine]. *Shenbao*, October 30, 1931.

"Shiye bu guanyu yufang meihuang si xiang banfa de ti'an" [A proposal by the Ministry of Industry regarding four measures to prevent a coal famine] (October 5, 1931). *MGDA*, ser. 5, no. 1, vol. 4, pt. 6, 466.

"Shots from the Firing Range." *Excavating Engineer* 11, no. 7 (April 1915): 268.

"The Significance of Saghalien." *China Weekly Review*, May 2, 1925.

"S.M.R. against Plan to Put Ban on Coal." *Japan Times*, July 12, 1932.

"The South Manchuria Railway Company." *Japan Times*, May 10, 1920.

"Strengthen Coal Washing Work, Raise Coking Coal Quality" (November 19, 1958). In *Chinese Communist Coal Mining Continues Technical Advance*. Translated from the Chinese. Joint Publications Research Services report no. 3686, August 11, 1960.

"Sulian renmin de xingfu shenghuo" [The happy lives of Soviet people]. *Fushun ribao*, November 1, 1952.

"Sulian zhuanjia zai wu nian lai dui Fushun meikuang geiyu juda de bangzhu" [Soviet experts have been a huge help to the Fushun colliery these past five years]. *Fushun ribao*, November 6, 1954.

"A Survey of the Coal Mining Industry in Manchuria." *Contemporary Manchuria* 1, no. 4 (November 1937): 66–76.

"Tech Man Who Moves Towns." *Technology Review* 24, no. 2 (May 1922): 189.

"Text of Kung Memorandum on Mine Development Given." *China Press*, March 29, 1931.

"Tiedao bu jiuji meihuang banfa" [Plan by the Ministry of Railways for coal famine relief]. *Shenbao*, November 28, 1931.

"Trade with Manchuria." *Japan Times*, April 30, 1910.

"Transfer of Mining Equipment from Japan to Manchuria Now Mooted to Increase Output." *Manchuria Daily News*, April 23, 1939.

"Treaty and Additional Agreement Relating to Manchuria, December 22, 1905." In *Treaties and Agreements with and Concerning China, 1894–1912*, compiled by John V. A. MacMurray, 1:549–54. New York: Oxford University Press, 1921.

"US Planes over New China." *China Monthly Review*, September 1, 1952, 293.

"Waga sekiyu mondai to Bujun san yubo ketsugan no kachi" [Fushun shale oil's value to our petroleum question]. *Manshū nichinichi shinbun*, March 10, 1924.

"Wang Keming huiyi cailiao" [Recollections of Wang Keming]. In *Zuixing, zuizheng, zuize*, vol. 1, *Erzhan shiqi Riben qinlüezhe zai woguo Dongbei canhai beifu renyuan zhuanti*, compiled by Fu Bo, 20–22. Shenyang: Liaoning renmin chubanshe, 1995.

"Wei Huakun jizhang Fukuang" [Wei Huakun, the next manager of the Fushun coal mine]. *Wudiao xunkan* 37 (February 1948): 20.

"World's Largest Shale-Oil Plant." *Far Eastern Review* 26, no. 2 (February 1930): 62–63, 79.

"Wu er bu zuo zhi Fushun meikuang" [No lows to which the Fushun colliery would not stoop]. *Kuangye zhoubao* 155 (August 1931): 3–4.

"Ximei shi zengchan gangtie de zhongyao huanjie" [Coal washing is an important connector for increasing steel output]. *Renmin ribao*, September 5, 1960.

"Xuexi Sulian xianjin jingyan wei zuguo gongyehua er fendou" [Learn from the advanced experience of the Soviet Union and fight for the industrialization of our country]. *Fushun ribao*, November 7, 1952.

"Yan Shuting huiyi cailiao" [Recollections of Yan Shuting]. In *Zuixing, zuizheng, zuize*, vol. 1, *Erzhan shiqi Riben qinlüezhe zai woguo Dongbei canhai beifu renyuan zhuanti*, compiled by Fu Bo, 23–25. Shenyang: Liaoning renmin chubanshe, 1995.

"Zhang Shenfu an xiongfan Mo Guangcheng de gongci" [Confession of Mo Guangcheng, murderer in the case of Zhang Shenfu]. *Qiaosheng bao*, August 30, 1946.

"Zhang Shenfu di er, lianjiao zhuanjia Yu Zailin zai Beipiao bei fei qiangsha" [The second Zhang Shenfu, coking expert Yu Zailin shot to death by bandits in Beipiao]. *Libao*, July 9, 1947.

"Zhang Shenfu gushi pingju yanchu" [Zhang Shenfu's story turned into a drama and performed]. *Haiyan* 2 (September 1946): 11.

Zhonghua renmin gongheguo fazhan guomin jingji de di yi ge wu nian jihua, 1953–1957 [The First Five-Year Plan to develop the national economy of the People's Republic of China, 1953–1957]. Beijing: Renmin chubanshe, 1955.

"Zonggonghui zhaokai chunjie jinglao dahui" [General union hosts lunar new year elderly appreciation mass meeting]. *Fushun ribao*, February 10, 1951.

Works of Known Authorship

Abe Isamu. *Bujun tan no hanro* [The market for Fushun coal]. Dairen: Minami Manshū tetsudō kabushiki gaisha shomubu chōsaka, 1925.

Abiko Kaoru and Shimeno Daiichi. *Manjin rōdōsha no eiyō* [The diet of Manchurian laborers]. Fushun: Mantetsu Bujun tankō, 1941. Liaoning Provincial Archives, file no. xingzheng 2851.

Adachi Masafusa. "Gaien bōshi undō no enkaku" [The development of the smoke abatement campaign]. *Nenryō kyōkai shi* 12, no. 12 (December 1933): 1467–71.

Akabane Katsumi. *Nihon no sekiyu mondai to Bujun san yubo ketsugan no kachi* [Japan's oil question and the value of Fushun's oil shale]. Dairen: Minami Manshū tetsudō kabushiki gaisha, 1924.

Allen, Michael. *Undermining the Japanese Miracle: Work and Conflict in a Coalmining Community*. Cambridge: Cambridge University Press, 1994.

Alvarez, Carlos Fernández. "Global Coal Demand Surpassed Pre-Covid Levels in Late

2020, Underlining the World's Emissions Challenge." IEA, March 23, 2021. https://www.iea.org/commentaries/global-coal-demand-surpassed-pre-covid-levels-in-late-2020-underlining-the-world-s-emissions-challenge.

Amrith, Sunil S. *Migration and Diaspora in Modern Asia*. New York: Cambridge University Press, 2011.

An Qifu. "Sulian zhuanjia bangzhu zamen zhen rexin" [Soviet experts are really sincere in helping us]. *Fushun ribao*, September 8, 1952.

Anderson, Warwick. *Colonial Pathologies: American Tropical Medicine, Race, and Hygiene in the Philippines*. Durham, NC: Duke University Press, 2006.

Anderson, Warwick. "From Subjugated Knowledge to Conjugated Subjects: Science and Globalisation, or Postcolonial Studies of Science?" *Postcolonial Studies* 12, no. 4 (December 2009): 389–400.

Andō Yoshio, ed. *Nihon keizai seisaku shi ron* [A history of Japanese economic policy]. Tokyo: Tōkyō daigaku shuppankai, 1976.

Andreas, Joel. *Disenfranchised: The Rise and Fall of Industrial Citizenship in China*. New York: Oxford University Press, 2019.

Andreas, Joel. *Rise of the Red Engineers: The Cultural Revolution and the Origins of China's New Class*. Stanford, CA: Stanford University Press, 2009.

Andrews, Thomas G. *Killing for Coal: America's Deadliest Labor War*. Cambridge, MA: Harvard University Press, 2008.

Andrews-Speed, Philip. *Energy Policy and Regulation in the People's Republic of China*. The Hague: Kluwer Law International, 2004.

Andrews-Speed, Philip, Guo Ma, Xunpeng Shi, and Bingjia Shao. "The Impact of, and Responses to, the Closure of Small-Scale Coal Mines in China: A Preliminary Analysis." In *The Socio-Economic Impacts of Artisanal and Small-Scale Mining in Developing Countries*, edited by Gavin M. Hilson, 486–503. Lisse: A. A. Balkema, 2003.

Andrews-Speed, Philip, and Xin Ma. "Energy Production and Social Marginalisation in China." In *China's Search for Energy Security: Domestic Sources and International Implications*, edited by Suisheng Zhao, 96–121. London: Routledge, 2013.

Appelbaum, Richard P., and Jeffrey Henderson, eds. *States and Development in the Asian Pacific Rim*. Newbury Park, CA: Sage, 1992.

Arase, David. "Public-Private Sector Interest Coordination in Japan's ODA." *Pacific Affairs* 67, no. 2 (Summer 1994): 171–99.

Arboleda, Martín. *Planetary Mine: Territories of Extraction under Late Capitalism*. New York: Verso, 2020.

Arimoto Kunitarō. "Toshi kūki no osen ni tsuite" [Regarding urban air pollution]. *Nenryō kyōkai shi* 9, no. 9 (September 1930): 981–96.

Aurelius. "Manchukuo—The World's Greatest Military Base." *China Weekly Review*, March 9, 1935.

Austin, Gareth, ed. *Economic Development and Environmental History in the Anthropocene: Perspectives on Asia and Africa*. London: Bloomsbury, 2017.

Bacon, C. A. "Coal Supplies of Shanghai: Actual and Potential." *Chinese Economic Journal* 6, no. 2 (February 1930): 195–218.

Bain, H. Foster. *Ores and Industry in the Far East: The Influence of Key Mineral Resources on the Development of Oriental Civilization*. Rev. and enl. ed. New York: Council on Foreign Relations, 1933.

Bank of Chosen. *Economic History of Chosen*. Seoul: Bank of Chosen, 1921.

Bao Min. "Fushun de shiming" [Fushun's mission]. *Fukuang xunkan, jinian te hao* (April 1948): 5.

Barak, On. *Powering Empire: How Coal Made the Middle East and Sparked Global Carbonization*. Oakland: University of California Press, 2020.

Barlow, Tani E. "Colonialism's Career in Postwar China Studies." *positions* 1, no. 1 (Spring 1993): 224–67.

Barnaby, Frank. "Effects of the Atomic Bombings of Hiroshima and Nagasaki." In *Hiroshima and Nagasaki: Retrospect and Prospect*, edited by Frank Barnaby and Douglas Holdstock, 1–7. New York: Routledge, 1995.

Barnaby, Frank, and Douglas Holdstock, eds. *Hiroshima and Nagasaki: Retrospect and Prospect*. New York: Routledge, 1995.

Barnes, Nicole Elizabeth. *Intimate Communities: Wartime Healthcare and the Birth of Modern China, 1937–1945*. Oakland: University of California Press, 2018.

Barnhart, Michael A. *Japan Prepares for Total War: The Search for Economic Security, 1919–1941*. Ithaca, NY: Cornell University Press, 1988.

Bassett, Ross. *The Technological Indian*. Cambridge, MA: Harvard University Press, 2016.

Bazhenov, I. I., I. A. Leonenko, and A. K. Kharohenko, "Organization of Production, Labor Productivity and Costs in the Coal Industry of the CPR." Translated from the Russian. Joint Publications Research Services report no. 5198, August 8, 1960.

Beard, Charles A., ed. *Whither Mankind: A Panorama of Modern Civilization*. New York: Longmans, Green, 1928.

Beasley, W. G. *Japanese Imperialism, 1894–1945*. Oxford: Oxford University Press, 1985.

Beer, John J. "Coal Tar Dye Manufacture and the Origins of the Modern Industrial Research Laboratory," *Isis* 49, no. 2 (June 1958): 123–31.

Bergère, Marie-Claire. *The Golden Age of the Chinese Bourgeoisie, 1911–1937*. Translated from the French by Janet Lloyd. Cambridge: Cambridge University Press, 1989.

Berglund, Abraham. "The Iron and Steel Industry of Japan and Japanese Continental Policies." *Journal of Political Economy* 30, no. 5 (October 1922): 623–54.

Bergsten, C. Fred, Charles Freeman, Nicholas R. Lardy, and Derek J. Mitchell. *China's Rise: Challenges and Opportunities*. Washington, DC: Peterson Institute for International Economics; Center for Strategic and International Studies, 2008.

Bernstein, Thomas P., and Hua-yu Li, eds. *China Learns from the Soviet Union, 1949–Present*. Lanham, MD: Lexington Books, 2010.

Bian, Morris L. "Building State Structure: Guomindang Institutional Rationalization during the Sino-Japanese War, 1937–1945." *Modern China* 31, no. 1 (January 2005): 35–71.

Bian, Morris L. *The Making of the State Enterprise System in Modern China: The Dynamics of Institutional Change*. Cambridge, MA: Harvard University Press, 2005.

Black, Brian. *Petrolia: The Landscape of America's First Oil Boom*. Baltimore: Johns Hopkins University Press, 2000.

Black, Megan. *The Global Interior: Mineral Frontiers and American Power*. Cambridge, MA: Harvard University Press, 2018.

Bloch, Kurt. "Coal and Power Shortage in Japan." *Far Eastern Survey* 9, no. 4 (February 1940): 39–45.

Bochove, Danielle, Aya Takada, and Aaron Clark. "Nine Years after Fukushima, Japan Can't Quit Its Coal Habit." *Japan Times*, March 17, 2020.

Boecking, Felix. *No Great Wall: Trade, Tariffs, and Nationalism in Republican China, 1927–1945*. Cambridge, MA: Harvard University Asia Center, 2017.

Bōeichō bōei kenshūjo senshishitsu. *Kaigun gunsenbi* [Naval war preparations]. Vol. 1. Tokyo: Asagumo shinbunsha, 1969.

Boyer, Dominic. "Energopower: An Introduction." *Anthropology Quarterly* 87, no. 2 (April 2014): 309–33.

Braden, Peter. "Serve the People: Bovine Experiences in China's Civil War and Revolution, 1935–1961." PhD diss., University of California, San Diego, 2020.

Bradley, John R., and Donald W. Smith. *Fuel and Power in Japan.* Washington, DC: United States Government Printing Office, 1935.

Brandt, Loren, and Thomas G. Rawski, eds. *China's Great Economic Transformation.* Cambridge: Cambridge University Press, 2008.

Braun, Bruce. "Producing Vertical Territory: Geology and Governmentality in Late Victorian Canada." *Ecumene* 7, no. 1 (January 2000): 7–46.

Braun, Lundy. *Breathing Race into the Machine: The Surprising Career of the Spirometer from Plantation to Genetics.* Minneapolis: University of Minnesota Press, 2014.

Braun, Stuart. "Is China's Five Year Plan a Decarbonization Blueprint?" Deutsche Welle, March 5, 2021. https://p.dw.com/p/3pfo9.

Bray, Francesca. "Only Connect: Comparative, National, and Global History as Frameworks for the History of Science and Technology in Asia." *East Asian Science, Technology and Society* 6, no. 2 (June 2012): 233–41.

Brazelton, Mary Augusta. *Mass Vaccination: Citizens' Bodies and State Power in Modern China.* Ithaca, NY: Cornell University Press, 2019.

Brooke, John L., and Julia C. Strauss. "Introduction: Approaches to State Formations." In *State Formations: Global Histories and Cultures of Statehood,* edited by John L. Brooke, Julia C. Strauss, and Greg Anderson, 1–21. Cambridge: Cambridge University Press, 2018.

Brooke, John L., Julia C. Strauss, and Greg Anderson, eds. *State Formations: Global Histories and Cultures of Statehood.* Cambridge: Cambridge University Press, 2018.

Brown, Jeremy. *City versus Countryside in Mao's China: Negotiating the Divide.* Cambridge: Cambridge University Press, 2012.

Brown, Jeremy. "When Things Go Wrong: Accidents and the Legacy of the Mao Era in Today's China." In *Restless China,* edited by Perry Link, Richard P. Madsen, and Paul G. Pickowicz, 11–35. Lanham, MD: Rowman and Littlefield, 2013.

Brown, Jeremy, and Paul G. Pickowicz, eds. *Dilemmas of Victory: The Early Years of the People's Republic of China.* Cambridge, MA: Harvard University Press, 2007.

Brown, Kate. *Plutopia: Nuclear Families, Atomic Cities, and the Great Soviet and American Plutonium Disasters.* Oxford: Oxford University Press, 2013.

Brown, Philip C. "Constructing Nature." In *Japan at Nature's Edge: The Environmental Context of a Global Power,* edited by Ian Jared Miller, Julia Adeney Thomas, and Brett C. Walker, 90–114. Honolulu: University of Hawai'i Press, 2013.

Browne, George Waldo. *China: The Country and Its People.* Boston: D. Estes, 1901.

Bryan, Ford R. *Beyond the Model T: The Other Ventures of Henry Ford.* Detroit: Wayne State University Press, 1990.

Bulman, Harrison Francis. *The Working of Coal and Other Stratified Minerals.* New York: Wiley, 1927.

Bunkichi Fujihirada. *Manshū ni okeru kōzan rōdōsha* [Miners in Manchuria]. Dairen: Minami Manshū tetsudō kōgyōbu chishitsuka, 1918.

Burton, W. Donald. *Coal-Mining Women in Japan: Heavy Burdens.* London: Routledge, 2014.

Cai Yunsheng. "Saitan kūrii no jihaku" [Confessions of a coal-mining coolie]. *Bujun* 12 (October 1913): 23–27. Fushun Mining Group Documents Room, Fushun.

Cameron, W. H. Morton, ed. *Present Day Impressions of Japan.* Chicago: Globe Encyclopedia, 1919.

Canavan, Gerry. "Addiction." In *Fueling Culture: 101 Words for Energy and Environment,* edited by Imre Szeman, Jennifer Wenzel, and Patricia Yaeger, 25–27. New York: Fordham University Press, 2017.

Cao Shengzhi. "Guanyu Zhang Shenfu zhongzhong" [Several matters regarding Zhang Shenfu]. *Qianxian ribao*, February 24, 1946.

Carlson, Ellsworth C. *The Kaiping Mines (1877–1912)*. Cambridge, MA: Harvard University Press, 1957.

Carr, Edward Hallett. *The Twenty Years' Crisis, 1919–1939: An Introduction to the Study of International Relations*. 2nd ed. London: Macmillan, 1956. First published 1946.

Case, Holly. *The Age of Questions: Or, A First Attempt at an Aggregate History of the Eastern, Social, Woman, American, Jewish, Polish, Bullion, Tuberculosis, and Many Other Questions over the Nineteenth Century, and Beyond*. Princeton, NJ: Princeton University Press, 2018.

Castells, Manuel. "Four Asian Tigers with a Dragon Head: A Comparative Analysis of the State, Economy, and Society in the Asian Pacific Rim." In *States and Development in the Asian Pacific Rim*, edited by Richard P. Appelbaum and Jeffrey Henderson, 33–70. Newbury Park, CA: Sage, 1992.

Cecil, Lamar J. R. "Coal for the Fleet That Had to Die." *American Historical Review* 69, no. 4 (July 1964): 990–1005.

Cederlöf, Gunnel. "The Agency of the Colonial Subject: Claims and Rights in Forestlands in the Early Nineteenth-Century Nilgiris." *Studies in History* 21, no. 2 (September 2005): 247–69.

Chakrabarty, Dipesh. "The Climate of History: Four Theses." *Critical Inquiry* 35, no. 2 (Winter 2009): 197–222.

Chang Ching-kiang. "A Reconstruction Program for 1930." *China Weekly Review*, March 1, 1930.

Chang, Leslie T. *Factory Girls: From Village to City in a Changing China*. New York: Siegel and Grau, 2008.

Chatterjee, Elizabeth. "The Asian Anthropocene: Electricity and Fossil Developmentalism." *Journal of Asian Studies* 79, no. 1 (February 2020): 3–24.

Chen Ciyu. *Riben zai Hua meiye touzi sishi nian* [Forty years of Japan's investment in the Chinese coal industry]. Taipei: Daoxiang chubanshe, 2004.

Chen Gongbo. *Si nian congzheng lu* [Account of four years in government]. Shanghai: Shangwu yinshuguan, 1936.

Chen Jiarang. "Zhang Shenfu xunguo sanshi nian" [Thirty years after Zhang Shenfu died for his country]. *Zhuanji wenxue* 28, no. 2 (February 1976): 39–49.

Chen, Tsu-yu. "The Development of the Coal Mining Industry in Taiwan during the Japanese Colonial Occupation, 1895–1945." In *Studies in the Economic History of the Pacific Rim*, edited by Sally M. Miller, A. J. H. Latham, and Dennis O. Flynn, 181–96. London: Routledge, 1997.

Cheng Linsun. "The Industrial Activities of the National Resources Commission and Their Legacies in Communist China." *American Journal of Chinese Studies* 12, no. 1 (April 2005): 45–64.

Cheng Yufeng and Cheng Yuhuang, comps. *Ziyuan weiyuanhui jishu renyuan fu Mei shixi shiliao: Minguo sanshiyi nian huipai* [Historical documents on the training trip to the United States by National Resources Commission engineers: Team sent in 1942]. 2 vols. Taipei: Guoshiguan, 1988.

Chinese Ministry of Information. *China Handbook, 1937–1943: A Comprehensive Survey of Major Developments in China in Six Years of War*. New York: Macmillan, 1943.

Christmas, Sakura. "Japanese Imperialism and Environmental Disease on a Soy Frontier, 1890–1940." *Journal of Asian Studies* 78, no. 4 (November 2019): 809–36.

Clancey, Gregory. *Earthquake Nation: The Cultural Politics of Japanese Seismicity, 1868–1930*. Berkeley: University of California Press, 2006.

Cleveland, Cutler J., ed. *Encyclopedia of Energy.* 6 vols. Amsterdam: Elsevier, 2004.

Coble, Parks M., Jr. *The Shanghai Capitalists and the Nationalist Government, 1927–1937.* 2nd ed. Cambridge, MA: Council on East Asian Studies, Harvard University, 1986. First published 1980.

Coen, Deborah R. *Climate in Motion: Science, Empire, and the Problem of Scale.* Chicago: University of Chicago Press, 2018.

Cohen, Aaron J., Michael Brauer, Richard Burnett, H. Ross Anderson, Joseph Frostad, Kara Estep, Kalpana Balakrishnan, et al. "Estimates and 25-Year Trends of the Global Burden of Disease, Attributable to Ambient Air Pollution: An Analysis of Data from the Global Burden of Diseases Study 2015." *Lancet* 389 (April 10, 2017): 1907–18.

Cohen, Lizabeth. *Making a New Deal: Industrial Workers in Chicago, 1919–1939.* 2nd ed. Cambridge: Cambridge University Press, 2008. First published 1990.

Cohen, Paul A. *Speaking to History: The Story of King Goujian in Twentieth-Century China.* Berkeley: University of California Press, 2010.

Colby, Frank Moore, ed. *The New International Year Book: A Compendium of the World's Progress for the Year 1921.* New York: Dodd, Mead, 1922.

Cole, Matthew A., Robert J. R. Elliott, and Toshihiro Okubo. "International Environmental Outsourcing." *Review of World Economics* 150, no. 4 (June 2014): 639–64.

Colliery Engineer Company. *The Elements of Mining Engineering.* Vol. 2. Scranton, PA: Colliery Engineer Company, 1900.

Cook, Eli. *The Pricing of Progress: Economic Indicators and the Capitalization of American Life.* Cambridge, MA: Harvard University Press, 2017.

Coopersmith, Jonathan C. *The Electrification of Russia, 1880–1926.* Ithaca, NY: Cornell University Press, 1992.

Coronil, Fernando. *The Magical State: Nature, Money, and Modernity in Venezuela.* Chicago: University of Chicago Press, 1997.

Coumbe, Albert T., Jr. *Petroleum in Japan.* Washington, DC: United States Government Printing Office, 1924.

Cowan, Ruth Schwartz. *More Work for Mother: The Ironies of Household Technology from the Open Hearth to the Microwave.* New York: Basic Books, 1983.

Crawcour, E. Sydney. "Industrialization and Technological Change, 1885–1920." In *The Cambridge History of Japan.* Vol. 6, *The Twentieth Century,* edited by Peter Duus, 385–450. New York: Cambridge University Press, 1998.

Crosby, Alfred W. *Children of the Sun: A History of Humanity's Unappeasable Appetite for Energy.* New York: W. W. Norton, 2008.

Cui Dongyuan. "Lutiankuang chang xianzai yu jianglai" [The open-pit mine's present and future]. *Fukuang xunkan* 1, no. 17 (September 1947): 205–8.

Culp, Robert, Eddy U, and Wen-hsin Yeh, eds. *Knowledge Acts in Modern China: Ideas, Institutions, and Identities.* Berkeley: Institute of East Asian Studies, University of California, Berkeley, 2016.

Culter, Suzanne. *Managing Decline: Japan's Coal Industry Restructuring and Community Response.* Honolulu: University of Hawai'i Press, 1999.

Curry, H. A., N. Jardine, J. A. Second, and E. C. Spary, eds. *Worlds of Natural History.* Cambridge: Cambridge University Press, 2018.

Daggert, Cara New. *The Birth of Energy: Fossil Fuels, Thermodynamics, and the Politics of Work.* Durham, NC: Duke University Press, 2019.

Dangdai Zhongguo congshu bianji bu. *Dangdai Zhongguo de meitan gongye* [Contemporary China's coal industry]. Beijing: Dangdai Zhongguo chubanshe, 1988.

Davenport, L. D. *Open Cut Mining by the Railway Approach System: Fushun Colliery.* Fushun: Technical Advisory Board, South Manchuria Railway Company, 1924.

de Bary, Wm. Theodore, Carol Gluck, and Arthur E. Tiedemann, eds. *Sources of Japanese Tradition.* Vol. 2, *1600–2000.* 2nd ed. New York: Columbia University Press, 2005. First published 1958.

Debeir, Jean-Claude, Jean-Paul Deléage, and Daniel Hémery. *In the Servitude of Power: Energy and Civilization through the Ages.* Translated from the French by John Barzman. London: Zed Books, 1990.

Deffeyes, Kenneth S. *Hubbert's Peak: The Impending World Oil Shortage.* Princeton, NJ: Princeton University Press, 2001.

Delahanty, Thomas W., and Charles C. Concannon. *Chemical Trade of Japan.* Washington, DC: United States Government Printing Office, 1924.

de Magalhães, Gabriel. *A New History of China: Containing a Description of the Most Considerable Particulars of that Vast Empire.* London: Thomas Newborough, 1688.

Demuth, Bathsheba. *Floating Coast: An Environmental History of the Arctic.* New York: W. W. Norton, 2019.

DeNovo, John A. "The Movement for an Aggressive American Oil Policy Abroad, 1918–1920." *American Historical Review* 61, no. 4 (July 1956): 854–76.

Denton, Kirk A. *Exhibiting the Past: Historical Memory and the Politics of Museums in Postsocialist China.* Honolulu: University of Hawai'i Press, 2014.

Denyer, Simon. "In China, a Ghost Town Points to Shifting Fortunes." *Washington Post,* August 24, 2015.

Department of Justice, War Division, Economic Warfare Section. "Report on Fushun, Part 1: Oil Shale Refinery and Dubbs Cracking Plant." June 11, 1943.

Department of Justice, War Division, Economic Warfare Section. "Report on Fushun, Part 3: Coal Hydrogenation Plant." December 16, 1943.

Derr, Jennifer L. *The Lived Nile: Environment, Disease, and Material Colonial Economy in Egypt.* Stanford, CA: Stanford University Press, 2019.

Devine, Warren D., Jr. "Coal Mining: Underground and Surface Mechanization." In *Electricity in the American Economy: Agent of Technological Progress,* edited by Sam H. Schurr, Calvin C. Burwell, Warren D. Devine Jr., and Sidney Sonenblum, 181–208. New York: Greenwood, 1991.

Dickinson, Frederick R. *World War I and the Triumph of New Japan, 1919–1930.* Cambridge: Cambridge University Press, 2013.

Dinmore, Eric G. "The Hydrocarbon Ring: Indonesia Fossil Fuel, Japanese 'Cooperation,' and US Cold War Order in Asia." In *Engineering Asia: Technology, Colonial Development, and the Cold War Order,* edited by Hiromi Mizuno, Aaron S. Moore, and John DiMoia, 113–36. London: Bloomsbury, 2018.

Dinmore, Eric Gordon. "A Small Island Nation Poor in Resources: Natural and Human Resource Anxieties in Trans-World War II Japan." PhD diss., Princeton University, 2006.

Dirlik, Arif. "Developmentalism: A Critique." *Interventions* 16, no. 1 (January 2014): 30–48.

Dong Wenqi. "Wangyou Zhang Shenfu sishi nian ji" [Fortieth-year memorial for my deceased friend Zhang Shenfu]. *Zhuanji wenxue* 50, no. 1 (January 1987): 22–28.

Dong, Yifu. "Coal, Which Built a Chinese City, Now Threatens to Bury It." *New York Times,* October 6, 2015.

Dower, John. *Embracing Defeat: Japan in the Wake of World War II.* New York: W. W. Norton, 1999.

Driscoll, Mark. *Absolute Erotic, Absolute Grotesque: The Living, Dead, and Undead in Japan's Imperialism, 1895–1945*. Durham, NC: Duke University Press, 2010.

Duara, Prasenjit. *The Crisis of Global Modernity: Asian Traditions and a Sustainable Future*. Cambridge: Cambridge University Press, 2015.

Duara, Prasenjit. *Sovereignty and Authenticity: Manchuria and the East Asia Modern*. Lanham, MD: Rowman and Littlefield, 2003.

Dudden, Alexis. *Japan's Colonization of Korea: Discourse and Power*. Honolulu: University of Hawai'i Press, 2005.

Duke, Benjamin C. *The History of Modern Japanese Education: Constructing the National School System, 1872–1890*. New Brunswick, NJ: Rutgers University Press, 2009.

Duus, Peter. *The Abacus and the Sword: The Japanese Penetration of Korea, 1895–1910*. Berkeley: University of California Press, 1998.

Duus, Peter, ed. *The Cambridge History of Japan*. Vol. 6, *The Twentieth Century*. Cambridge: Cambridge University Press, 1988.

Dyer, Henry. *Dai Nippon: The Britain of the East*. London: Blackie, 1904.

Eastman, Lloyd E. *The Abortive Revolution: China under Nationalist Rule, 1927–1937*. Cambridge, MA: Council on East Asian Studies, Harvard University, 1974.

Eastman, Lloyd E. "Nationalist China during the Sino-Japanese War, 1937–1945." In *The Cambridge History of China*. Vol. 13, *Republican China, 1912–1949*, pt. 2, edited by John K. Fairbank and Albert Feuerwerker, 547–608. Cambridge: Cambridge University Press, 1986.

Eastman, Lloyd E. *Seeds of Destruction: Nationalist China in War and Revolution, 1937–1949*. Stanford, CA: Stanford University Press, 1984.

Eckert, Carter J. *Park Chung Hee and Modern Korea: The Roots of Militarism, 1866–1945*. Cambridge, MA: Harvard University Press, 2016.

Edgerton, David. *The Shock of the Old: Technology and Global History since 1900*. London: Profile Books, 2006.

Elleman, Bruce A. *Moscow and the Emergence of Communist Power in China, 1925–30: The Nanchang Uprising and the Birth of the Red Army*. New York: Routledge, 2009.

Elleman, Bruce A. "Soviet Policy on Outer Mongolia and the Chinese Communist Party." *Journal of Asian History* 28, no. 2 (January 1994): 108–23.

Elleman, Bruce A., and Stephen Kotkin, eds. *Manchurian Railways and the Opening of China: An International History*. Armonk, NY: M. E. Sharpe, 2010.

Elliott, Mark C. "The Limits of Tartary: Manchuria in Imperial and National Geographies." *Journal of Asian Studies* 59, no. 3 (August 2000): 603–46.

Elliott, Mark C. *The Manchu Way: The Eight Banners and Ethnic Identity in Late Imperial China*. Stanford, CA: Stanford University Press, 2001.

Elman, Benjamin A. *On Their Own Terms: Science in China, 1550–1900*. Cambridge, MA: Harvard University Press, 2005.

Engels, Friedrich. *The Condition of the Working Class in England*. Edited by David McLellan. Oxford: Oxford University Press, 1999. First published in English in 1892.

Esarey, Ashley, Mary Alice Haddad, Joanna I. Lewis, and Stevan Harrell, eds. *Greening East Asia: The Rise of the Eco-developmental State*. Seattle: University of Washington Press, 2020.

Evans, David C., and Mark R. Peattie. *Kaigun: Strategy, Tactics, and Technology in the Imperial Japanese Navy, 1887–1941*. Annapolis, MD: Naval Institute Press, 1997.

Evans, Peter B., Dietrich Rueschemeyer, and Theda Skocpol, eds. *Bringing the State Back In*. Cambridge: Cambridge University Press, 1985.

Eyferth, Jacob. "Women's Work and the Politics of Homespun in Socialist China, 1949–1980." *International Review of Social History* 57, no. 3 (December 2012): 365–91.

Fairbank, John K., and Albert Feuerwerker, eds. *The Cambridge History of China*. Vol. 13, *Republican China, 1912–1949*, pt. 2. Cambridge: Cambridge University Press, 1986.

Fan, Fa-ti. "Can Animals Predict Earthquakes? Bio-sentinels as Seismic Sensors in Communist China and Beyond." *Studies in the History and Philosophy of Science* 70 (August 2018): 58–69.

Fan, Fa-ti. "'Collective Monitoring, Collective Defense': Science, Earthquakes, and Politics in Communist China." *Science in Context* 25, no. 1 (2012): 127–54.

Fan, Fa-ti. "Modernity, Region, and Technoscience: One Small Cheer for Asia as Method." *Cultural Sociology* 10, no. 3 (September 2016): 352–68.

Fan Yincang. "Lutiankuang 'mei' de zishu" [Autobiography of coal from the open pit]. *Fushun ribao*, July 28, 1951.

Fedman, David. *Seeds of Control: Japan's Empire of Forestry in Colonial Korea*. Seattle: University of Washington Press, 2020.

Fedman, David. "Wartime Forestry and the 'Low Temperature Lifestyle' in Late Colonial Korea, 1937–1945." *Journal of Asian Studies* 77, no. 2 (May 2018): 333–50.

Feng Hao and Tom Baxter. "China's Coal Consumption on the Rise." *China Dialogue*, March 1, 2019. https://chinadialogue.net/en/energy/11107-china-s-coal-consumption -on-the-rise/.

Finamore, Barbara. *Will China Save the Planet?* Cambridge: Polity, 2018.

Fischer, Frank. *Technocracy and the Politics of Expertise*. Newbury Park, CA: Sage, 1990.

Fogel, Joshua A. "Yosano Akiko and Her China Travelogue of 1928." In *Travels in Manchuria and Mongolia: A Feminist Poet from Japan Encounters Prewar China*, by Yosano Akiko, translated from the Japanese and edited by Joshua A. Fogel, 1–10. New York: Columbia University Press, 2001.

Foucault, Michel. *Discipline and Punish: The Birth of the Prison*. Translated from the French by Alan Sheridan. New York: Pantheon Books, 1977.

Frank, Alison Fleig. *Oil Empire: Visions of Prosperity in Austrian Galicia*. Cambridge, MA: Harvard University Press, 2007.

Fraser, Nancy, and Rahel Jaeggi. *Capitalism: A Conversation in Critical Theory*. Edited by Brian Milstein. Cambridge: Polity, 2018.

Frazier, Mark W. *The Making of the Chinese Industrial Workplace: State, Revolution, and Labor Management*. Cambridge: Cambridge University Press, 2002.

Freeman, Joshua B. *Behemoth: A History of the Factory and the Making of the Modern World*. New York: W. W. Norton, 2018.

French, Thomas, ed. *The Economic and Business History of Occupied Japan: New Perspectives*. London: Routledge, 2017.

Friedman, Jeffrey. *Power without Knowledge: A Critique of Technocracy*. New York: Oxford University Press, 2019.

Fu Bo. "Duli shi shi jiefang Fushun" [Tenth Independent Squad liberates Fushun]. In *Fushun minguo wangshi* [Former happenings in Republican Fushun], edited by Fushun shi zhengxie wenhua he wenshi ziliao weiyuanhui, 251–58. Shenyang: Liaoning renmin chubanshe, 2014.

Fu Bo. "'Jiu-yi-ba' qian de Fushun kuanggong" [Fushun miners before the September 18 incident]. In *Fushun wenshi ziliao xuanji* [Selection of historical materials on Fushun], vol. 3, edited by Zhonghua renmin zhengzhi xieshang huiyi Liaoning sheng Fushun shi weiyuanhui wenshi weiyuanhui, 73–86. Fushun: Liaoning sheng Fushun shi weiyuanhui wenshi weiyuanhui, 1984.

Fu Bo. "Zaixian de lishi yu lishi de qishi" [Reproduced history and history's revelation]. In *Zuixing, zuizheng, zuize*, vol. 1, *Erzhan shiqi Riben qinlüezhe zai woguo Dongbei canhai beifu renyuan zhuanti,* compiled by Fu Bo, 4–22. Shenyang: Liaoning renmin chubanshe, 1995.

Fu Bo. *Zhong-Ri Fushun meikuang an jiaoshe shimo* [The ins and outs of the Sino-Japanese negotiations over the Fushun coal mines]. Ha'erbin: Heilongjiang renmin chubanshe, 1987.

Fu Bo, comp. *Zuixing, zuizheng, zuize* [Crime, proof, guilt]. Vol. 1, *Erzhan shiqi Riben qinlüezhe zai woguo Dongbei canhai beifu renyuan zhuanti* [Special topic on the Japanese invaders' cruel treatment of captured prisoners in our country's northeast during World War II]. Shenyang: Liaoning renmin chubanshe, 1995.

Fu Bo, Liu Chang, and Wang Pinglu, eds. *Fushun difang shi gailan* [A general overview of Fushun's local history]. Fushun: Fushun shi shehui kexueyuan, 2001.

Fu Bo and Xiao Jingquan, comps. *Zuixing, zuizheng, zuize* [Crime, proof, guilt]. Vol. 2, *Riben qinlüezhe zhizao Pingdingshan can'an zhuanti* [Special topic on the Pingdingshan atrocity committed by the Japanese invaders]. Shenyang: Liaoning renmin chubanshe, 1998.

Fuchs, J. R. "Oral History Interview with Edwin W. Pauley." Los Angeles, March 3, 4, and 9, 1971. Harry S. Truman Library and Museum. https://www.trumanlibrary.gov/library/oral-histories/pauleye#transcript.

Fuess, Harald, ed. *The Japanese Empire in East Asia and Its Postwar Legacy.* Munich: Iudicium, 1998.

Fukushima, S. "Japanese Gas Works." *Japan Magazine* 6, no. 8 (December 1915): 489–90.

Fukuzawa Yukichi. "Datsu-A ron" [Leaving Asia]. *Jiji shimpō,* March 16, 1885.

Fushun kuangwu ju. *Canguan shouce* [Observational record]. Fushun: Fushun yinshuasuo, 1949.

Fushun kuangwu ju meitan zhi bianzuan weiyuanhui. *Fushun kuangqu dashiji, 1901–1985* [Chronology of the Fushun coalfield, 1901–1985]. Fushun: Fushun kuangwu ju, 1987.

Fushun shi difang zhi bangongshi. *Fushun shi zhi* [Gazetteer of Fushun city]. Shenyang: Liaoning renmin chubanshe, 1993.

Fushun shi renmin zhengfu difang zhi bangongshi and Fushun shi shehui kexue yanjiusuo. *Fushun dashiji* [A record of major events in Fushun]. Shenyang: Liaoning renmin chubanshe, 1987.

Fushun shi shehui kexueyuan. *Zhonggong Fushun difang shi* [Local history of the Chinese Communist Party in Fushun]. Shenyang: Liaoning minzu chubanshe, 1999.

Fushun shi shehui kexueyuan and Fushun shi renmin zhengfu difang zhi bangongshi. *Fushun shi zhi* [Fushun city gazetteer], *Gongye juan* [volume on industry]. Shenyang: Liaoning minzu chubanshe, 2003.

Fushun shi tongji ju. *Fushun sanshi nian: 1949–1978 jingji he wenhua jianshe chengjiu de tongji* [Fushun's thirty years: Statistics of the accomplishments in economic and cultural construction, 1949–1978]. Shenyang: Shenyang shi di yi yinshuachang, n.d.

Gamsa, Mark. "The Epidemic of Pneumonic Plague in Manchuria, 1910–1911." *Past and Present* 190, no. 1 (February 2006): 147–83.

Gao, Bai. *Economic Ideology and Japanese Industrial Policy.* Cambridge: Cambridge University Press, 1997.

Gao Baiyu. "China Relaxes Restrictions on Coal Power Expansion for Third Year Running." *China Dialogue,* April 17, 2020. https://chinadialogue.net/en/energy/11966-china-relaxes-restrictions-on-coal-power-expansion-for-third-year-running/.

Gao Kai. "Woguo meikuang gongye de zhanwang" [Prospects for our country's coal-mining industry]. *Renmin ribao*, December 2, 1949.

Gao Luming. "Ping–Han yanxian zhi meichan ji qi yunshu maoyi zhi gaikuang" [A survey of coal produced along the Ping–Han railway and its transport and trade]. *Jiaotong jingji huikan* 3, no. 1 (April 1930): 1–17.

Garon, Sheldon. "Fashioning a Culture of Diligence and Thrift: Savings and Frugality Campaigns in Japan, 1900–1931." In *Japan's Competing Modernities: Issues in Culture and Democracy, 1900–1930*, edited by Sharon A. Minichiello, 312–34. Honolulu: University of Hawai'i Press, 1998.

Garon, Sheldon M. *The State and Labor in Modern Japan*. Berkeley: University of California Press, 1987.

Gavin, Martin J. *Oil Shale: An Historical, Technical, and Economic Study*. Washington, DC: United States Government Printing Office, 1924.

Gerth, Karl. *China Made: Consumer Culture and the Creation of the Nation*. Cambridge, MA: Harvard University Asia Center, 2003.

Gerth, Karl. *Unending Capitalism: How Consumerism Negated China's Communist Revolution*. Cambridge: Cambridge University Press, 2020.

Ghosh, Amitav. *The Great Derangement: Climate Change and the Unthinkable*. Chicago: University of Chicago Press, 2016.

Ghosh, Arunabh. *Making It Count: Statistics and Statecraft in the Early People's Republic of China*. Princeton, NJ: Princeton University Press, 2020.

Giersch, C. Patterson. *Corporate Conquests: Business, the State, and the Origins of Ethnic Inequality in Southwest China*. Stanford, CA: Stanford University Press, 2020.

Gillin, Donald G., and Ramon H. Myers. "Introduction." In *Last Chance in Manchuria: The Diary of Chang Kia-ngau*, edited by Donald G. Gillin and Ramon H. Myers and translated from the Chinese by Dolores Zen, 1–58. Stanford, CA: Hoover Institution Press, 1989.

Gillin, Donald G., and Ramon H. Myers, eds. *Last Chance in Manchuria: The Diary of Chang Kia-ngau*. Translated from the Chinese by Dolores Zen. Stanford, CA: Hoover Institution Press, 1989.

Gilpin, Alan. *Environmental Impact Assessment: Cutting Edge for the Twenty-First Century*. Cambridge: Cambridge University Press, 1995.

Goforth, Paul. "The Battle of the Coal: Fire in Open Pit Adds Drama, Inspires Visitor." *Manchuria Daily News*, October 31, 1933.

Golas, Peter J. *Science and Civilisation in China*. Vol. 5, *Chemistry and Chemical Technology*, pt. 13, *Mining*. Cambridge: Cambridge University Press, 1999.

Goldman, Merle. *Literary Dissent in Communist China*. Cambridge, MA: Harvard University Press, 1967.

Gong Yadong. "Fushun meikuang zhi yan'ge ji zuzhi" [The Fushun colliery's evolution and organization]. *Fukuang xunkan: qingzhu zongtong jiuzhi zhuankan* (May 1948): 24–31.

Gooday, Graeme J. N., and Morris F. Low. "Technology Transfer and Cultural Exchange: Western Scientists and Engineers Encounter Lake Tokugawa and Meiji Japan." "Beyond Joseph Needham: Science, Technology, and Medicine in East and Southeast Asia," *Osiris* 13 (1998): 99–128.

Goodman, Bryna, and David S. G. Goodman. "Introduction: Colonialism and China." *Twentieth-Century Colonialism and China: Localities, the Everyday and the World*, edited by Bryna Goodman and David S. G. Goodman, 1–22. London: Routledge, 2012.

Goodman, Bryna, and David S. G. Goodman, eds. *Twentieth-Century Colonialism and China: Localities, the Everyday and the World*. London: Routledge, 2012.

Goodrich, Carter. *The Miner's Freedom: A Study of the Working Life in a Changing Industry*. Boston: Marshall Jones, 1925.

Goossaert, Vincent, Jan Kiely, and John Lagerwey, eds. *Modern Chinese Religion II: 1850–2015*. Vol. 1. Leiden: Brill, 2015.

Gordon, Andrew. *Labor and Imperial Democracy in Prewar Japan*. Berkeley: University of California Press, 1991.

Gorz, André. *Ecology as Politics*. Boston: South End Press, 1980.

Gotō Shinpei. "Jinsei to nenryō mondai" [Human life and the fuel question]. *Nenryō kyōkai shi* 5, no. 1 (January 1926): 28–37.

Gottschang, Thomas R., and Diana Lary. *Swallows and Settlers: The Great Migration from North China to Manchuria*. Ann Arbor: Center for Chinese Studies, University of Michigan, 2000.

Graham, Maurice. "Round the World—And Some Gas-Works." *Journal of Gas Lighting, Water Supply and Sanitary Improvement* 104, no. 2380 (December 22, 1908): 820–25.

Great Britain Foreign Office, Historical Section. *Sakhalin*. London: H. M. Stationery Office, 1920.

Greenberg, Dolores. "Energy, Power, and Perceptions of Social Change in the Early Nineteenth Century." *American Historical Review* 95, no. 3 (June 1990): 693–714.

Greene, J. Megan. "Looking toward the Future: State Standardization and Professionalization of Science in Wartime China." In *Knowledge Acts in Modern China: Ideas, Institutions, and Identities*, edited by Robert Culp, Eddy U, and Wen-hsin Yeh, 275–303. Berkeley: Institute of East Asian Studies, University of California, Berkeley, 2016.

Greene, J. Megan. *The Origins of the Developmental State in Taiwan: Science Policy and the Quest for Modernization*. Cambridge, MA: Harvard University Press, 2008.

Greenhalgh, Susan. "Governing through Science: The Anthropology of Science and Technology in Contemporary China." In *Can Science and Technology Save China?*, edited by Susan Greenhalgh and Li Zhang, 1–24. Ithaca, NY: Cornell University Press, 2020.

Greenhalgh, Susan, and Li Zhang, eds. *Can Science and Technology Save China?* Ithaca, NY: Cornell University Press, 2020.

Grunden, Walter F. *Secret Weapons and World War II: Japan in the Shadow of Big Science*. Lawrence: University Press of Kansas, 2005.

Guojia tongji ju. *Weida de shi nian* [Ten great years]. Beijing: Renmin chubanshe, 1959.

Hama, Katsuhiko. "The Daqing Oil Field: A Model in China's Struggle for Rapid Industrialization." *Developing Economies* 18, no. 2 (June 1980): 180–205.

Hamauzu, Tetsuo. "The Changing Structure of Oil Connections." In *Japan in the Contemporary Middle East*, edited by Kaoru Sugihara and J. A. Allan, 50–76. London: Routledge, 1993.

Hamlin, Christopher. *Cholera: The Biography*. Oxford: Oxford University Press, 2009.

Handler, Sandra. *Austere Luminosity of Chinese Classical Furniture*. Berkeley: University of California Press, 2001.

Hanley, Susan B. *Everyday Things in Premodern Japan: The Hidden Legacy of Material Culture*. Berkeley: University of California Press, 1997.

Hara Akira. "1930-nendai no Manshū keizai tōsei seisaku" [Manchurian economic control policy in the 1930s]. In *Nihon teikokushugi ka no Manshū: "Manshūkoku" seiritsu zengo no keizai kenkyū*, edited by Manshū shi kenkyūkai, 1–114. Tokyo: Ochanomizu shobō, 1972.

Hara Akira. "'Manshū' ni okeru keizai tōsei seisaku no tenkai: Mantetsu kaisō to Mangyō setsuritsu o megutte" [The development of policies of the managed economy in Manchuria: A glance at the reorganization of Mantetsu and the establishment of Mangyō]. In *Nihon keizai seisaku shi ron*, edited by Andō Yoshio, 209–96. Tokyo: Tōkyō daigaku shuppankai, 1976.

Hara Seiji. *Harukanari Mantetsu: tsuioku no Mantetsu Bujun tankōchō Kubo Tōru* ["Faraway" Mantetsu: Looking back on Mantetsu's Fushun colliery's head Kubo Tōru]. Tokyo: Shin jinbutsu ōraisha, 2000.

Hara Teruyuki. "Japan Moves North: The Japanese Occupation of Northern Sakhalin (1920s)." In *Rediscovering Russia in Asia: Siberia and the Russian Far East*, edited by Stephen Kotkin and David Wolff, 55–67. Armonk, NY: M. E. Sharpe, 1995.

Harrell, Stevan. "The Eco-developmental State and the Environmental Kuznets Curve." In *Greening East Asia: The Rise of the Eco-developmental State*, edited by Ashley Esarey, Mary Alice Haddad, Joanna I. Lewis, and Stevan Harrell, 241–66. Seattle: University of Washington Press, 2020.

Harrison, Henrietta. *The Man Awakened from Dreams: One Man's Life in a North China Village, 1857–1942*. Stanford, CA: Stanford University Press, 2005.

Hartman, Howard L., and Jan M. Mutmansky. *Introductory Mining Engineering*. 2nd ed. Hoboken, NJ: John Wiley and Sons, 2002. First published 1987.

Hartwell, Robert. "A Cycle of Economic Change in Imperial China: Coal and Iron in Northeast China, 750–1350." *Journal of Economic and Social History of the Orient* 10, no. 1 (July 1967): 102–59.

Hartwell, Robert. "A Revolution in the Chinese Iron and Coal Industries during the Northern Sung, 960–1126 A.D." *Journal of Asian Studies* 21, no. 2 (February 1962): 153–62.

Hashimoto, Michio. "History of Air Pollution Control in Japan." In *How to Conquer Air Pollution: A Japanese Experience*, edited by H. Nishimura, 1–93. Amsterdam: Elsevier, 1989.

Hayakawa, T. "New Foreign Minister." *Japan Magazine* 9, no. 4 (August 1918): 189–96.

Hayase, Yukiko. "The Career of Gotō Shinpei: Japan's Statesman of Research, 1857–1929." PhD diss., Florida State University, 1974.

He Suhua. "Kangzhan shiqi guoying meikuangye de fazhan" [The development of the state coal-mining sector during the War of Resistance]. MA thesis, National Taiwan University, 1990.

He, Y. G. "Mining and Utilization of Chinese Fushun Oil Shale." *Oil Shale* 21, no. 3 (September 2004): 259–64.

He Zhen. "San nian lai Fushun shi de funü gongzuo" [Women's work in Fushun over the past three years]. *Fushun ribao*, October 1, 1952.

Hecht, Gabrielle. *The Radiance of France: Nuclear Power and National Identity after World War II*. Cambridge, MA: MIT Press, 1998.

Heenan, Patrick, ed. *The Japan Handbook*. London: Fitzroy Dearborn, 1998.

Hein, Laura E. *Fueling Growth: The Energy Revolution and Economic Policy in Postwar Japan*. Cambridge, MA: Council on East Asian Studies, Harvard University, 1990.

Heroy, W. B. "Petroleum." In *Ores and Industry in the Far East: The Influence of Key Mineral Resources on the Development of Oriental Civilization*, by H. Foster Bain, rev. and enl. ed., 118–43. New York: Council on Foreign Relations, 1933. First published 1927.

Hershatter, Gail. *The Gender of Memory: Rural Women and China's Collective Past*. Berkeley: University of California Press, 2011.

Hess, Christian A. "Big Brother Is Watching: Local Sino-Soviet Relations and the Building of New Dalian, 1945–55." In *Dilemmas of Victory: The Early Years of the People's Repub-*

lic of China, edited by Jeremy Brown and Paul G. Pickowicz, 160–83. Cambridge, MA: Harvard University Press, 2007.

Hevia, James. *English Lessons: The Pedagogy of Imperialism in Nineteenth-Century China.* Durham, NC: Duke University Press, 2003.

Heynen, Robert, and Emily van der Meulen, eds. *Making Surveillance States: Transnational Histories.* Toronto: University of Toronto Press, 2019.

Hilson, Gavin M., ed. *The Socio-Economic Impacts of Artisanal and Small-Scale Mining in Developing Countries.* Lisse: A. A. Balkema, 2003.

Hilton, Isabel. "How China's Big Overseas Initiative Threatens Global Climate Progress." *Yale Environment 360,* January 3, 2019.

Hirata, Koji. "From the Ashes of Empire: The Reconstruction of Manchukuo's Enterprises and the Making of China's Northeastern Industrial Base, 1948–1952." In *Overcoming Empire in Post-Imperial East Asia: Repatriation, Redress and Rebuilding,* edited by Barak Kushner and Sherzod Muminov, 147–62. London: Bloomsbury, 2019.

Hirata, Koji. "Steel Metropolis: Industrial Manchuria and the Making of Chinese Socialism, 1916–1964." PhD diss., Stanford University, 2018.

Hirzel, Thomas, and Nanny Kim, eds. *Metals, Monies, and Markets in Early Modern Societies: East Asian and Global Perspectives.* Berlin: LIT, 2008.

Hisada Keita. "Chikuhō tankō hatten ki ni okeru gijutsusha no jitsuzō: kōgyōsha Matsuda Buichirō no keieishateki sokumen" [A true picture of technical experts in the development of the Chikuhō coal mines: Mining industrialist Matsuda Buichirō's managerial profile]. Graduate research paper, Kyushu University, 2012.

Hoar, H. M. *The Coal Industry of the World.* Washington, DC: United States Government Printing Office, 1930.

Hokimoto Hiromi. "Bujun saishū zanryū ki" [Recollections on being left behind in Fushun at the end]. In *Mantetsu shain shūsen kiroku,* edited by Mantetsu kai, 675–81. Tokyo: Mantetsu kai, 1996.

Hokimoto Hiromi. "Saishū zanryūsha no kiroku" [Records of one left behind at the end]. In *Bujun tankō shūsen no ki,* edited by Mantetsu Tōkyō Bujun kai, 259–68. Tokyo: Mantetsu Tōkyō Bujun kai, 1973.

Honda sei. "Santō kūrii ni tsuite" [Regarding the coolies from Shandong]. *Bujun 11* (September 1913): 26–29. Fushun Mining Group Documents Room, Fushun.

Hornibrook, Jeff. *A Great Undertaking: Mechanization and Social Change in a Late Imperial Chinese Coalmining Community.* Albany: State University of New York Press, 2015.

Hou Defeng, ed. *Zhongguo kuangye jiyao, di si ci* [General statement of China's mining industry, 4th issue]. Beiping: Shiye bu Dizhi diaocha suo, 1932.

Hou Defeng, ed. *Zhongguo kuangye jiyao, di wu ci* [General statement of China's mining industry, 5th issue]. Beiping: Shiye bu Dizhi diaocha suo, 1935.

Hou Li. *Building for Oil: Daqing and the Formation of the Chinese Socialist State.* Cambridge, MA: Harvard University Asia Center, 2018.

Hsia, Ching-Lin. *The Status of Shanghai: A Historical Review of the International Settlement, Its Future Development and Possibilities through Sino-Foreign Co-operation.* Shanghai: Kelly and Walsh, 1929.

Hsia, C. T. *A History of Modern Chinese Fiction.* 3rd ed. Introduction by David Der-wei Wang. Bloomington: Indiana University Press, 1999. First published 1961.

Hu, Cheng. "Quarantine Sovereignty during the Pneumonic Plague in Northeast China (November 1910–April 1911)." *Frontiers of History in China* 5, no. 2 (June 2010): 294–339.

Hu Shengsan, Liu Xiuyuan, and Cheng Yuqi. "Caimei shi shang de jishu geming—woguo zongcai fazhan 40 nian" [A technical revolution in coal-mining history: The development of fully mechanized mining in our country over forty years]. *Meitan xuebao* 35, no. 11 (November 2010): 1769–71.

Hu Shih. "The Civilizations of the East and the West." In *Whither Mankind: A Panorama of Modern Civilization*, edited by Charles A. Beard, 25–41. New York: Longmans, Green, 1928.

Huagong. "Ximei" [Coal washing]. *Renmin ribao*, January 24, 1959.

Huang, Yasheng. *Capitalism with Chinese Characteristics: Entrepreneurship and the State.* Cambridge: Cambridge University Press, 2008.

Huber, Matthew T. *Lifeblood: Oil, Freedom, and the Forces of Capital.* Minneapolis: University of Minnesota Press, 2013.

Human Rights Watch. *Paying the Price: Worker Unrest in Northeast China.* New York: Human Rights Watch, 2002.

Hustrulid, William A. *Blasting Principles for Open Pit Mining.* Rotterdam: A. A. Balkema, 1999.

Iguchi, Haruo. *Unfinished Business: Ayukawa Yoshisuke and U.S.-Japan Relations, 1937–1953.* Cambridge, MA: Harvard University Asia Center, 2003.

Iiduka Yasushi. "Mantetsu Bujun oirushēru jigyō no kigyōka to sono hatten" [The commercialization of Mantetsu's Fushun shale oil industry and its development]. *Ajia keizai* 44, no. 8 (August 2003): 2–32.

Ikonnikov, A. B. *The Coal Industry of China.* Canberra: Australian National University, 1977.

Immerwahr, Daniel. *How to Hide an Empire: A History of the Greater United States.* New York: Farrar, Straus and Giroux, 2019.

Imperial Japanese Government Railways. *An Unofficial Guide to Eastern Asia: Trans-Continental Connections between Europe and Asia.* Vol. 1, *Manchuria and Chōsen.* Tokyo: n.p., 1913.

Imura. "Reconstruction of Manchuria and Mongolia with Their Natural Resources." *Manchuria Daily News Monthly Supplement*, February 1, 1932, 10–11.

Imura, Hidefumi, and Miranda A. Schreurs, eds. *Environmental Policy in Japan.* Cheltenham: Edward Elgar, 2005.

Imura, Hidefumi, and Miranda A. Schreurs. "Learning from Japanese Environmental Management Experiences." In *Environmental Policy in Japan*, edited by Hidefumi Imura and Miranda A. Schreurs, 1–14. Cheltenham: Edward Elgar, 2005.

Inglis, Sam. "Tibetan Headwaters of the Yangtze under Threat." GlacierHub, July 12, 2016. http://glacierhub.org/2016/07/12/tibetan-headwaters-of-the-yangtze-under -threat/.

Inhara, K., ed. *The Japan Year Book, 1933.* Tokyo: Kenkyusha, 1933.

Inhara, K., ed. *The Japan Year Book, 1935.* Tokyo: Foreign Affairs Association of Japan, 1935.

Inoue Tadashirō. "Manshū no tankō" [The coal mines of Manchuria]. *Ōsaka mainichi shinbun*, August 7, 1919.

Inoue Toshirō. *Bujun tankō hōkokusho* [Report on the Fushun coal pits] (1913). Unpublished report, Earth Systems Library, Department of Engineering, University of Tokyo, Tokyo.

Inouye, Kinosuke. "Geology of the Southern Part of the Province of Hsinking, China." *Bulletin of the Imperial Geological Survey of Japan* 18, no. 2 (December 1905): 1–49.

Inouye, Kinosuke. "Japan's Position in the World of Coal." *Trans-Pacific* 4, no. 2 (February 1921): 51–55.

Inskeep, Steve, and Ashley Westerman. "Why Is China Placing a Global Bet on Coal?" *NPR*, April 29, 2019. https://www.npr.org/2019/04/29/716347646/why-is-china -placing-a-global-bet-on-coal.

Irfan, Umair. "Why the World's Third-Largest Economy Is Still Betting on Coal." *Vox*, February 18, 2020. https://www.vox.com/2020/2/18/21128205/climate-change-japan -coal-energy-emissions-pikachu.

Iriye, Akira, ed. *The Chinese and the Japanese: Essays in Political and Cultural Interactions.* Princeton, NJ: Princeton University Press, 1980.

Iriye, Akira, and William Wei, eds. *Essays in the History of the Chinese Republic.* Urbana, IL: University of Illinois Center for Asian Studies, 1983.

Isett, Christopher Mills. *State, Peasant, and Merchant in Qing Manchuria, 1644–1862.* Stanford, CA: Stanford University Press, 2006.

Ishibashi, Koki. "Present Status of Properties and Refining of Fushun Shale Oil." In *World Petroleum Congress Proceedings*, edited by World Petroleum Congress, 183–212. London: Applied Science, 1937.

Ishiwata Nobutarō. "Bujun tankō" [The Fushun coal pits]. *Chikuhō sekitan kōgyō kumiai geppō* 2, no. 29 (November 1906): 33–40.

Ishiwata Nobutarō. "Bujun tankō ni tsuite no jokan" [Impressions of the Fushun coal pits]. *Chikuhō sekitan kōgyō kumiai geppō* 3, no. 33 (March 1907): 4–9.

Ishiwata Nobutarō. "Nenryō jūōdan (sono jū)" [The warp and weft of fuel (number ten)]. *Nenryō kyōkai shi* 12, no. 10 (October 1933): 1286–90.

Ishiwata Nobutarō. "Waga kuni shōrai no sekitan mondai" [Our country's future coal question]. *Nenryō kyōkai shi* 3, no. 5 (May 1924): 287–312.

Islam, S. Nazrul, and John Winkel. "Climate Change and Social Inequality." DESA Working Paper 152, October 2017.

Itō Kobun. *Caitan jishu rumen* [An introduction to coal-mining technology]. Translated from the Japanese by Zhang Dounan. Fushun: Fushun kuangwu ju mishuchu shiliao ke, 1949.

Jakes, Aaron G. "Boom, Bugs, Bust: Egypt's Ecology of Interest, 1882–1914." *Antipode* 49, no. 4 (September 2017): 1035–59.

Japanese Chamber of Commerce of New York. *Manchukuo: The Founding of the New State in Manchuria.* New York: Japanese Chamber of Commerce, 1933.

Japan Statistical Association. "Ichiji enerugii kokunai kyōkū (Shōwa 28-nendo—Heisei 12-nendo)" [Domestic supply of primary energy (1953–2003)]. *Nihon chōki tōkei sōran* [Overview of Japan's long-term statistics]. *JapanKnowledge Lib.* 2020. http://japanknowledge.com.

Jasanoff, Sheila. "The Idiom of Co-production." In *States of Knowledge: The Co-production of Science and Social Order*, edited by Sheila Jasanoff, 1–12. London: Routledge, 2004.

Jasanoff, Sheila, ed. *States of Knowledge: The Co-production of Science and Social Order.* London: Routledge, 2004.

Jensen, Lionel M., and Timothy B. Weston, eds. *China's Transformations: The Stories Beyond the Headlines.* Lanham, MD: Rowman and Littlefield, 2007.

Jevons, William Stanley. *The Coal Question: An Inquiry Concerning the Progress of the Nation, and the Probable Exhaustion of Our Coal Mines.* 3rd ed. London: Macmillan, 1906. First published 1865.

Ji Min. "Zhongguo kuangzhang he Riben gongchengshi de youyi" [The friendship between a Chinese colliery manager and a Japanese engineer]. *Wenshi jinghua* 213 (February 2008): 38–44.

Jiang, Lijing. "The Socialist Origins of Artificial Carp Reproduction in Maoist China." *Science, Technology, and Society* 22, no. 1 (March 2017): 59–77.

Jiang Xing. "Laohutai kuang fangsong baoan jiaoyu, dai yanhuo ru jing xianxiang zhuyue zengduo" [Laohutai mine slackened safety education, the practice of bringing cigarettes down into the pits increasing month by month]. *Fushun ribao*, October 7, 1954.

Jin Huihua, ed. *Zhongguo kuangye jiyao, di liu ci, Xi'nan qu* [General statement of China's mining industry, 6th issue, the Southwest]. Chongqing: Jingji bu Zhongyang dizhi diaocha suo, 1941.

Johnson, Bob. "Energy Slaves: Carbon Technologies, Climate Change, and the Stratified History of the Fossil Economy." *American Quarterly* 68, no. 4 (December 2016): 955–79.

Johnson, Chalmers. *MITI and the Japanese Miracle: The Growth of Industrial Policy, 1925–1975*. Stanford, CA: Stanford University Press, 1982.

Johnson, Edward A. "Geology of the Fushun Coalfield, Liaoning Province, Republic of China." *International Journal of Coal Geology* 14, no. 3 (March 1990): 217–36.

Jones, Christopher F. *Routes of Power: Energy and Modern America*. Cambridge, MA: Harvard University Press, 2014.

Jonsson, Fredrik Albritton. "The Origins of Cornucopianism: A Preliminary Genealogy." *Critical Historical Studies* 1, no. 1 (March 2014): 151–68.

Jonsson, Fredrik Albritton, John Brewer, Neil Fromer, and Frank Trentmann. "Introduction." In *Scarcity in the Modern World: History, Politics, Society and Sustainability, 1800–2075*, edited by Fredrik Albritton Jonsson, John Brewer, Neil Fromer, and Frank Trentmann, 1–17. London: Bloomsbury, 2019.

Jonsson, Fredrik Albritton, John Brewer, Neil Fromer, and Frank Trentmann, eds. *Scarcity in the Modern World: History, Politics, Society and Sustainability, 1800–2075*. London: Bloomsbury, 2019.

Jordan, Donald A. *China's Trial by Fire: The Shanghai War of 1932*. Ann Arbor: University of Michigan Press, 2001.

Jung, Moon-Ho. *Coolies and Cane: Race, Labor, and Sugar in the Age of Emancipation*. Baltimore: Johns Hopkins University Press, 2006.

Kado Itsumi. *Bujun tankō chōsa hōkoku* [Report on the Fushun colliery] (1914). Unpublished report, Archives of the Department of Resources and Environmental Engineering, School of Creative Science and Engineering, Waseda University.

Kajima, Morinosuke. *The Diplomacy of Japan, 1894–1922*. Vol. 3, *First World War, Paris Peace Conference, Washington Conference*. Tokyo: Kajima Institute for International Peace, 1980.

Kantō totokufu. *Manshū sangyō chōsa shiryō (kōzan)* [Survey documents on Manchurian industry (mining)]. Dairen: Kantō totokufu, 1906.

Kantō totokufu. *Manshū sangyō chōsa shiryō (shōgyō seizōgyō)* [Survey documents on Manchurian industry (commerce and manufacturing)]. Dairen: Kantō totokufu, 1906.

Kaple, Deborah A. *Dream of a Red Factory: The Legacy of High Stalinism in China*. New York: Oxford University Press, 1994.

Karl, Rebecca E. *The Magic of Concepts: History and the Economic in Twentieth-Century China*. Durham, NC: Duke University Press, 2019.

Kashima, K. "Japan Letter." *Journal of Industrial and Engineering Chemistry* 14, no. 10 (October 1922): 988.

Kaske, Elisabeth. "The Price of an Office: Venality, the Individual and the State in 19th Century China." In *Metals, Monies, and Markets in Early Modern Societies: East Asian and Global Perspectives*, edited by Thomas Hirzel and Nanny Kim, 279–304. Berlin: LIT, 2008.

Katō Kiyofumi. *Manekarezaru kokuhin: Abe Ryōnosuke* [An unwelcomed national treasure: Abe Ryōnosuke]. Tokyo: Yumani shobō, 2011.

Kawashima, Ken C. *The Proletarian Gamble: Korean Workers in Interwar Japan.* Durham, NC: Duke University Press, 2009.

Kelkar, Govind S. "The Chinese Experience of Political Campaigns and Mass Mobilization." *Social Scientist* 7, no. 5 (December 1978): 45–63.

Kharohenko, A. K. "Development of the Coal Industry in the Chinese People's Republic." Translated from the Russian. Joint Publications Research Services report no. DC-373, November 18, 1958.

Kido Chūtarō. "Bujun tandan chishitsu chōsa hōkoku" [Geological survey report of the Fushun coalfield]. *Chikuhō sekitan kōgyō kumiai geppō* 8, no. 101 (November 1912): 9–17.

King, Amy. "Reconstructing China: Japanese Technicians and Industrialization in the Early Years of the People's Republic of China." *Modern Asian Studies* 50, no. 1 (January 2016): 141–74.

Kingsberg, Miriam. *Moral Nation: Modern Japan and Narcotics in Global History.* Berkeley: University of California Press, 2014.

Kingston, Ewan. "Climate Justice and Temporally Remote Emissions." *Social Theory and Practice* 40, no. 2 (April 2014): 281–303.

Kinzley, Judd C. *Natural Resources and the New Frontier: Constructing Modern China's Borderlands.* Chicago: University of Chicago Press, 2018.

Kinzley, William Dean. "Japan in the World of Welfare Capitalism: Imperial Railroad Experiments with Welfare Work." *Labor History* 47, no. 2 (May 2006): 189–212.

Kirby, William C. "Engineering China: Birth of the Developmental State, 1928–1937." In *Becoming Chinese: Passages to Modernity and Beyond*, edited by Wen-hsin Yeh, 137–60. Berkeley: University of California Press, 2000.

Kirby, William C. *Germany and Republican China.* Stanford, CA: Stanford University Press, 1984.

Kirby, William C. "Kuomintang China's 'Great Leap Outward': The 1936 Three Year Plan for Industrial Development." In *Essays in the History of the Chinese Republic*, edited by Akira Iriye and William Wei, 43–66. Urbana, IL: University of Illinois Center for Asian Studies, 1983.

Kirby, William C., ed. *The People's Republic of China at 60: An International Assessment.* Cambridge, MA: Harvard University Asia Center, 2011.

Kirby, William C. "Planning Postwar China: China, the United States and Postwar Economic Strategies, 1941–1948." In *Proceedings of Conference on Dr. Sun Yat-sen and Modern China*, 3:216–48. Taipei: Compilation Committee, 1986.

Kirby, William C. "Technocratic Organization and Technological Development in China: The Nationalist Experience and Legacy, 1928–1953." In *Science and Technology in Post-Mao China*, edited by Denis Fred Simon and Merle Goldman, 23–43. Cambridge, MA: Council on East Asian Studies, Harvard University, 1989.

Kita. "Sekitan kai no jukyū kan" [A look at the supply and demand of the coal sector]. *Ōsaka mainichi shinbun*, August 20–23, 1919.

Kitamura Yoshio. "Bujun tankō kara kaette Chūkyō kōgyō no gen dankai" [Returning from the Fushun colliery; the current level of the Chinese Communists' mining industry]. *Taiheiyō mondai* 11, no. 1/2 (February 1955): 1–9.

Kobayashi Hideo. *Mantetsu Chōsabu: "ganso shinku tanku" no tanjō to hōkai* [Mantetsu's Research Section: The rise and fall of "the first think tank"]. Tokyo: Heibonsha, 2005.

Kobayashi Kyūhei. "Gyoyu yori sekiyu seizō shigen hōkoku narabi ni sekiyu no seiin nit suite" [The manufacture of petroleum from fish oil and the origins of petroleum]. *Kōgyō kagaku zasshi* 24, no. 1 (January 1921): 1–26.

Kobori, Satoru. "The Development of Energy-Conservation Technology in Japan, 1920–70: An Analysis of Energy-Intensive Industries and Energy-Conservation Policies." In *Economic Development and Environmental History in the Anthropocene: Perspectives on Asia and Africa*, edited by Gareth Austin, 219–43. London: Bloomsbury, 2017.

Kobori Satoru. *Nihon no enerugii kakumei: shigen shōkoku no kingendai* [Japan's energy revolution: A resource-weak country in the modern era]. Nagoya: Nagoya daigaku shuppankai, 2010.

Koga, Yukiko. *The Inheritance of Loss: China, Japan, and the Political Economy of Redemption after Empire*. Chicago: University of Chicago Press, 2016.

Kojima Seiichi. *Nenryō dōryoku keizai dokuhon* [A reader on the economy of fuel and power]. Tokyo: Chikura shobō, 1937.

Kokugakuin daigaku toshokan chōsaka. *Inoue Tadashirō monjo mokuroku* [Index to the Inoue Tadashirō archives]. Tokyo: Kokugakuin daigaku toshokan, 1992.

Köll, Elisabeth. *Railroads and the Transformation of China*. Cambridge, MA: Harvard University Press, 2019.

Kopp, Otto C. "Coal." In *Encyclopaedia Britannica*, November 13, 2020. https://www.britannica.com/science/coal-fossil-fuel.

Koshizawa Akira. *Zhongguo Dongbei dushi jihua shi* [A history of urban planning in China's Northeast]. Translated from the Japanese by Huang Shimeng. Taipei: Dajia chubanshe, 1987. First published 1982.

Kotkin, Stephen, and David Wolff, eds. *Rediscovering Russia in Asia: Siberia and the Russian Far East*. Armonk, NY: M. E. Sharpe, 1995.

Koyama Shizuko. *Ryōsai Kenbo: The Educational Ideal of 'Good Wife, Wise Mother' in Modern Japan*. Translated from the Japanese by Stephen Filler. Leiden: Brill, 2015.

Kratoska, Paul H., ed. *Asian Labor in the Wartime Japanese Empire: Unknown Histories*. Armonk, NY: M. E. Sharpe, 2005.

Kratoska, Paul H. "Labor Mobilization in Japan and the Japanese Empire." In *Asian Labor in the Wartime Japanese Empire: Unknown Histories*, edited by Paul H. Kratoska, 1–22. Armonk, NY: M. E. Sharpe, 2005.

Kubo, Tohru. "The Hydraulic Stowage Mining System in Fushun Colliery, South Manchuria." In *Proceedings, World Engineering Congress, Tokyo 1929*, vol. 37, *Mining and Metallurgy*, pt. 5, *Mineral Resources and Mining*, edited by World Engineering Congress, 221–28. Tokyo: Kogakkai, 1931.

Kubo Tōru. *Bujun tankō hōkoku* [A report on the Fushun colliery] (1912). Unpublished report, Earth Systems Library, Department of Engineering, University of Tokyo, Tokyo.

Kubo Tōru. "Bujun tankō ni hattatsu seru shase jūten tankutsu hō" [The hydraulic stowage method developed at the Fushun colliery]. PhD diss., Tokyo Imperial University, 1932.

Kubo Tōru. "Bujun tankō ni okeru saitanki ni tsuite" [Coal getting machines used in Fushun colliery, South Manchuria]. *Nihon kōgyōkai shi* 50, no. 587 (March 1934): 264–79.

Kubo Tōru. "Bujun tankō no saitan jigyō" [The coal-mining business of the Fushun colliery]. *Kōsei* 119 (October 1929): 27–33.

Kubo Tōru. "Saitan no gōrika" [The rationalization of coal mining]. *Sekitan jihō* 7, no. 1 (January 1932): 4–6.

Kubo Tōru. *TōA no sekitan hōsaku* [General plan for East Asia's coal]. Tokyo: TōA shinsho, 1941.

Kuboyama Yuzo. *Saishin tankō kōgaku* [The latest in mining engineering]. 3 vols. Tokyo: Kōronsha.

Kume Kanae. "Heichōzan jiken to sono shūmatsu" [The Pingdingshan incident and its conclusion]. In *Bujun tankō shūsen no ki,* edited by Mantetsu Tōkyō Bujun kai, 72–82. Tokyo: Mantetsu Tōkyō Bujun kai, 1973.

Kunisawa Kiyoko. *Katei ni okeru nenryō no setsuyaku* [Fuel economy in the home]. Tokyo: Ōkura shoten, 1920.

Kuribayashi Kurata. "Konwakai ni shimon jisshi ni tsuite" [Discussion council on the implementation of fingerprinting]. In *Mantie midang: Mantie yu laogong* [Mantetsu's secret archives: Mantetsu and laborers], vol. 1, compiled by Liaoning sheng dang'anguan, 50–53. Guilin: Guixi shifan daxue chubanshe, 2003.

Kurita, J. "Oil Consumption Index to Japan's Industry." *Trans-Pacific* 7, no. 1 (July 1922): 43–50.

Kushner, Barak. *Men to Devils, Devils to Men: Japanese War Crimes and Chinese Justice.* Cambridge, MA: Harvard University Press, 2015.

Kushner, Barak, and Sherzod Muminov, eds. *Overcoming Empire in Post-Imperial East Asia: Repatriation, Redress and Rebuilding.* London: Bloomsbury, 2019.

Lattimore, Owen. *Manchuria: Cradle of Conflict.* New York: Macmillan, 1932.

Lean, Eugenia. *Vernacular Industrialism in China: Local Innovation and Translated Technologies in the Making of a Cosmetics Empire, 1900–1940.* New York: Columbia University Press, 2020.

LeCain, Timothy J. *Mass Destruction: The Men and the Giant Mines That Wired America and Scarred the Planet.* New Brunswick, NJ: Rutgers University Press, 2009.

Lee, Ching Kwan. *Against the Law: Labor Protests in China's Rustbelt and Sunbelt.* Berkeley: University of California Press, 2007.

Lee, Chong-Sik. *Revolutionary Struggle in Manchuria: Chinese Communism and Soviet Interest, 1922–1945.* Berkeley: University of California Press, 1983.

Lee, En-Han. "China's Response to Foreign Investment in Her Mining Industry." *Journal of Asian Studies* 28, no. 1 (November 1968): 55–76.

Lee, Leo Ou-fan. *The Romantic Generation of Modern Chinese Writers.* Cambridge, MA: Harvard University Press, 1973.

Leech, Brian James. *The City That Ate Itself: Butte, Montana and Its Expanding Berkeley Pit.* Reno: University of Nevada Press, 2018.

Lei Lifang, Fang Yibing, and Qian Wei. "Kangzhan shiqi houfang yejin ranliao de yanjiu: yi Jingji bu Kuangye yanjiusuo weili" [Metallurgical research in the interior during the War of Resistance: The example of the Ministry of Economic Affairs' Mining and Metallurgy Research Bureau]. *Zhongguo keji shi zazhi* 37, no. 3 (September 2016): 271–83.

Lei, Sean Hsiang-lin. "Sovereignty and the Microscope: Constituting Notifiable Infectious Disease and Containing the Manchurian Plague (1910–11)." In *Health and Hygiene in Chinese East Asia: Politics and Publics in the Long Twentieth Century,* edited by Angela Ki Che Leung and Charlotte Furth, 73–106. Durham, NC: Duke University Press, 2010.

Leiyu. "Minguo 19 niandu Dongbei meikuangye niaokan" [A bird's eye view of the coal-mining industry in the Northeast in 1930]. *Zhongdong banyuekan* 2, no. 18 (October 1931): 5–13.

Leland Stanford Junior University. *Graduate Study, 1915–16.* Stanford, CA: Stanford University Press, 1915.

Lenin, Vladimir I. *Collected Works.* 4th English ed. Vol. 30. Moscow: Progress, 1965.

Leung, Angela Ki Che, and Charlotte Furth, eds. *Health and Hygiene in Chinese East Asia:*

Politics and Publics in the Long Twentieth Century. Durham, NC: Duke University Press, 2010.

Levine, Steven I. *Anvil of Victory: The Communist Revolution in Manchuria, 1945–1948*. New York: Columbia University Press, 1987.

Li, Cheng. *China's Leaders: The New Generation*. Lanham, MD: Rowman and Littlefield, 2001.

Li Cheng and Lynn White. "The Fifteenth Central Committee of the Chinese Communist Party: Full-Fledged Technocratic Leadership with Partial Control by Jiang Zemin." *Asian Survey* 38, no. 3 (March 1998): 231–64.

Li Cheng and Lynn White. "The Thirteenth Central Committee of the Chinese Communist Party: From Mobilizers to Managers." *Asian Survey* 28, no. 4 (April 1988): 371–99.

Li, Danke. *Echoes of Chongqing: Women in Wartime China*. Urbana, IL: University of Illinois Press, 2010.

Li Enhan. *WanQing de shouhui kuangquan yundong* [Late-Qing mining rights recovery movement]. Taipei: Zhongyang yanjiuyuan Jindaishi yanjiusuo, 1963.

Li Fuchun. "Guanyu fazhan guomin jingji de di yi ge wu nian jihua de baogao" [Report regarding the First Five-Year Plan to develop the national economy] (July 5–6, 1955). In *Zhonghua renmin gongheguo fazhan guomin jingji de di yi ge wu nian jihua, 1953–1957*, 157–238. Beijing: Renmin chubanshe, 1955.

Li, Huaiyin. *Village China under Socialism and Reform: A Micro-History, 1948–2008*. Stanford, CA: Stanford University Press, 2009.

Li Minghe. "Shi nian lai zhi meikuangye" [The coal-mining industry over the past decade]. In *Shi nian lai zhi Zhongguo jingji*, edited by Tan Xihong, sec. 1. Shanghai: Zhonghua shuju, 1948.

Li, Yifei, and Judith Shapiro. *China Goes Green: Coercive Environmentalism for a Troubled Planet*. Cambridge: Polity, 2020.

Li Zixiang. "Wai mei qingqiao yu woguo mei gongye zhi qiantu" [The dumping of foreign coal and the future of our country's coal industry]. *Shenbao yuekan* 2, no. 11 (November 1933): 41–50.

Lindee, M. Susan. *Rational Fog: Science and Technology in Modern War*. Cambridge, MA: Harvard University Press, 2020.

Link, Perry, Richard P. Madsen, and Paul G. Pickowicz, eds. *Restless China*. Lanham, MD: Rowman and Littlefield, 2013.

Link, Stefan J. *Forging Global Fordism: Nazi Germany, Soviet Russia, and the Contest over the Industrial Order*. Princeton, NJ: Princeton University Press, 2020.

Liu, Andrew B. *Tea War: A History of Capitalism in China and India*. New Haven, CT: Yale University Press, 2020.

Liu Dianxin. "Fushun meikuang kaicai fangfa de yanbian" [Evolution of mining methods at the Fushun colliery]. *FMBN*, 67–77.

Liu Shaoxin. "Wei Xijiu chuangzao xiaoyinji" [Wei Xijiu creates noise-canceling device]. *Fushun ribao*, January 6, 1953.

Lohmann, Larry. "Marketing and Making Carbon Dumps: Commodification, Calculation and Counterfactuals in Climate Change Mitigation." *Science as Culture* 14, no. 3 (September 2005): 203–35.

Long, Priscilla. *Where the Sun Never Shines: A History of America's Bloody Coal Industry*. New York: Paragon House, 1989.

Low, Morris F. "Mapping the Japanese Empire: Colonial Science in Shanghai and Manchuria." *Papers of the British Association for Korean Studies* 6 (1996): 134–49.

Lü Shifang. "Sulian zhuanjia gei Fushun meikuang bumen de bangzhu" [The assistance that Soviet experts have been giving the departments at the Fushun colliery]. *Fushun ribao*, February 14, 1951.

Lu, Sidney Xu. *The Making of Japanese Settler Colonialism: Malthusianism and Trans-Pacific Migration, 1868–1961.* Cambridge: Cambridge University Press, 2019.

Lucier, Paul. *Scientists and Swindlers: Consulting on Coal and Oil in America, 1820–1880.* Baltimore: Johns Hopkins University Press, 2008.

Luo Fujun. "Kong Xiangrui xiaozu zenyang zuo dao san nian duo anquan shengchan" [Kong Xiangrui's small group able to accomplish more than three years of safe production]. *Fushun ribao*, March 15, 1952.

Lynch, Catherine, Robert B. Marks, and Paul G. Pickowicz, eds. *Radicalism, Revolution, and Reform in Modern China: Essays in Honor of Maurice Meisner.* Lanham, MD: Lexington Books, 2011.

MacFarquhar, Roderick. *The Origins of the Cultural Revolution.* Vol. 2, *The Great Leap Forward, 1958–1960.* New York: Columbia University Press, 1983.

MacMurray, John V. A., comp. *Treaties and Agreements with and Concerning China, 1894–1912.* Vol. 1. New York: Oxford University Press, 1921.

Maier, Charles S. *Among Empires: American Ascendency and Its Predecessors.* Cambridge, MA: Harvard University Press, 2006.

Maier, Charles S. "Between Taylorism and Technocracy: European Ideologies and the Vision of Industrial Productivity in the 1920s." *Journal of Contemporary History* 5, no. 2 (April 1970): 27–61.

Maier, Charles S. *Leviathan 2.0: Inventing Modern Statehood.* Cambridge, MA: Harvard University Press, 2012.

Malm, Andreas. *Fossil Capital: The Rise of Steam Power and the Roots of Global Warming.* New York: Verso, 2016.

Manhattan Engineer District. "The Atomic Bombings of Hiroshima and Nagasaki" (June 29, 1946). Atomic Archive. http://www.atomicarchive.com/Docs/MED/index.shtml.

Manshikai. *Manshū kaihatsu yonjūnen shi* [A forty-year history of Manchuria's development]. Vol. 2. Tokyo: Manshū kaihatsu yonjūnen shi kankōkai, 1965.

Manshū shi kenkyūkai, ed. *Nihon teikokushugi ka no Manshū: "Manshūkoku" seiritsu zengo no keizai kenkyū* [Manchuria under Japanese imperialism: research on the economy before and after the establishment of "Manchukuo"]. Tokyo: Ochanomizu shobō, 1972.

Manshū tankō kabushiki gaisha. *Manshū tankō kabushiki gaisha annai* [A guide to the Manchurian Coal Company]. N.p.: Manshū tankō kabushiki gaisha, n.d.

Mantetsu. "Tōkei nenpō" [Annual statistics] (1937). In *MSZ*, no. 2, 398–99.

Mantetsukai, ed. *Mantetsu shain shūsen kiroku* [Records of Mantetsu employees at the end of the war]. Tokyo: Mantetsukai, 1996.

Mantetsu rōmuka. *Minami Manshū kōzan rōdō jijō* [Mining labor conditions in southern Manchuria]. Dairen: Minami Manshū tetsudō kabushiki gakisha, 1931.

Mantetsu Tōkyō Bujun kai, ed. *Bujun tankō shūsen no ki* [Recollections of the Fushun colliery at the end of the war]. Tokyo: Mantetsu Tōkyō Bujun kai, 1973.

Manuel, Jeffrey T. *Taconite Dreams: The Struggle to Sustain Mining on Minnesota's Iron Range, 1915–2000.* Minneapolis: University of Minnesota Press, 2015.

Mao zhuxi de hao zhanshi Lei Feng jinianguan, comp. *Lei Feng riji* [The diary of Lei Feng]. N.p.: Mao zhuxi de hao zhanshi Lei Feng jinianguan, 1968. First published 1963.

Marks, Robert B. "Chinese Communists and the Environment." In *Radicalism, Revolution, and Reform in Modern China: Essays in Honor of Maurice Meisner*, edited by Catherine

Lynch, Robert B. Marks, and Paul G. Pickowicz, 105–32. Lanham, MD: Lexington Books, 2011.

Marx, Leo. *The Machine in the Garden: Technology and the Pastoral Ideal in America.* New York: Oxford University Press, 1964.

Masuda Hajimu. *Cold War Crucible: The Korean Conflict and the Postwar World.* Cambridge, MA: Harvard University Press, 2015.

Matoba Chū. "Bujun tankō ni tsuki" [Regarding the Fushun coal pits]. *Chikuhō sekitan kōgyō kumiai geppō* 3, no. 36 (June 1907): 1–8.

Matsuda Buichirō. "Shisui" [Prospecting]. In *Sekitan kōgyō ronshū* [A collection of papers on coal mining], edited by Takanoe Mototarō, 3.1–3.3. Fukuoka: Hatsubaijo sekizenkan shiten, 1910.

Matsuda Junkichi. "Matsuda Buichirō shōden" [A biographical sketch of Matsuda Buichirō] (1996). Typescript, Matsuda ka shiryō, E 11, Manuscript Library, Kyushu University.

Matsuda Kamezō. *Mantetsu chishitsu chōsajo shiki* [Private records of Mantetsu's Geological Research Institute]. Tokyo: Hakueisha, 1990.

Matsumoto Toshirō. *"Manshūkoku" kara shin Chūgoku e: Anzan tekkōgyō kara mita Chūgoku tōhoku no saihen katei, 1940–1954* [From "Manchukuo" to New China: Perspectives on the process of Northeast Asia's reorganization from Anshan's iron and steel industry]. Nagoya: Nagoya daigaku shuppankai, 2000.

Matsumura Takao. "Fushun meikuang gongren shitai" [Facts about workers at the Fushun colliery]. In *Mantie yu Zhongguo laogong* [Mantetsu and Chinese labor], edited by Xie Xueshi and Matsumura Takao, 316–62. Beijing: Shehui kexue wenxian chubanshe, 2003.

Matsusaka, Yoshihisa Tak. *The Making of Japanese Manchuria, 1904–1932.* Cambridge, MA: Harvard University Asia Center, 2001.

Matsusaka, Y. Tak. "Japan's South Manchuria Railway Company in Northeast China, 1906–34." In *Manchurian Railways and the Opening of China: An International History*, edited by Bruce A. Elleman and Stephen Kotkin, 37–58. Armonk, NY: M. E. Sharpe, 2010.

Matsuzawa Dentarō. *Kokubōjō oyobi sangyōjō yori mitaru kakkoku no sekiyu seisaku* [The petroleum policy of various countries from the perspective of national defense and industry]. Tokyo: Ikkyōsha, 1922.

McCormack, Gavan. *Chang Tso-lin in Northeast China, 1911–1928: China, Japan, and the Manchurian Idea.* Stanford, CA: Stanford University Press, 1977.

McDonald, Kate. *Placing Empire: Travel and the Social Imagination in Imperial Japan.* Oakland: University of California Press, 2017.

McElroy, Michael B. *Energy: Perspectives, Problems, and Prospects.* Oxford: Oxford University Press, 2010.

McInnes, William, D. B. Dowling, and W. W. Leach, eds. *The Coal Resources of the World.* Vol. 1. Toronto: Morang, 1913.

McIvor, Arthur, and Ronald Johnston. *Miners' Lung: A History of Dust Disease in British Coal Mining.* Aldershot: Ashgate, 2007.

McNeill, J. R. *Something New under the Sun: An Environmental History of the Twentieth-Century World.* New York: W. W. Norton, 2000.

McNeill, J. R., and Peter Engelke. *The Great Acceleration: An Environmental History of the Anthropocene since 1945.* Cambridge, MA: Harvard University Press, 2014.

McNeill, J. R., and George Vrtis, eds. *Mining North America: An Environmental History since 1522.* Oakland: University of California Press, 2017.

Meidu Fushun bianxie xiaozu. *Meidu Fushun* [Coal capital Fushun]. Shenyang: Liaoning renmin chubanshe, 1959.

Meisner, Maurice. *Mao's China and After: A History of the People's Republic.* 3rd ed. New York: Free Press, 1999. First published 1977.

Melosi, Martin V. *Coping with Abundance: Energy and Environment in Industrial America.* New York: Knopf, 1985.

Meng, C. U. W. "More Evidence of Japanese Brutalities at Fushun." *China Weekly Review,* January 7, 1933.

Meng Yue. "Hybrid Science versus Modernity: The Practice of the Jiangnan Arsenal, 1864–1897." *East Asian Science, Technology, and Medicine* 16 (1999): 13–52

Merkel-Hess, Kate. *The Rural Modern: Reconstructing the Self and State in Republican China.* Chicago: University of Chicago Press, 2016.

Metzler, Mark. "Japan's Postwar Social Metabolic Crisis." In *The Economic and Business History of Occupied Japan: New Perspectives,* edited by Thomas French, 31–52. London: Routledge, 2017.

Metzler, Mark. *Lever of Empire: The International Gold Standard and the Crisis of Liberalism in Prewar Japan.* Berkeley: University of California Press, 2006.

Meyskens, Covell F. *Mao's Third Front: The Militarization of Cold War China.* Cambridge: Cambridge University Press, 2020.

Miao Shutang. "Woguo meikuang he wasi zuo douzheng de chengjiu" [Achievements in our country's coal mines' struggle against gas]. *Mei* 73 (May 1954): 27–30.

Miller, Ian Jared. *The Nature of the Beasts: Empire and Exhibition at the Tokyo Imperial Zoo.* Berkeley: University of California Press, 2013.

Miller, Ian Jared, Julia Adeney Thomas, and Brett L. Walker, eds. *Japan at Nature's Edge: The Environmental Context of a Global Power.* Honolulu: University of Hawai'i Press, 2013.

Miller, Ian Jared, and Paul Warde. "Energy Transitions as Environmental Events." *Environmental History* 24, no. 3 (May 2019): 464–71.

Miller, Sally M., A. J. H. Latham, and Dennis O. Flynn, eds. *Studies in the Economic History of the Pacific Rim.* London: Routledge, 1997.

Milward, Alan S. *War, Economy, and Society, 1939–1945.* London: Allen Lane, 1977.

Mimura, Janis. *Planning for Empire: Reform Bureaucrats and the Japanese Wartime State.* Ithaca, NY: Cornell University Press, 2011.

Minami Manshū tetsudō kabushiki gaisha. *Bujun yuboketsugan jigyō rengō kyōgikai kiroku* [Records of the combined meeting on the Fushun shale oil industry]. Dairen: Minami Manshū tetsudō kabushiki gaisha, 1925.

Minami Manshū tetsudō kabushiki gaisha. *Dainijūhachi-kai eigyō hōkokusho* [Twenty-eighth company report]. Dairen: Minami Manshū tetsudō kabushiki gaisha, 1928.

Minami Manshū tetsudō kabushiki gaisha. *Economic Construction Program of Manchukuo.* New York: South Manchuria Railway Company, 1933.

Minami Manshū tetsudō kabushiki gaisha. *Katei ni okeru sekitan no takikata* [Ways of burning coal in the home]. Dairen: Minami Manshū tetsudō kabushiki gaisha, n.d.

Minami Manshū tetsudō kabushiki gaisha. *Minami Manshū tetsudō kabushiki gaisha dainiji jūnen shi* [Second ten-year history of Mantetsu]. Dairen: Minami Manshū tetsudō kabushiki gaisha, 1928.

Minami Manshū tetsudō kabushiki gaisha. *Minami Manshū tetsudō kabushiki gaisha daisanji jūnen shi* [Third ten-year history of Mantetsu]. Tokyo: Ryūkei shosha, 1976.

Minami Manshū tetsudō kabushiki gaisha. *Report on Progress in Manchuria, 1907–1928.* Dairen: South Manchuria Railway, 1929.

Minami Manshū tetsudō kabushiki gaisha Bujun tankō. *Tankō dokuhon* [A coal-mining reader]. Fushun: Bujun tankō, 1939.

Minichiello, Sharon A., ed. *Japan's Competing Modernities: Issues in Culture and Democracy, 1900–1930*. Honolulu: University of Hawai'i Press, 1998.

Ministry of Information of the Republic of China. *China after Five Years of War*. New York: Chinese News Service, 1942.

Mitchell, Kate L. *Industrialization of the Western Pacific*. New York: Institute of Pacific Relations, 1942.

Mitchell, Kate L. *Japan's Industrial Strength*. New York: Alfred A. Knopf, 1942.

Mitchell, Timothy. *Carbon Democracy: Political Power in the Age of Oil*. New York: Verso, 2011.

Mitchell, Timothy. "The Limits of the State: Beyond Statist Approaches and Their Critics." *American Political Science Review* 85, no. 1 (March 1991): 77–96.

Mitchell, Timothy. *Rule of Experts: Egypt, Techno-Politics, Modernity*. Berkeley: University of California Press, 2002.

Mitchell, Timothy. "Society, Economy, and the State Effect." In *State/Culture: State-Formation after the Cultural Turn*, edited by George Steinmetz, 76–97. Ithaca, NY: Cornell University Press, 1999.

Mitsuki Rokurō. "Jūshūnen no omoide (sono go)" [Memories on the tenth anniversary (number five)]. *Nenryō kyōkai shi* 11, no. 10 (October 1932): 1315.

Mitter, Rana. *China's War with Japan, 1937–1945: A Struggle for Survival*. London: Allen Lane, 2013.

Mitter, Rana. *The Manchurian Myth: Nationalism, Resistance, and Collaboration in Modern China*. Berkeley: University of California Press, 2000.

Miyakuchi Katashi. *Bujun tankō Ryūhō kō chōsa hōkokusho* [Report on the Longfeng mine at the Fushun colliery] (1923). Unpublished report, Archives of the Department of Resources and Environmental Engineering, School of Creative Science and Engineering, Waseda University, Tokyo.

Mizuno, Hiromi. "Introduction: A Kula Ring for the Flying Geese; Japan's Technology Aid and Postwar Asia." In *Engineering Asia: Technology, Colonial Development, and the Cold War Order*, edited by Hiromi Mizuno, Aaron S. Moore, and John DiMoia, 1–40. London: Bloomsbury, 2018.

Mizuno, Hiromi. *Science for the Empire: Scientific Nationalism in Modern Japan*. Stanford, CA: Stanford University Press, 2009.

Mizuno, Hiromi, Aaron S. Moore, and John DiMoia, eds. *Engineering Asia: Technology, Colonial Development, and the Cold War Order*. London: Bloomsbury, 2018.

Mizutani Kōtarō. "Gunbi seigen to sekiyu mondai" [The limiting of naval vessels and the oil question]. *Sekiyu jihō* 517 (January 1922): 11–14.

Mizutani Kōtarō. "Liquid Fuel and Value of Manchuria." *Manchuria Daily News*, October 19, 1930.

Mizutani Kōtarō. "Manshū ni okeru ekitai nenryō no kaiko to tenbō" [Recollections and outlooks for liquid fuels in Manchuria] (1938). In *MSZ*, no. 3, 819–23, 825–26.

Mokyr, Joel, ed. *The British Industrial Revolution: An Economic Perspective*. 2nd ed. Boulder, CO: Westview, 1999. First published 1993.

Mokyr, Joel. "Editor's Introduction: The New Economic History and the Industrial Revolution." In *The British Industrial Revolution: An Economic Perspective*, 2nd ed., edited by Joel Mokyr, 1–127. Boulder, CO: Westview, 1999. First published 1993.

Molony, Barbara. *Technology and Investment: The Prewar Japanese Chemical Industry*. Cambridge, MA: Council on East Asian Studies, Harvard University, 1990.

Mommsen, Wolfgang J., and Jürgen Osterhammel, eds. *Imperialism and After: Continuities and Discontinuities.* London: Allen and Unwin, 1986.

Moore, Aaron S. "From 'Constructing' to 'Developing' Asia: Japanese Engineers and the Formation of the Postcolonial, Cold War Discourse of Development in Asia." In *Engineering Asia: Technology, Colonial Development, and the Cold War Order,* edited by Hiromi Mizuno, Aaron S. Moore, and John DiMoia, 85–112. London: Bloomsbury, 2018.

Moore, Aaron Stephen. *Constructing East Asia: Technology, Ideology, and Empire in Japan's Wartime Era, 1931–1945.* Stanford, CA: Stanford University Press, 2013.

Moore, Aaron Stephen. "'The Yalu River Era of Developing Asia': Japanese Expertise, Colonial Power, and the Construction of the Sup'ung Dam." *Journal of Asian Studies* 72, no. 1 (February 2013): 115–39.

Moore, Jason W., ed. *Anthropocene or Capitalocene? Nature, History, and the Crisis of Capitalism.* Oakland, CA: PM Press, 2016.

Moore, Jason W. *Capitalism in the Web of Life: Ecology and Accumulation of Capital.* New York: Verso, 2015.

Morris-Suzuki, Tessa. *The Technological Transformation of Japan: From the Seventeenth to the Twenty-First Century.* Cambridge: Cambridge University Press, 1994.

Mosley, Stephen. *Chimney of the World: A History of Smoke Pollution in Victorian and Edwardian Manchester.* Cambridge: White Horse, 2001.

Mougey, Thomas. "Needham at the Crossroads: History, Politics and International Science in Wartime China (1942–1946)." *British Journal for the History of Science* 50, no. 1 (March 2017): 83–109.

Mullaney, Thomas S. "The Moveable Typewriter: How Chinese Typists Developed Predictive Text during the Height of Maoism." *Technology and Culture* 53, no. 4 (October 2012): 777–814.

Mumford, Lewis. *Technics and Civilization.* Chicago: University of Chicago Press, 2010. First published 1934.

Murakushi, Nisaburo. *Technology and Labour in Japanese Coal Mining.* Tokyo: United Nations University, 1980.

Murakushi, Nisaburo. *The Transfer of Coal-Mining Technology from Japan to Manchuria and Manpower Problems: Focusing on the Development of the Fushun Coal Mines.* Tokyo: United Nations University, 1981.

Muscolino, Micah. "Global Dimensions of Modern China's Environmental History." *World History Connected* 6, no. 1 (March 2009). http://worldhistoryconnected.press.uillinois.edu/6.1/muscolino.html.

Muscolino, Micah S. *The Ecology of War in China: Henan Province, the Yellow River, and Beyond, 1938–1950.* New York: Cambridge University Press, 2015.

Myers, Ramon H., and Herbert Bix. "Economic Development in Manchuria under Japanese Imperialism: A Dissenting View." *China Quarterly* 55 (July–September 1973): 547–59.

Naitō Yū. "Kōjō keizai yori mita sekitan no riyō" [Viewing coal use from the perspective of the factory economy]. *Nenryō kyōkai shi* 3, no. 5 (May 1924): 334–39.

Naitō Yū. "Nenryō kenkyū no igi" [The significance of fuel research]. *Nenryō kyōkai shi* 2, no. 1 (January 1923): 3–7.

Naitou Hisako. "Korean Forced Labor in Japan's Wartime Empire." In *Asian Labor in the Wartime Japanese Empire: Unknown Histories,* edited by Paul H. Kratoska, 90–98. Armonk, NY: M. E. Sharpe, 2005.

Nakagane Katsuji. "Manchukuo and Economic Development." In *The Japanese Informal*

Empire in China, 1895–1937, edited by Peter Duus, Ramon H. Myers, and Mark R. Peattie, 73–97. Princeton, NJ: Princeton University Press, 1989.

Nakamura, Takafusa. "Depression, Recovery, and War, 1920–1945." Translated from the French by Jacqueline Kaminsky. In *The Cambridge History of Japan*. Vol. 6, *The Twentieth Century*, edited by Peter Duus, 451–93. Cambridge: Cambridge University Press, 1988.

Nakamura, Takafusa. "Japan's Economic Thrust into North China, 1933–1938: Formation of the North China Development Corporation." In *The Chinese and the Japanese: Essays in Political and Cultural Interactions*, edited by Akira Iriye, 220–53. Princeton, NJ: Princeton University Press, 1980.

Nakamura, Takafusa, and Kōnosuke Odaka, eds. *The Economic History of Japan: 1600–1990*. Vol. 3, *Economic History of Japan, 1914–1955: A Dual Structure*. Translated from the Japanese by Noah S. Brannen. Oxford: Oxford University Press, 2002.

Nakamura, Takafusa, and Kōnosuke Odaka. "The Inter-war Period: 1914–1937, an Overview." In *The Economic History of Japan: 1600–1990*. Vol. 3, *Economic History of Japan, 1914–1955: A Dual Structure*, edited by Takafusa Nakamura and Kōnosuke Odaka and translated from the Japanese by Noah S. Brannen, 1–54. Oxford: Oxford University Press, 2002.

Nakazato Shigeji. "Kaigun to sekiyu" [The navy and oil]. *Sekiyu jihō* 535 (July 1923): 8–19.

Nathan, Carl F. *Plague Prevention and Politics in Manchuria, 1910–1931*. Cambridge, MA: East Asian Research Center, Harvard University, 1967.

National Bureau of Statistics of China. "6-5 Coal Balance Sheet." *China Statistical Yearbook 2008*. http://www.stats.gov.cn/tjsj/ndsj/2008/indexeh.htm.

National Bureau of Statistics of China. "7-4 Coal Balance Sheet." In *Yearly Data 1999*. http://www.stats.gov.cn/english/statisticaldata/yearlydata/YB1999e/g04e.htm.

National Bureau of Statistics of China. "7-5 Coal Balance Sheet." In *China Statistical Yearbook 2001*. http://www.stats.gov.cn/english/statisticaldata/yearlydata/YB2001e/htm/g0705e.htm.

National Bureau of Statistics of China. "7-5 Coal Balance Sheet." In *China Statistical Yearbook 2003*. http://www.stats.gov.cn/english/statisticaldata/yearlydata/yarbook2003_e.pdf.

National Bureau of Statistics of China. "7-5 Coal Balance Sheet." In *China Statistical Yearbook 2012*. http://www.stats.gov.cn/tjsj/ndsj/2012/indexeh.htm.

National Bureau of Statistics of China. "9-5 Coal Balance Sheet." In *China Statistical Yearbook 2015*. http://www.stats.gov.cn/tjsj/ndsj/2015/indexeh.htm.

Naughton, Barry. *The Chinese Economy: Transitions and Growth*. Cambridge, MA: MIT Press, 2007.

Naughton, Barry. *Growing Out of the Plan: Chinese Economic Reform, 1978–1993*. Cambridge: Cambridge University Press, 1995.

Needham, Andrew. *Power Lines: Phoenix and the Making of the Modern Southwest*. Princeton, NJ: Princeton University Press, 2014.

Needham, Joseph. "Chinese Scientists Go to War." *China at War* 14, no. 3 (March 1945): 56–60.

Needham, Joseph, and Dorothy Needham, eds. *Science Outpost: Papers of the Sino-British Science Co-operation Office (British Council Scientific Office in China), 1942–1946*. London: Pilot, 1948.

Nemet, Gregory F. *How Solar Energy Became Cheap: A Model for Low-Carbon Innovation*. London: Routledge, 2019.

Neushul, Peter, and Zuoyue Wang. "Between the Devil and the Deep Sea: C. K. Tseng, Mariculture, and the Politics of Science in Modern China." *Isis* 91, no. 1 (March 2000): 59–88.

Nishikawa Tadashi. "Bujun tankō no genjō (ni)" [Conditions at the Fushun colliery (part two)]. *Sekitan hyōron* 4, no. 11 (November 1953): 700–704.

Nishimura, H., ed. *How to Conquer Air Pollution: A Japanese Experience.* Amsterdam: Elsevier, 1989.

Nishiyama, Takashi. *Engineering War and Peace in Modern Japan, 1868–1964.* Baltimore: Johns Hopkins University Press, 2014.

Noble, David F. *Progress without People: New Technology, Unemployment, and the Message of Resistance.* Toronto: Between the Lines, 1995.

Nye, David E. *American Technological Sublime.* Cambridge, MA: MIT Press, 1994.

Nye, David E. *America's Assembly Line.* Cambridge, MA: MIT Press, 2013.

Nystrom, Eric C. *Seeing Underground: Maps, Models, and Mining Engineering in America.* Reno: University of Nevada Press, 2014.

O'Dwyer, Emer. *Significant Soil: Settler Colonialism and Japan's Urban Empire in Manchuria.* Cambridge, MA: Harvard University Asia Center, 2015.

O'Dwyer, Emer Sinéad. "People's Empire: Democratic Imperialism in Japanese Manchuria." PhD diss., Harvard University, 2007.

Ogasawara, Midori. "Bodies as Risky Resources: Japan's Colonial Identification Systems in Northeast China." In *Making Surveillance States: Transnational Histories,* edited by Robert Heynen and Emily van der Meulen, 163–85. Toronto: University of Toronto Press, 2019.

Oka, Y. "Electric Power Development in Fushun." *Far Eastern Review* 27, no. 5 (May 1931): 307–14.

Okamura Kinzō. "Bujun yuboketsugan jigyō no genjō." *Nenryō kyōkai shi* 9, no. 11 (November 1930): 1265–83.

Okunaka, K. "Fuel Problem in Japan." In *The Japan Year Book, 1927,* edited by Y. Takenobu, supplement, 30–36. Tokyo: Japan Year Book Office, 1927.

Olson, Richard G. *Scientism and Technocracy in the Twentieth Century: The Legacy of Scientific Management.* Lanham, MD: Lexington Books, 2016.

Orbach, Danny. *Curse on This Country: The Rebellious Army of Imperial Japan.* Ithaca, NY: Cornell University Press, 2017.

Oreskes, Naomi. *Why Trust Science?* Princeton, NJ: Princeton University Press, 2019.

Oreskes, Naomi, and Eric M. Conway. *The Collapse of Western Civilization: A View from the Future.* New York: Columbia University Press, 2014.

Orwell, George. *The Road to Wigan Pier.* London: Victor Gollancz, 1937.

Ōshima Yoshikiyo. "Katei nenryō ni tsuite" [Regarding household fuel]. *Nenryō kyōkai shi* 8, no. 6 (June 1926): 642–44.

Ōshima Yoshikiyo. "Ko Yonekura Kiyotsugu hakase ryakureki" [A brief personal record of the late Dr. Yonekura Kiyotsugu]. *Nenryō kyōkai shi* 10, no. 109 (October 1931), unnumbered back matter.

Osterhammel, Jürgen. "Semi-colonialism and Informal Empire in Twentieth-Century China: Towards a Framework of Analysis." In *Imperialism and After: Continuities and Discontinuities,* edited by Wolfgang J. Mommsen and Jürgen Osterhammel, 290–314. London: Allen and Unwin, 1986.

Oswald, Yannick, Anne Owen, and Julia K. Steinberger. "Large Inequality in International and Intranational Energy Footprints between Income Groups and across Consumption Categories." *Nature Energy* 5, no. 3 (March 2020): 231–39.

Paine, S. C. M. "The Chinese Eastern Railway." In *Manchurian Railways and the Opening of China: An International History*, edited by Bruce A. Elleman and Stephen Kotkin, 13–36. Armonk, NY: M. E. Sharpe, 2010.

Paine, S. C. M. *Imperial Rivals: China, Russia, and Their Disputed Frontier*. Armonk, NY: M. E. Sharpe, 1996.

Paine, S. C. M. *The Wars for Asia, 1911–1949*. New York: Cambridge University Press, 2012.

Pan Huanjing. "Zhang Shenfu furen tan Shenfu" [Madam Zhang Shenfu talks about Shenfu]. *Zhoubo* 2 (March 1946): 7.

Parenti, Christian. "Environment-Making in the Capitalocene: Political Ecology of the State." In *Anthropocene or Capitalocene? Nature, History, and the Crisis of Capitalism*, edited by Jason W. Moore, 166–84. Oakland, CA: PM Press, 2016.

Park, Hyun Ok. *Two Dreams in One Bed: Empire, Social Life, and the Origins of the North Korean Revolution in Manchuria*. Durham, NC: Duke University Press, 2005.

Park, Soon-Won. *Colonial Industrialization and Labor in Korea: The Onoda Cement Factory*. Cambridge, MA: Harvard University Asia Center, 1999.

Parthasarathi, Prasannan. *Why Europe Grew Rich and Asia Did Not: Global Economic Divergence, 1600–1850*. Cambridge: Cambridge University Press, 2011.

Patel, Kiran Klaus. *The New Deal: A Global History*. Princeton, NJ: Princeton University Press, 2016.

Patrikeeff, Felix. "Railway as Political Catalyst: The Chinese Eastern Railway and the 1929 Sino-Soviet Conflict." In *Manchurian Railways and the Opening of China: An International History*, edited by Bruce A. Elleman and Stephen Kotkin, 81–102. Armonk, NY: M. E. Sharpe, 2010.

Pauley, Edwin W. *Report on Japanese Assets in Manchuria to the President of the United States, July 1946*. Washington, DC: United States Government Printing Office, 1946.

Peattie, Mark R. "Japanese Attitudes toward Colonization, 1895–1945." In *The Japanese Colonial Empire, 1895–1945*, edited by Ramon H. Myers and Mark R. Peattie, 80–127. Princeton, NJ: Princeton University Press, 1984.

Peluso, Nancy Lee, and Michael Watts, eds. *Violent Environments*. Ithaca, NY: Cornell University Press, 2001.

Peng Zuocheng. "Zhong-Su youhao tongmeng huzhu tiaoyue dingli yi nian lai Sulian gei women na xie juti bangzhu?" [After one year of the Sino-Soviet Treaty of Friendship, Alliance, and Mutual Assistance, what specific assistance has the Soviet Union given us?]. *Fushun ribao*, February 11, 1951.

Pepper, Suzanne. "The KMT-CCP Conflict, 1945–1949." In *The Cambridge History of China*. Vol. 13, *Republican China, 1912–1949*, pt. 2, edited by John K. Fairbank and Albert Feuerwerker, 723–88. Cambridge: Cambridge University Press, 1986.

Perry, Elizabeth J. *Anyuan: Mining China's Revolutionary Tradition*. Berkeley: University of California Press, 2012.

Perry, Elizabeth J. "Masters of the Country? Shanghai Workers in the Early People's Republic." In *Dilemmas of Victory: The Early Years of the People's Republic of China*, edited by Jeremy Brown and Paul G. Pickowicz, 59–79. Cambridge, MA: Harvard University Press, 2007.

Perry, Elizabeth J. *Rebels and Revolutionaries in North China, 1845–1945*. Stanford, CA: Stanford University Press, 1980.

Perry, Elizabeth J. "The Red Spears Reconsidered: An Introduction." In *The Red Spears, 1916–1949*, by Tai Hsüan-chih, translated from the Chinese by Ronald Suleski, 1–15. Ann Arbor: Center for Chinese Studies, University of Michigan, 1985.

Pietz, David A. *Engineering the State: The Huai River and Reconstruction in Nationalist China, 1927–1937.* New York: Routledge, 2002.

Piper, Liza. *The Industrial Transformation of Subarctic Canada.* Vancouver: UBC Press, 2009.

Podobnik, Bruce. *Global Energy Shifts: Fostering Sustainability in a Turbulent Age.* Philadelphia: Temple University Press, 2005.

Pomeranz, Kenneth. *The Great Divergence: China, Europe, and the Making of the Modern World Economy.* Princeton, NJ: Princeton University Press, 2000.

Pomeranz, Kenneth. "The Great Himalayan Watershed: Agrarian Crisis, Mega-Dams and the Environment." *New Left Review* 58 (July/August 2009): 5–39.

Pong, David, and Edmund S. K. Fung, eds. *Ideal and Reality: Social and Political Change in Modern China.* Lanham, MD: University Press of America, 1985.

Powell, J. B. "The Marvellous Fushun Colliery." *China Weekly Review,* November 7, 1931.

Priest, Tyler. "Hubbert's Peak: The Great Debate over the End of Oil." *Historical Studies in the Natural Sciences* 44, no. 1 (February 2014): 37–79.

Pritchard, Sara B. *Confluence: The Nature of Technology and the Remaking of Rhône.* Cambridge, MA: Harvard University Press, 2011.

Putnam, Robert D. "Elite Transformation in Advanced Industrial Societies: An Empirical Assessment of the Theory of Technocracy." *Comparative Political Studies* 10, no. 3 (October 1977): 383–412.

Qi Kechang. "Yingjie hong wuyue, jinyibu gaohao laodong jingsai" [Welcome Red May, do a better job with labor competitions]. *Fushun ribao,* May 1, 1951.

Qi Weiping and Wang Jun. "Guanyu Mao Zedong 'chao Ying gan Mei' sixiang yanbian jieduan de lishi kaocha" [A historical inquiry into the development of Mao Zedong's idea of "surpassing Britain and catching up with the United States"]. *Shixue yuekan* (February 2002): 66–71.

Qian Nanjing Guomin zhengfu Sifa xingzheng bu. *Minshi xiguan diaocha baogao lu* [An investigative report on civil customs]. Vol. 1. Beijing: Zhongguo zhengfa daxue chubanshe, 2000. First published 1930.

Quackenbush, William M., and Quentin E. Singewald. *Fushun Coal Field, Manchuria.* Tokyo: General Headquarters, Supreme Commander for the Allied Powers, Natural Resources Section, 1947.

Quanguo zhengxie wenshi he xuexi weiyuanhui, ed. *Huiyi Guomindang zhengfu Ziyuan weiyuanhui* [Remembering the Nationalist government's National Resources Commission]. Beijing: Zhongguo wenshi chubanshe, 2015.

Raask, Eric. *Mineral Impurities in Coal Combustion: Behavior, Problems, and Remedial Measures.* Washington, DC: Hemisphere, 1985.

Rabinbach, Anson. *The Human Motor: Energy, Fatigue, and the Origins of Modernity.* Berkeley: University of California Press, 1992.

Raj, Kapil. "Beyond Postcolonialism . . . and Postpositivism: Circulation and the Global History of Science." *Isis* 104, no. 2 (June 2013): 337–47.

Rawski, Evelyn S. *Early Modern China and Northeast Asia: Cross-Border Perspectives.* Cambridge: Cambridge University Press, 2015.

Read, Thomas T. "The Earliest Industrial Use of Coal." *Transactions of the Newcomen Society for the Study of the History of Engineering and Technology* 20, no. 1 (January 1939): 119–33.

Reardon-Anderson, James. *The Study of Change: Chemistry in China, 1840–1949.* New York: Cambridge University Press, 1991.

Renn, Jürgen. *The Evolution of Knowledge: Rethinking Science for the Anthropocene.* Princeton, NJ: Princeton University Press, 2020.

Richthofen, Ferdinand von. *Baron Richthofen's Letters: 1870–1872*. Shanghai: North China Herald, n.d.

Ritchie, Hannah. "Energy Mix." *Our World in Data*. https://ourworldindata.org/energy-mix.

Ritchie, Hannah, and Max Roser. "Fossil Fuels." *Our World in Data*. https://ourworldindata.org/fossil-fuels.

Rithmire, Meg E. *Land Bargains and Chinese Capitalism: The Politics of Property Rights under Reform*. New York: Cambridge University Press, 2015.

Rivett, Rohan D. *Behind Bamboo: An Inside Story of the Japanese Prison Camps*. Sydney: Angus and Robertson, 1946.

Roberts, Lissa. "Situating Science in Global History: Local Exchanges and Networks of Circulation." *Itinerario* 33, no. 1 (March 2009): 9–30.

Rogaski, Ruth. *Hygienic Modernity: Meanings of Health and Disease in Treaty-Port China*. Berkeley: University of California Press, 2004.

Rogaski, Ruth. "Vampires in Plagueland: The Multiple Meanings of *Weisheng* in Manchuria." In *Health and Hygiene in Chinese East Asia: Politics and Publics in the Long Twentieth Century*, edited by Angela Ki Che Leung and Charlotte Furth, 132–59. Durham, NC: Duke University Press, 2010.

Ross, Corey. *Ecology and Power in the Age of Empire: Europe and the Transformation of the Tropical World*. Oxford: Oxford University Press, 2017.

Rossabi, Morris. "The Ming and Inner Asia." In *The Cambridge History of China*, vol. 8, *The Ming Dynasty, 1368–1644*, pt. 2, edited by Denis Twitchett and Frederick W. Mote, 221–71. Cambridge: Cambridge University Press, 1998.

Roumasset, James, Kimberly Burnett, and Hua Wang. "Environmental Resources and Economic Growth." In *China's Great Economic Transformation*, edited by Loren Brandt and Thomas G. Rawski, 250–85. Cambridge: Cambridge University Press, 2008.

Russell, Edmund, James Allison, Thomas Finger, John K. Brown, Brian Balogh, and W. Bernard Carlson. "The Nature of Power: Synthesizing the History of Technology and Environmental History." *Technology and Culture* 52, no. 2 (April 2011): 246–59.

Rutter, Frank R. *The Electrical Industry of Japan*. Washington, DC: United States Government Printing Office, 1922.

Sabin, Paul. *Crude Politics: The California Oil Market, 1900–1940*. Berkeley: University of California Press, 2005.

Sakamoto Toshiatsu. "Nentō no kotoba" [New Year's address]. *Nenryō kyōkai shi* 11, no. 1 (January 1932): 1–3.

Samuels, Richard J. *The Business of the Japanese State: Energy Markets in Comparative and Historical Perspective*. Ithaca, NY: Cornell University Press, 1987.

Samuels, Richard J. *"Rich Nation, Strong Army": National Security and the Technological Transformation of Japan*. Ithaca, NY: Cornell University Press, 1994.

Samuels, Richard J. "Sources and Uses of Energy." In *The Japan Handbook*, edited by Patrick Heenan, 48–58. London: Fitzroy Dearborn, 1998.

Sand, Jordan. *House and Home in Modern Japan: Architecture, Domestic Space, and Bourgeois Culture, 1880–1930*. Cambridge, MA: Harvard University Asia Center, 2003.

Santiago, Myrna I. *The Ecology of Oil: Environment, Labor, and the Mexican Revolution, 1900–1938*. New York: Cambridge University Press, 2006.

Saraiva, Tiago. *Fascist Pigs: Technoscientific Organisms and the History of Fascism*. Cambridge, MA: MIT Press, 2016.

Satō Jin. *'Motazaru kuni' no shigen ron: jizoku kanō na kokudo o meguru mō hitotsu no chi* [Resource thinking of the 'have-nots': Sustainability of land and alternative vision in Japan]. Tokyo: University of Tokyo Press, 2011.

Schäfer, Dagmar. *The Crafting of the 10,000 Things: Knowledge and Technology in Seventeenth-Century China*. Chicago: University of Chicago Press, 2011.

Schencking, J. Charles. *The Great Kantō Earthquake and the Chimera of National Reconstruction in Japan*. New York: Columbia University Press, 2013.

Scherer, James A. B. "Manchoukuo Down to Date." *Japan Times and Mail*, January 14, 1934.

Schiltz, Michael. *The Money Doctors from Japan: Finance, Imperialism, and the Building of the Yen Bloc, 1895–1937*. Cambridge, MA: Harvard University Asia Center, 2012.

Schlesinger, Jonathan. *A World Trimmed with Fur: Wild Things, Pristine Places, and the Natural Fringes of Qing Rule*. Stanford, CA: Stanford University Press, 2016.

Schmalzer, Sigrid. "On the Appropriate Use of Rose-Colored Glasses: Reflections on Science in Socialist China." *Isis* 98, no. 3 (September 2007): 571–83.

Schmalzer, Sigrid. *The People's Peking Man: Popular Science and Human Identity in Twentieth-Century China*. Chicago: University of Chicago Press, 2008.

Schmalzer, Sigrid. "Red and Expert." In *Afterlives of Chinese Communism: Political Concepts from Mao to Xi*, edited by Christian Sorace, Ivan Franceschini, and Nicholas Loubere, 215–20. Acton: Australian National University Press, 2019.

Schmalzer, Sigrid. *Red Revolution, Green Revolution: Scientific Farming in Socialist China*. Chicago: University of Chicago Press, 2016.

Schneider, Laurence. *Biology and Revolution in Twentieth-Century China*. Lanham, MD: Rowman and Littlefield, 2003.

Schoenhals, Michael. "Elite Information in China." *Problems of Communism* 34 (September/October 1985): 65–71.

Schumpeter, E. B., ed. *The Industrialization of Japan and Manchukuo, 1930–1940*. New York: Macmillan, 1940.

Schumpeter, E. B. "Japan, Korea, and Manchukuo, 1936–1940." In *The Industrialization of Japan and Manchukuo, 1930–1940*, edited by E. B. Schumpeter, 271–474. New York: Macmillan, 1940.

Schurr, Sam H., Calvin C. Burwell, Warren D. Devine Jr., and Sidney Sonenblum, eds. *Electricity in the American Economy: Agent of Technological Progress*. New York: Greenwood Press, 1991.

Schwartz, Benjamin I. *In Search of Wealth and Power: Yen Fu and the West*. Cambridge, MA: Belknap Press of Harvard University Press, 1964.

Scott, A. P. "The History of Oil in Japan," In *Present Day Impressions of Japan*, edited by W. H. Morton Cameron, 483–94. Chicago: Globe Encyclopedia, 1919.

Scott, James C. *Seeing Like a State: How Certain Schemes to Improve the Human Condition Have Failed*. New Haven, CT: Yale University Press, 1998.

Scott, J. H. "Changes and Still More Changes in Japan." *Japan Review: A Herald of the Pacific Era* 4, no. 1 (November 1919): 109–11.

Sengoopta, Chandak. *Imprint of the Raj: How Fingerprinting Was Born in Colonial India*. London: Macmillan, 2003.

Seow, Victor. "*The Coal Question* in the Age of Carbon." Joint Center for History and Economics, December 31, 2016. https://histecon.fas.harvard.edu/energyhistory/seow.html.

Seta Hidetoshi. *Bujun tankō Rōkudai kō hōkokusho* [A report on the Laohutai mine at the Fushun colliery] (1923). Unpublished report, Library of the Department of Metallurgical Engineering, Osaka University, Osaka.

Shanghai shangye chuxu yinhang diaocha bu. *Mei yu meiye* [Coal and the coal industry]. Shanghai: Shanghai shangye chuxu yinhang diaocha bu, 1935.

Shao, Dan. *Remote Homeland, Recovered Borderland: Manchus, Manchoukuo, and Manchuria, 1907–1985*. Honolulu: University of Hawai'i Press, 2011.

Shapiro, Judith. *Mao's War against Nature: Politics and the Environment in Revolutionary China*. New York: Cambridge University Press, 2001.

Shen, Grace Yen. "Scientism in the Twentieth Century." In *Modern Chinese Religion II: 1850–2015*, edited by Vincent Goossaert, Jan Kiely, and John Lagerwey, 1:91–140. Leiden: Brill, 2015.

Shen, Grace Yen. *Unearthing the Nation: Modern Geology and Nationalism in Republican China*. Chicago: University of Chicago Press, 2014.

Shen Yue. "Nanwang '2.13': huiyi Mao Zedong shicha Fushun" [Unforgettable February 13: Remembering Mao Zedong's survey of Fushun]. In *Fushun gongye bai nian huimou*, edited by Zhengxie Fushun shi weiyuanhui, Fushun kuangye jituan youxian zeren gongsi, and Zhongguo shiyou Fushun shihua gongsi, 1:155–58. Shenyang: Shenyang chubanshe, 2008.

Shen Zhihua. *Mao, Stalin, and the Korean War: Trilateral Communist Relations in the 1950s*. Translated from the Chinese by Neil Silver. London: Routledge, 2012.

Shigeto Tsuru. *The Political Economy of the Environment: The Case of Japan*. London: Athlone, 1999.

Shinozaki Hikoji. *Bujun tankō Ōyama saitanjo hōkoku* [Report on the Ōyama mine at the Fushun colliery] (1920). Unpublished report, Kyushu University Archives, Fukuoka.

Shirato Inchū. "Mokutan gasu jidōsha no shinpo ni tsuite" [The progress of the charcoal gas automobile]. *Nenryō kyōkai shi* 10, no. 12 (December 1931): 1406–22.

Shiroyama, Tomoko. *China during the Great Depression: Market, State, and the World Economy, 1929–1937*. Cambridge, MA: Harvard University Asia Center, 2009.

Shulman, Peter A. *Coal and Empire: The Birth of Energy Security in Industrial America*. Baltimore: Johns Hopkins University Press, 2015.

Sieferle, Rolf Peter. *The Subterranean Forest: Energy Systems and the Industrial Revolution*. Cambridge: White Horse, 2001.

Simon, Denis Fred, and Merle Goldman, eds. *Science and Technology in Post-Mao China*. Cambridge, MA: Council on East Asian Studies, Harvard University, 1989.

Sivasundaram, Sujit. "The Oils of Empire." In *Worlds of Natural History*, edited by H. A. Curry, N. Jardine, J. A. Second, and E. C. Spary, 378–98. Cambridge: Cambridge University Press, 2018.

Smil, Vaclav. *Energy and Civilization: A History*. Cambridge, MA: MIT Press, 2017.

Smil, Vaclav. "War and Energy." In *Encyclopedia of Energy*, edited by Cutler J. Cleveland, 6:363–71. Amsterdam: Elsevier, 2004.

Smith, Benjamin Lyman. *Geological Survey of the Oil Lands of Japan*. Tokei: Public Works Department, 1877.

Smith, Crosbie. *The Science of Energy: A Cultural History of Energy Physics in Victorian Britain*. London: Athlone, 1998.

Smith, Norman, ed. *Empire and Environment in the Making of Manchuria*. Vancouver: UBC Press, 2017.

Smith, Norman. "'Hibernate No More!': Winter, Health, and the Great Outdoors." In *Empire and Environment in the Making of Manchuria*, edited by Norman Smith, 130–51. Vancouver: UBC Press, 2017.

Smith, Norman. *Intoxicating Manchuria: Alcohol, Opium, and Culture in China's Northeast*. Vancouver: UBC Press, 2012.

Smyth, Helen. "China's Petroleum Industry." *Far Eastern Survey* 15, no. 12 (June 1946): 187–90.

Smyth, Russell, Zhai Qingguo, and Wang Jing. "Labour Market Reform in China's State-Owned Enterprises: A Case Study of Post-Deng Fushun in Liaoning Province." *New Zealand Journal of Asian Studies* 3, no. 2 (December 2001): 42–72.

Sohn, Yul. *Japanese Industrial Governance: Protectionism and the Licensing State*. London: RoutledgeCurzon, 2005.

Song Yingxing. *Tiangong kaiwu* [The works of heaven and the inception of things]. Taipei: Zhonghua congshu weiyuanhui, 1955. First published 1637.

Sorace, Christian, Ivan Franceschini, and Nicholas Loubere, eds. *Afterlives of Chinese Communism: Political Concepts from Mao to Xi*. Acton: Australian National University Press, 2019.

South Manchuria Railway Company. *Fifth Report on Progress in Manchuria to 1936*. Dairen: South Manchuria Railway Company, 1936.

South Manchuria Railway Company. *Sixth Report on Progress in Manchuria, to 1939*. Dairen: South Manchuria Railway Company, 1939.

Speight, J. G. *The Chemistry and Technology of Coal*. 2nd ed., rev. and enl. New York: Marcel Dekker, 1994. First published 1983.

State Statistical Bureau. *Major Aspects of the Chinese Economy through 1956*. Beijing: T'ung-chi ch'u-pan-she, 1958.

Steffen, Will, Jacques Grinevald, Paul Crutzen, and John McNeill. "The Anthropocene: Conceptual and Historical Perspectives." *Philosophical Transactions: Mathematics, Physical and Engineering Sciences* 369, no. 1938 (March 2011): 842–67.

Steinmetz, George, ed. *State/Culture: State-Formation after the Cultural Turn*. Ithaca, NY: Cornell University Press, 1999.

Stranges, Anthony N. "Friedrich Bergius and the German Synthetic Fuel Industry." *Isis* 75, no. 4 (December 1984): 642–67.

Stranges, Anthony N. "Synthetic Fuel Production in Prewar and World War II Japan: A Case Study of Technological Failure." *Annals of Science* 50 (1993): 229–65.

Strauss, Julia C. *Strong Institutions in Weak Polities: State Building in Republican China, 1927–1940*. Oxford: Clarendon, 1998.

Suleski, Ronald. *Civil Government in Warlord China: Tradition, Modernization and Manchuria*. New York: P. Lang, 2002.

Summers, William C. *The Great Manchurian Plague of 1910–1911: The Geopolitics of an Epidemic Disease*. New Haven, CT: Yale University Press, 2012.

Sun Yat-sen. *The International Development of China*. Shanghai: Commercial Press, 1920.

Sun Yueqi. "Chuanmei chanxiao zhi huigu yu qianzhan" [Reviewing and forecasting the production and marketing of Sichuanese coal]. *Xi'nan shiye tongxun* 2, no. 2 (August 1940): 41–45.

Sun Yueqi. "Wo he Ziyuan weiyuanhui" [The National Resources Commission and me]. In *Huiyi Guomindang zhengfu Ziyuan weiyuanhui*, edited by Quanguo zhengxie wenshi he xuexi weiyuanhui, 14–67. Beijing: Zhongguo wenshi chubanshe, 2015.

Sun Yueqi. "Zhongguo gongye de qiantu" [The future of Chinese industry]. *Xi'nan shiye tongxun* 12, no. 3/4 (October 1945): 42–43.

Sun Yueqi keji jiaoyu jijin guanweihui. *Sun Yueqi zhuan* [Biography of Sun Yueqi]. Beijing: Shiyou gongye chubanshe, 1994.

Sun Zhongshan. *Shiye jihua* [Industrial Plan]. Shanghai: Shanghai guomin shuju, 1926. First published 1920.

Swope, Kenneth M. *The Military Collapse of China's Ming Dynasty, 1618–44*. London: Routledge, 2014.

Szeman, Imre. "Literature and Energy Futures." *PLMA* 126, no. 2 (March 2011): 323–25.

Szeman, Imre, Jennifer Wenzel, and Patricia Yaeger, eds. *Fueling Culture: 101 Words for Energy and Environment*. New York: Fordham University Press, 2017.

Tai Hsüan-chih. *The Red Spears, 1916–1949*. Translated from the Chinese by Ronald Suleski. Ann Arbor: Center for Chinese Studies, University of Michigan, 1985.

Takahashi, Kazuo. "The Iran-Japan Petrochemical Project: A Complex Issue." In *Japan in the Contemporary Middle East*, edited by Kaoru Sugihara and J. A. Allan, 77–86. London: Routledge, 1993.

Takano Asako. *Shimon to kindai: idō suru shintai no kanri to tōchi no gihō* [Fingerprinting and modernity: The management of moving bodies and the techniques of governance]. Tokyo: Misuzu shobō, 2016.

Takanoe Mototarō. *Nihon tankō shi* [Gazetteer of Japan's coal mines]. Tokyo: Meiji bunken shiryō kankōkai, 1970. First published 1908.

Takanoe Mototarō, ed. *Sekitan kōgyō ronshū* [A collection of papers on coal mining]. Fukuoka: Hatsubaijo sekizenkan shiten, 1910.

Takenob, Y., ed. *The Japan Year Book, 1919–20*. Tokyo: Japan Year Book Office, 1920.

Takenob, Y., ed. *The Japan Year Book, 1920–21*. Tokyo: Japan Year Book Office, 1921.

Takenob, Y., ed. *The Japan Year Book, 1921–22*. Tokyo: Japan Year Book Office, 1922.

Takenob, Y., ed. *The Japan Year Book, 1923*. Tokyo: Japan Year Book Office, 1923.

Takenobu, Y., ed. *The Japan Year Book, 1927*. Tokyo: Japan Year Book Office, 1927.

Tamanoi, Mariko Asano, ed. *Crossed Histories: Manchuria in the Age of Empire*. Honolulu: University of Hawai'i Press, 2005.

Tan Beizhan. "Nanjing Guomin zhengfu shiqi guoying meikuang shiye jingying de dianxing: yi Jianshe weiyuanhui yu Huainan meikuang weili de kaocha" [A model of a state-owned coal-mining enterprise during the Nanjing government period: An inquiry through the case of the National Reconstruction Commission's Huainan colliery]. *Anhui shixue* 2010, no. 2 (February 2010): 106–14.

Tan, Isaac C. K. "Science and Empire: Tracing the Imprint of Dactylography in Manchuria, 1924–1945." *Japan Forum* (June 2019): 1–24.

Tan Xihong, ed. *Shi nian lai zhi Zhongguo jingji* [China's economy over the past decade]. Shanghai: Zhonghua shuju, 1948.

Tan, Ying Jia. "Revolutionary Current: Electricity and the Formation of the Party State in China and Taiwan." PhD diss., Yale University, 2015.

Tanaka Hiroshi. "Shimon ōnatsu no genten: Chūgoku tōhokubu (kyū Manshū) o aruite" [The origins of fingerprinting: Walking through China's Northeast (former Manchuria)]. *Asahi jaanaru* 29 (1987): 21–23.

Tanaka, Shoko. *Post-War Japanese Resource Policies and Strategies: The Case of Southeast Asia*. Ithaca, NY: Cornell East Asia Series, 1986.

Tanaka, Stefan. *New Times in Modern Japan*. Princeton, NJ: Princeton University Press, 2004.

Tanner, Harold M. *The Battle for Manchuria and the Fate of China: Siping, 1946*. Bloomington: Indiana University Press, 2013.

Tanner, Harold M. *Where Chiang Kai-shek Lost China: The Liao-Shen Campaign, 1948*. Bloomington: Indiana University Press, 2015.

Taylor, Frederick Winslow. *The Principles of Scientific Management*. New York: W. W. Norton, 1947. First published 1911.

Taylor, Jay. *The Generalissimo's Son: Chiang Ching-kuo and the Revolutions in China and Taiwan*. Cambridge, MA: Harvard University Press, 2000.

Teh, Limin. "Labor Control and Mobility in Japanese-Controlled Fushun Coalmine (China), 1907–1932." *International Review of Social History* 60, no. S1 (August 2015): 95–119.

Teh, Limin. "Mining for Difference: Race, Chinese Labor, and Japanese Colonialism in Fushun Coalmine, 1907–1945." PhD diss., University of Chicago, 2014.

Thai, Philip. *China's War on Smuggling: Law, Economic Life, and the Making of the Modern State, 1842–1965*. New York: Columbia University Press, 2018.

Thomas, Julia Adeney. "Reclaiming Ground: Japan's Great Convergence." *Japanese Studies* 34, no. 3 (September 2014): 253–63.

Thomas, Julia Adeney, Prasannan Parthasarathi, Rob Linrothe, Fa-ti Fan, Kenneth Pomeranz, and Amitav Ghosh. "*JAS* Round Table on Amitav Ghosh, *The Great Derangement: Climate Change and the Unthinkable*." *Journal of Asian Studies* 75, no. 4 (November 2016): 929–55.

Thomson, Elspeth. *The Chinese Coal Industry: An Economic History*. London: Routledge-Curzon, 2003.

Tian, Xiaofei. "The Making of a Hero: Lei Feng and Some Issues of Historiography." In *The People's Republic of China at 60: An International Assessment*, edited by William C. Kirby, 293–305. Cambridge, MA: Harvard University Asia Center, 2011.

Tiedao bu quanguo tielu shangyun huiyi banshichu, comp. *Quanguo tielu shangyun huiyi huikan* [Collected documents of the National Railway Commercial Transport Meeting]. Nanjing: Tiedao bu quanguo tielu shangyun huiyi banshichu, 1931.

Ting, Leonard G. *The Coal Industry in China*. Tianjin: Nankai Institute of Economics, Nankai University, 1937.

Ting, V. K. "Mining Legislation and Development in China." *North-China Daily News*, May 26, 1917.

Tōa keizai chōsakyoku. *Honpō o chūshin to seru sekitan jukyū* [The supply and demand of coal in our country]. Tokyo: Tōa keizai chōsakyoku, 1933.

Tocqueville, Alexis de. *Journeys to England and Ireland*. Translated from the French by George Lawrence and K. P. Mayer. New Haven, CT: Yale University Press, 1958.

Tooze, Adam. *The Wages of Destruction: The Making and Breaking of the Nazi Economy*. New York: Viking, 2006.

Torgasheff, Boris P. "Mining Labor in China." *Chinese Economic Journal and Bulletin* 6, no. 4 (April 1930): 392–417.

Tsuboguchi Ikuo. "Tan yanghuai zaolin" [Regarding afforestation with acacia]. *Fukuang xunkan* 2, no. 1 (October 1947): 235.

Tsutsui, William M. "Landscapes in the Dark Valley: Toward an Environmental History of Wartime Japan." *Environmental History* 8, no. 2. (April 2003): 294–311.

Tsutsui, William M. *Manufacturing Ideology: Scientific Management in Twentieth-Century Japan*. Princeton, NJ: Princeton University Press, 1998.

Tucker, David. "Labor Policy and the Construction Industry in Manchukuo: Systems of Recruitment, Management, and Control." In *Asian Labor in the Wartime Japanese Empire: Unknown Histories*, edited by Paul H. Kratoska, 25–57. Armonk, NY: M. E. Sharpe, 2005.

Twitchett, Denis, and Frederick W. Mote, eds. *The Cambridge History of China*. Vol. 8, *The Ming Dynasty, 1368–1644*, pt. 2. Cambridge: Cambridge University Press, 1998.

United States Strategic Bombing Survey, Electric Power Division. *The Electric Power Industry of Japan*. Washington, DC: United States Strategic Bombing Survey, Electric Power Division, 1946.

van de Ven, Hans J. *War and Nationalism in China, 1925–1945*. London: Routledge-Curzon, 2003.

Vogel, Ezra F. *The Four Little Dragons: The Spread of Industrialization in East Asia*. Cambridge, MA: Harvard University Press, 1991.

Vrtis, George, and J. R. McNeill. "Introduction: Of Mines, Minerals, and North American Environmental History." In *Mining North America: An Environmental History since 1522*, edited by J. R. McNeill and George Vrtis, 1–16. Oakland: University of California Press, 2017.

Wakahara Tomiko. "Ryōri to nenryō" [Cooking and fuel]. *Nenryō kyōkai shi* 9, no. 10 (October 1930): 1114–22.

Wakeman, Frederick, Jr. *The Great Enterprise: The Manchu Reconstruction of Imperial Order in Seventeenth-Century China*. Vol. 1. Berkeley: University of California Press, 1985.

Walker, Brett L. *Toxic Archipelago: A History of Industrial Disease in Japan*. Seattle: University of Washington Press, 2010.

Wang Buyi. *Bujun rotenbori hōkoku* [Report on the Fushun open-pit mine] (1923). Unpublished report, Library of the Department of Metallurgical Engineering, Osaka University, Osaka.

Wang Daoping. "Kuangye zhuanjia Cheng Zongyang xiansheng xiaozhuan" [Biographical sketch of mining and metallurgy specialist Mr. Cheng Zongyang]. *Zhuanji wenxue* 31, no. 5 (November 1977): 110.

Wang, Dong. *China's Unequal Treaties: Narrating National History*. Lanham, MD: Lexington Books, 2005.

Wang Hongyan. *"Manshūkoku" rōkō no shiteki kenkyū: Kahoku chiku kara no nyūMan rōkō* [Historical research on laborers in "Manchukuo": Migrant laborers from North China]. Tokyo: Nihon keizai hyōron sha, 2015.

Wang, Jianhua. "Development and Prospect on Fully Mechanized Mining in Chinese Coal Mines." *International Journal of Coal Science and Technology* 1, no. 3 (October 2014): 253–60.

Wang Lianzhong. "Fushun meikuang bowuguan canguan ji" [Record of a visit to the Fushun Coal Mine Museum]. *Fushun qiqian nian*, April 13, 2018. http://www.fs7000.com/wap/news/?12296.html.

Wang, Mark, Zhiming Chen, Pingyu Zhu, Lianjun Tong, and Yanji Ma, eds. *Old Industrial Cities Seeking New Road of Industrialization: Models of Revitalizing Northeast China*. Singapore: World Scientific, 2013.

Wang Xinqi. "Mei zhi yongtu ji qi gongye zhizao" [Coal's use and its industrial manufacturing]. *Huabei gongye jikan* 1, no. 2 (December 1930): 11–28.

Wang Xinsan. "Di er ci lai Fushun jieshou kuangwu ju" [Second time coming to Fushun to take over the Mining Affairs Bureau]. In *FKZ*, 280–84.

Wang Xinsan. "Zai di yi ci jun, kuang lianxi huiyi shang de jianghua zhalu" [Excerpts from the speech at the first joint meeting between the military and the colliery]. Fushun Mining Affairs Bureau Archives, file no. 3.1. In *FKZ*, 57.

Wang Yusheng. "Ming xiu Fushun cheng yu Fushun de ming" [The construction of Fushun city and the naming of Fushun in the Ming]. In *Fushun difang shi gailan* [A general overview of Fushun's local history], edited by Fu Bo, Liu Chang, and Wang Pinglu, 75–77. Fushun: Fushun shi shehui kexueyuan, 2001.

Wang, Zuoyue. "Saving China through Science: The Science Society of China, Scientific Nationalism, and Civil Society in Republican China." "Science and Civil Society," *Osiris*, 2nd ser., 17 (2002): 291–322.

Wasserstrom, Jeffrey N. *Student Protests in Twentieth-Century China: The View from Shanghai*. Stanford, CA: Stanford University Press, 1991.

Watt, Lori. *When Empire Comes Home: Repatriation and Reintegration in Postwar Japan*. Cambridge, MA: Harvard University Asia Center, 2010.

Watts, Michael. "Petro-Violence: Community, Extraction, and Political Economy in a Mythical Community." In *Violent Environments*, edited by Nancy Lee Peluso and Michael Watts, 189–212. Ithaca, NY: Cornell University Press, 2001.

Weart, Spencer R. *The Discovery of Global Warming*. Cambridge, MA: Harvard University Press, 2003.

Wei Ping. "Chang mofan Li Yuming" [Yi Yuming, factory model]. *Fushun ribao*, January 27, 1951.

Wei, Yi-Ming, Qiao-Mei Liang, Gang Wu, and Hua Liao. *Energy Economics: Understanding Energy Security in China*. Bingley: Emerald Publishing, 2019.

Wells, Richard Evan. "The Manchurian Bean: How the Soybean Shaped the Modern History of China's Northeast, 1862–1945." PhD diss., University of Wisconsin–Madison, 2018.

Wemheuer, Felix. *Famine Politics in Maoist China and the Soviet Union*. New Haven, CT: Yale University Press, 2014.

Weng Wenhao. "Zhanhou Zhongguo gongyehua wenti" [The problem of China's postwar industrialization]. *Gangtie jie* 1, no. 3 (June 1943): 5–7.

Weng Wenhao. "Zhongguo dixia fuyuan de guji" [Estimates of China's subsurface natural resources]. *Duli pinglun* 17 (September 1932): 6–10.

Weng Wenhao. "Zhongguo meikuangye de eyun: jingji zhan de yi ge li" [The misfortune of China's coal-mining industry: An example of economic war]. *Duli pinglun* 23 (October 1932): 4–7.

Wenzel, Jennifer. "Introduction." In *Fueling Culture: 101 Words for Energy and Environment*, edited by Imre Szeman, Jennifer Wenzel, and Patricia Yaeger, 1–15. New York: Fordham University Press, 2017.

Weston, Timothy B. "Fueling China's Capitalist Transformation." In *China's Transformations: The Stories beyond the Headlines*, edited by Lionel M. Jensen and Timothy B. Weston, 68–89. Lanham, MD: Rowman and Littlefield, 2007.

White, Richard. *The Organic Machine: The Remaking of the Columbia River*. New York: Hill and Wang, 1995.

Whitt, Laurelyn. *Science, Colonialism, and Indigenous Peoples: The Cultural Politics of Law and Knowledge*. New York: Cambridge University Press, 2009.

Whitten, David O., and Bessie E. Whitten. *Handbook of American Business History*. Vol. 2, *Extractives, Manufacturing, and Services: A Historiographical and Bibliographical Guide*. Westport, CT: Greenwood, 1997.

Williams, Rosalind H. *Notes on the Underground: An Essay on Technology, Society, and the Imagination*. Cambridge, MA: MIT Press, 1990.

Wilson, Sandra. *The Manchurian Crisis and Japanese Society, 1931–33*. London: Routledge, 2002.

Wittfogel, Karl. *Oriental Despotism: A Comparative Study of Total Power*. New Haven, CT: Yale University Press, 1957.

Wong Wen-hao, "China's Economic Reconstruction in War Time." *China Critic*, September 19, 1940.

Woo-Cumings, Meredith, ed. *The Developmental State*. Ithaca, NY: Cornell University Press, 1999.

Woodhead, H. G. W. *A Visit to Manchukuo*. Shanghai: Mercury Press, 1932.

World Bank Group. *Managing Coal Mine Closure: Achieving a Just Transition for All.* Washington, DC: World Bank, 2018.

World Engineering Congress, ed. *Proceedings, World Engineering Congress, Tokyo 1929,* vol. 37, *Mining and Metallurgy,* pt. 5, *Mineral Resources and Mining.* Tokyo: Kogakkai, 1931.

World Petroleum Congress, ed. *World Petroleum Congress Proceedings.* London: Applied Science, 1937.

Wray, Harry, and Hilary Conroy, eds. *Japan Examined: Perspectives on Modern Japanese History.* Honolulu: University of Hawai'i Press, 1984.

Wright, Tim. *Coal Mining in China's Economy and Society, 1895–1937.* Cambridge: Cambridge University Press, 1984.

Wright, Tim. "'A Method of Evading Management': Contract Labor in Chinese Coal Mines before 1937." *Comparative Studies in Society and History* 23, no. 4 (October 1981): 656–78.

Wright, Tim. "The Nationalist State and the Regulation of Chinese Industry during the Nanjing Decade: Competition and Control in Coal Mining." In *Ideal and Reality: Social and Political Change in Modern China,* edited by David Pong and Edmund S. K. Fung, 127–52. Lanham, MD: University Press of America, 1985.

Wright, Tim. *The Political Economy of the Chinese Coal Industry: Black Gold and Blood-Stained Coal.* Abingdon: Routledge, 2012.

Wright, Tim. "Sino-Japanese Business in China: The Luda Company, 1923–1937." *Journal of Asian Studies* 39, no. 4 (August 1980): 711–27.

Wrigley, E. A. *Continuity, Chance and Change: The Character of the Industrial Revolution in England.* Cambridge: Cambridge University Press, 1988.

Wrigley, E. A. *Energy and the English Industrial Revolution.* Cambridge: Cambridge University Press, 2010.

Wu Hung. "Tiananmen Square: A Political History of Monuments." *Representations* 35 (Summer 1991): 84–117.

Wu Keyi. "Kangzhan qijian guoying meikuang zhi kaifa ji zengchan liyong" [The development, increase in production, and use of state-run coal mines during the War of Resistance]. *Ziyuan weiyuanhui jikan* 5, no. 3 (September 1945): 126–49.

Wu Mingyao and Gu Yiyou. "Gongren de shenghuo bian le yang" [Workers' lives have become different]. *Fushun ribao,* September 22, 1952.

Wu, Shellen Xiao. *Empires of Coal: Fueling China's Entry into the Modern World Order, 1860–1920.* Stanford, CA: Stanford University Press, 2015.

Wu Yong and Liu Ce. "Eight Cities Partner for Shenyang Economic Zone." *China Daily,* December 3, 2010.

Wu, Yuan-li. *Economic Development and the Use of Energy Resources in Communist China.* New York: Frederick A. Praeger, 1963.

Xian Min. "Benju jiwu chu gaikuang" [Situation at our colliery's machine manufacturing and maintenance department]. *Fukuang xunkan* 2, no. 6 (May 1947): 290–91.

Xiao Jingquan and Jin Hui. *Wangshi jiu ying: lao zhaopian zhong de Fushun lishi* [Old shadows of the past: Fushun's history in old photographs]. Shenyang: Liaoning renmin chubanshe, 2008.

Xiao Jun. *Dongbei riji, 1946–1950* [Diary from the Northeast, 1946–1950]. Hong Kong: Oxford University Press, 2014.

Xiao Jun. *Wuyue de kuangshan* [Coal mines in May]. Harbin: Heilongjiang renmin chubanshe, 1982. First published 1954.

Xie Jiarong, ed. *Zhongguo kuangye jiyao, di er ci* [General statement of China's mining industry, 2nd issue]. Beiping: Shiye bu Dizhi diaocha suo, 1926.

Xie Shuying. "Cong jianku zhong fendou" [Striving in the midst of adversity]. *Fukuang xunkan* 2, no. 1 (October 1947): 229–30.

Xie Xueshi, and Matsumura Takao, eds. *Mantie yu Zhongguo laogong* [Mantetsu and Chinese labor]. Beijing: Shehui kexue wenxian chubanshe, 2003.

Xu, Yan. *The Soldier Image and State-Building in Modern China, 1924–1949*. Lexington, KY: University of Kentucky Press, 2018.

Xue Yi. *Gongkuang taidou Sun Yueqi* [Sun Yueqi, eminent figure in industry and mining]. Beijing: Zhongguo wenshi chubanshe, 1997.

Xue Yi. "Sun Yueqi yu 20 shiji Zhongguo meitan gongye" [Sun Yueqi and the twentieth-century Chinese coal industry]. *Zhongguo kuangye daxue xuebao* 2011, no. 3 (March 2011): 102–9.

Yadian. "Youguan Fushun di yi ren kuangzhang Songtian Wuyilang de ziliao" [Information regarding Fushun's first colliery manager Matsuda Buichirō]. *Fushun qiqian nian*. February 25, 2014. http://www.fs7000.com/news/?9102.html.

Yamaguchi Yoshiko and Fujiwara Sakuya. *Fragrant Orchid: The Story of My Early Life*. Translated from the Japanese by Chia-ning Chang. Honolulu: University of Hawai'i Press, 2015.

Yamaguchi Yoshiko and Fujiwara Sakuya. *Ri Kōran: watashi no hansei* [Li Xianglan: half of my lifetime]. Tokyo: Shinchōsha, 1990.

Yamamoto Yū. "Manshū oirushēru jigyō, 1909–31-nen" [Manchuria's shale oil industry, 1909–1931]. *Mita gakkai zasshi* 95, no. 2 (January 2003): 177–98.

Yamamoto Yūzō. *"Manshūkoku" keizai shi kenkyū* [Research on the economic history of "Manchukuo"]. Nagoya: Nagoya daigaku shuppankai, 2003.

Yamamuro Shin'ichi. *Manchuria under Japanese Domination*. Translated from the Japanese by Joshua A. Fogel. Philadelphia: University of Pennsylvania Press, 2006.

Yamaoka, N. "Manchurian Plant with Seam in Places Four Hundred Feet Thick Uses Sand Filling." *Coal Age* 19, no. 20 (May 19, 1921): 897–902.

Yamaoka, N. "Two Strip Pits and Much Allied Industry Operated by Fushun Colliery." *Coal Age* 19, no. 21 (May 26, 1921): 945–49.

Yamazaki Motoki. *Bujun tankō shucchō hōkokusho* [Written report of official trip to the Fushun coal mine] (September 6, 1946). Box no. 2, Kia-ngau Chang Papers, Hoover Institution Archives, Stanford, CA.

Yang, Daqing. "Resurrecting the Empire: Japanese Technicians in Postwar China, 1945–49." In *The Japanese Empire in East Asia and Its Postwar Legacy*, edited by Harald Fuess, 185–205. Munich: Iudicium, 1998.

Yang Naicang. "Fushun meikuang kaicai bai nian huigu" [A review of a century of coal mining in Fushun]. In *FMBN*, 2–15.

Yeh, Wen-hsin, ed. *Becoming Chinese: Passages to Modernity and Beyond*. Berkeley: University of California Press, 2000.

Yeh, Wen-hsin. *Shanghai Splendor: Economic Sentiments and the Making of Modern China, 1843–1949*. Berkeley: University of California Press, 2007.

Yellen, Jeremy A. *The Greater East Asia Co-Prosperity Sphere: When Total Empire Met Total War*. Ithaca, NY: Cornell University Press, 2019.

Yergin, Daniel. *The Prize: The Epic Quest for Oil, Money, and Power*. New York: Simon and Schuster, 1991.

Yiu, Angela. "Beach Boys in Manchuria: An Examination of Sōseki's *Here and There in Manchuria and Korea*, 1909." *Review of Japanese Culture and Society* 29 (2017): 109–25.

Yosano Akiko. *Travels in Manchuria and Mongolia: A Feminist Poet from Japan Encounters Prewar China.* Translated from the Japanese and edited by Joshua A. Fogel. New York: Columbia University Press, 2001.

Yosano Tekkan and Yosano Akiko. "Man-Mō yūki" [Travels in Manchuria and Mongolia]. In *Tekkan Akiko zenshū* [Complete works of Tekkan and Akiko], 26:3–238. Tokyo: Bensei suppan, 2001.

Yosano Tekkan and Yosano Akiko. *Tekkan Akiko zenshū* [Complete works of Tekkan and Akiko]. Vol. 26. Tokyo: Bensei suppan, 2001.

Yoshida, Phyllis Genther. "Japan's Energy Conundrum." In *Japan's Energy Conundrum: A Discussion of Japan's Energy Circumstances and U.S.-Japan Energy Relations,* edited by Phyllis Genther Yoshida, 1–9. Washington: Sasakawa Peace Foundation USA, 2018.

Yoshida, Phyllis Genther, ed. *Japan's Energy Conundrum: A Discussion of Japan's Energy Circumstances and U.S.-Japan Energy Relations.* Washington, DC: Sasakawa Peace Foundation USA, 2018.

Yoshimi Shun'ya. "Radioactive Rain and the American Umbrella." Translated from the Japanese by Shi-Lin Loh. *Journal of Asian Studies* 71, no. 2 (May 2012): 319–31.

Yoshimura Manji. "Nenryō kyōkai jūnen shi" [Ten-year history of the Fuel Society]. *Nenryō kyōkai shi* 11, no. 10 (October 1932): 1194–1210.

Yoshimura Manji. "Shunjitsu nenryō dansō" [Brief commentary regarding fuels on a spring day]. *Nenryō kyōkai shi* 15, no. 1 (January 1936): 88–91.

Yoshimura Manji. "Waga kuni ni okeru nenryō mondai gaisetsu" [Overview of our country's fuel question]. *Nenryō kyōkai shi* 1, no. 1 (August 1922): 1–13.

You Byoung Boo. *Mantetsu Bujun tankō no rōmu kanri shi* [A history of labor management in Mantetsu's Fushun colliery]. Fukuoka: Kyūshū daigaku shuppankai, 2004.

Young, Arthur N. *China's Nation Building Effort, 1927–1937: The Financial and Economic Record.* Stanford, CA: Hoover Institution Press, 1971.

Young, Arthur N. *China's Wartime Finance and Inflation, 1937–1945.* Cambridge, MA: Harvard University Press, 1965.

Young, C. Walter. *Japanese Jurisdiction in the South Manchuria Railway Areas.* Baltimore: Johns Hopkins Press, 1931.

Young, John. *The Research Activities of the South Manchurian Railway Company, 1907–1945.* New York: East Asian Institute, Columbia University, 1966.

Young, Louise. *Japan's Total Empire: Manchuria and the Culture of Wartime Imperialism.* Berkeley: University of California Press, 1998.

Young, Louise. "When Fascism Met Empire in Japanese-Occupied Manchuria." *Journal of Global History* 12, no. 2 (July 2017): 274–96.

Yu Ge. "Feiyue qianjin de Fushun shi funü men" [Women of Fushun leaping forward]. *Fushun ribao,* August 15, 1952.

Yu Heyin. *Fushun meikuang baogao* [Report on the Fushun colliery]. Beijing: Nongshang bu kuangzheng si, 1926.

Yu Heyin. "Liaoyang Yantai meikuang" [The Yantai coal mines of Liaoyang]. *Kuangye* 4, no. 14 (November 1930): 66–98.

Yu, Miin-ling. "'Labor Is Glorious': Model Laborers in the People's Republic of China." In *China Learns from the Soviet Union, 1949–Present,* edited by Thomas P. Bernstein and Hua-yu Li, 231–58. Lanham, MD: Lexington Books, 2010.

Yu Zailin. "Bensuo liang nian lai ximei lianjiao shiyan baogao" [Report of our bureau's experiments on coal washing and coke production over the past two years]. *Kuangye banyuekan* 3, no. 5/6 (April 1940): 8–12.

Yu Zailin. "Ximei gongye" [The coal-washing industry] (n.d.). In *Ziyuan weiyuanhui jishu renyuan fu Mei shixi shiliao: Minguo sanshiyi nian huipai*, vol. 2, compiled by Cheng Yufeng and Cheng Yuhuang, 1998–2003. Taipei: Guoshiguan, 1988.

Yun Yan. *Jindai Kailuan meikuang yanjiu* [A study of the Kailuan coal mines in modern times]. Beijing: Renmin chubanshe, 2015.

Zallen, Jeremy. *American Lucifers: The Dark History of Artificial Light, 1750–1865.* Chapel Hill: University of North Carolina Press, 2019.

Zanasi, Margherita. *Economic Thought in Modern China: Market and Consumption, c.1500–1937.* Cambridge: Cambridge University Press, 2020.

Zanasi, Margherita. *Saving the Nation: Economic Modernity in Republican China.* Chicago: University of Chicago Press, 2010.

Zhang Dan and Lin Yan. "Wei Xijiu chuangzao xiaoyinqi" [Wei Xijiu creates noise-canceling device], *Fushun ribao*, September 10, 1952.

Zhang Kexiang, and Zhou Zhizhen, *Fushun xian zhi* [Gazetteer of Fushun county]. N.p.: n.p., 1931.

Zhang, Lawrence. "Legacy of Success: Office Purchase and State-Elite Relations in Qing China." *Harvard Journal of Asiatic Studies* 73, no. 2 (December 2013): 259–97.

Zhang Linzhi. "Wei gao sudu fazhan meikuang gongye er fendou" [Fighting for the rapid development of the coal industry]. *Meikuang jishu* 1959, nos. 19/20 (October 1959): 4–9.

Zhang Shenfu. "Woguo de guofang yu zhonggongye" [Our country's national defense and heavy industry]. *Xingjian yuekan* 2, no. 1 (January 1933): 127–61.

Zhang Weibao. *Shiye jihua yu Guomin zhengfu: Zhongguo jindai jingji shi lunwenji* [The Industrial Plan and the Nationalist government: Essays in modern Chinese economic history]. Taipei: Tiangong, 2001.

Zhang Weibao, Zhao Shanxuan, and Luo Zhiqiang. *Jingji yu zhengzhi zhi jian: Zhongguo jingji shi zhuanti yanjiu* [Between economics and politics: Special topics in Chinese economic history]. Xiamen: Xiamen daxue chubanshe, 2010.

Zhang Zifu. "Jiefang liao de caimei gongren" [The coal miners who were liberated]. In *FKZ*, 287–88.

Zhao, Hai. "Manchurian Atlas: Competitive Geopolitics, Planned Industrialization, and the Rise of Heavy Industrial State in Northeast China, 1918–1954." PhD diss., University of Chicago, 2015.

Zhao, Suisheng, ed. *China's Search for Energy Security: Domestic Sources and International Implications.* London: Routledge, 2013.

Zhao Yuhang, Cheng Tingheng, and Li Jingrong. *Fushun xian zhi lüe* [Draft gazetteer of Fushun county]. N.p.: n.p., 1911.

Zheng Youkui, Cheng Linsun, and Zhang Chuanhong. *Jiu Zhongguo de Ziyuan weiyuanhui: shishi yu pingjia* [Old China's National Resources Commission: Historical fact and assessments]. Shanghai: Shanghai shehui kexueyuan chubanshe, 1991.

Zhengxie Fushun shi weiyuanhui, Fushun kuangye jituan youxian zeren gongsi, and Zhongguo shiyou Fushun shihua gongsi, eds. *Fushun gongye bai nian huimou* [A glance back on a century of Fushun's industries]. Vol. 1. Shenyang: Shenyang chubanshe, 2008.

Zhongguo Guomindang zhongyang weiyuanhui dangshi weiyuanhui, comp. *Zhang Jingjiang xiansheng wenji* [Collected works of Mr. Zhang Jingjiang]. Taipei: Zhongyang wenwu jingxiao, 1982.

Zhongguo jindai meikuang shi bianxie zu. *Zhongguo jindai meikuang shi* [Modern history of coal mining in China]. Beijing: Meitan gongye chubanshe, 1990.

Zhonghua renmin gongheguo Guojia tongji ju gongye tongji si. *Woguo gangtie, dianli, meitan, jiqi, fangzhi, zaozhi gongye de jinxi* [The past and present of our country's iron and steel, electric power, coal, textile, and papermaking industries]. Beijing: Tongji chubanshe, 1958.

Zhonghua renmin zhengzhi xieshang huiyi Liaoning sheng Fushun shi weiyuanhui wenshi weiyuanhui, ed. *Fushun wenshi ziliao xuanji* [Selection of historical materials on Fushun]. Vol. 3. Fushun: Liaoning sheng Fushun shi weiyuanhui wenshi weiyuanhui, 1984.

Zhong Lian. "Da ke zhuyi zhi meihuang wenti" [The coal famine problem that is much deserving of attention]. *Yinhang zhoubao* 15, no. 42 (November 1931): 3–4.

Zhou Enlai. "Zengchan meitan he jieyue yongmei tongshi bingju" [The concurrent increasing of coal production and saving on coal consumption]. *Renmin ribao*, November 2, 1957.

Zhu Dajing. "Shi nian lai zhi dianli shiye" [The electrical power industry over the past decade]. In *Shi nian lai zhi Zhongguo jingji*, edited by Tan Xihong, sec. J. Shanghai: Zhonghua shuju, 1948.

Zhu, Xiaoqi. "Re-examining the Conflict between Japanese Coal and Fushun Coal in 1932: Conflict Mediation and Policy Motivation of the Japanese Government." *Waseda University Journal of the Graduate School of Asia-Pacific Studies* 21 (April 2011): 229–56.

Zinda, John Aloysius, Yifei Li, and John Chung-En Liu. "China's Summons for Environmental Sociology." *Current Sociology Review* 66, no. 6 (October 2018): 867–85.

INDEX